Biological
Bases
of Sexual
Behavior

Animal Behavior Series Under the Editorship
of Vincent G. Dethier and H. Philip Zeigler

Biological Bases of Sexual Behavior

Gordon Bermant
Battelle Seattle Research Center

Julian M. Davidson
Stanford University

HARPER & ROW, PUBLISHERS
New York, Evanston, San Francisco, London

Sponsoring Editor: George A. Middendorf
Project Editor: David Nickol
Designer: June Negrycz
Production Supervisor: Bernice Krawczyk

Biological Bases of Sexual Behavior

Copyright © 1974 by Gordon Bermant and Julian M. Davidson

Library of Congress Cataloging in Publication Data

Bermant, Gordon.
    Biological bases of sexual behavior.
    (Animal behavior series)
    Bibliography: p.
    1. Sexual behavior in animals.   2. Sex.   3. Sex
(Biology)   I. Davidson, Julian M., joint author.
II. Title.   [DNLM: 1. Sex behavior.   HQ21–B516b–1974]
QL761.B45   591.5′6   73–21325
ISBN 0–06–040647–X

This book is dedicated
to Frank A. Beach

Contents

Preface, ix
Introduction, xi

CHAPTER 1   Understanding the Biological Psychology of Sex   1
Biological Psychology, 1
The Basic Concept of Sex, 9
The Plan of the Book, 10

CHAPTER 2   Sex, Reproduction, and Evolution   12
Reproductive Isolation, 13
Genetic Correlates of Sexual Behavior—Experimental
    Analyses, 27
Sexual and Nonsexual Reproduction, 38

CHAPTER 3   Environmental Determinants of Sexual Behavior   53
Varieties of Environmental Events, 54
Varieties of Effects on Sexual Behavior, 54
Plan of the Chapter, 55
Reproductive Cycles, 56

The Roles of Stimuli in Courtship and Mating, 67
Social Control of Mating Opportunities, 84
Social Control of Extended Copulatory Performance, 88
Environmental and Experiential Determinants of Abnormal
    Sexual Behavior, 92

CHAPTER 4   Neural Determinants of Sexual Behavior   97

Methodologies in Neural Psychology, 99
The Logic of Neural Organization, 102
Inhibition and Disinhibition of Sexual Reflexes, 103
Summary, 121

CHAPTER 5   Hormonal Determinants of Sexual Behavior   123

Effects of Hormones, 123
Endocrine Concepts, 125
The Relationship Between Steroid Hormones and Sexual
    Behavior, 132

CHAPTER 6   Neuroendocrine Mechanisms and Integration   151

Actions of Hormones on the Nervous System, 152
Effects of Social Stimuli on Hormone Production, 165

CHAPTER 7   Sexual Differentiation   178

The Extent of Sexual Differences, 178
The Causes of Sexual Differences, 186
Behavioral Differentiation, 193

CHAPTER 8   Approaches to the Study of Human Sexual Behavior   206

Direct Observation and Experimentation: Anatomy and
    Physiology of Human Sexual Response, 207
The Survey Approach: Anthropological and Sociological Perspectives, 214
The Clinical Medical Approach to Sex and Sexual Behavior, 232
The Psychoanalytic Approach: The Concept of Instinct (Drive),
    the Incest Taboo, and the Oedipus Complex, 250

References, 264
Index of Names, 295
Index of Subjects, 300

Preface

This book was conceived in the spring of 1963 while the authors were postdoctoral research fellows in the laboratory of Frank A. Beach in the Department of Psychology, University of California, Berkeley. Now, ten years later, it appears to us that the volume is rather slim to have demanded such a long period of gestation. However, there may have been some advantages to this protracted production time. We find that a substantial portion of our references were published after 1963. Many of the facts reported here were not available when the idea of this book first came up. The basic endocrinology was there, to be sure, and some awareness of the role of the hypothalamus in sexual neuroendocrine integration, as well as much of the material included here in Chapter 2 under the heading of sex, evolution, and reproduction. But many experimental data confirming the basic assumptions of the neuroendocrine model had not yet been collected. Nor, in the field of human sexual behavior, had Masters and Johnson published their first book; their work was known only by the sexual cognoscenti.

In several ways the decade between our commitment to the idea and the book's completion has been a very *sexual decade*. Some observers decry this fact, others rejoice in it. We simply point out, as far as the scientific aspect of this development goes, that we have been able to provide more information than we would have done had we completed the book sooner.

During the writing of the book we have had the benefit of much good advice. So many colleagues and students read so many drafts of the various chapters that a complete list would be prohibitively long. But one of our critics must not remain anonymous—our academic editor, Philip Zeigler. We feel very fortunate to have had the benefit of his objective and enormously detailed comments on several drafts of the manuscript.

Much of the typing, retyping, reference chasing, and other related tasks were accomplished at the Seattle Research Center of the Battelle Memorial Institute. We thank the officers of the Institute, in particular T. W. Ambrose, E. R. Irish, and R. S. Paul, for their support and encouragement. Throughout this task Barbara Hawley performed the secretarial work admirably. Mark Starr and Mary McGuire worked hard and well on the illustrations, permissions, and bibliography. Our wives, Roberta and Ann, have shown admirable restraint during these ten years.

Although this is primarily a text aimed at advanced undergraduates and graduate students in biology and psychology, we hope we have written simply enough for the layman to understand much of the material and deeply enough to interest our professional colleagues.

Introduction

Mark Twain once said of the weather that everybody talked about it but nobody did anything about it. For many years in polite society, the converse was true about sexual behavior. There have been times in this country when public references to bodily parts or functions were couched in euphemisms; for example, one spoke of "limbs" instead of "legs," and "breast" and "thigh" were avoided even in reference to chicken meat. People, and in particular "ladies," simply did not talk about sexual behavior.

Obviously, those times are over. It is no exaggeration to claim that in today's educated society frank talk about human sexual behavior, at least in the abstract, is admissible and in some cases even commonplace. By and large we do not make detailed references in public to our own sexual behavior, but we may feel quite free to discuss topics like the pros and cons of premarital and extramarital intercourse, contraceptive techniques and abortion, the status of homosexuals, and the social consequences of legalized and widely disseminated pornography. It is no accident that these and related topics may all be called *issues*, for it is characteristic of discussions of issues that competing or conflicting systems of values are brought to bear upon what are, presumably, the facts of the matter. Many people seem to be much clearer about their values concerning sexual behavior than they are about the facts. Most of us have faith that a clear and comprehensive understanding of the facts will facilitate resolutions of differences in values or ameliorate conflicts that arise because of them.

The social climate now allows wide public dissemination of facts about human sexual behavior, and large numbers of people are responding by buying books about sex and reading them. It has not been very many years since a consideration of human sexual behavior could be considered a "taboo topic" in psychology (Faberow, 1963). But at the date of this writing at least two books about sexual behavior have reached high positions on national best-seller lists (Masters and Johnson, 1966; Reuben, 1970). There are courses in human sexual behavior now being taught and well attended on numerous college campuses around the country, and textbooks are being prepared specifically for these courses (e.g., McCary, 1967; Katchadourian and Lunde, 1972). There is a magazine called *Medical Aspects of Human Sexuality* and a quarterly *Journal of Sex Research*. Add to these sources articles in daily, weekly, and monthly publications, not to mention special coverage of certain topics on radio and television—for example prostitution, venereal disease, and homosexuality—and the sum is an enormous public discussion of virtually every aspects of sexual behavior, all of it aimed at clarification and understanding. These days it is easy to be surrounded by talk about sex.

One might inquire, from the midst of all this sexually oriented publication, about the necessity or wisdom of providing still another book about sexual behavior. There are several features of this book that justify its entrance into an apparently overcrowded field.

First, we are concerned here with biological foundations and fundamentals, not as these are typically portrayed in manuals about sexual behavior (e.g., in cut-away diagrams of male and female genitalia), but as we understand them from the results of modern experimental physiology and psychology. The available knowledge about these fundamentals is substantial, but so is the remaining ignorance. From this point of view we still have a lot to learn about sex.

Second, people are not on center stage in this book. We believe that a deep comprehensive understanding of human sexual behavior develops most surely from a correct view of our species' place in the animal kingdom. Acquiring this view requires comparative study. In addition there is substantial intrinsic interest in the sexual behavior of nonhuman species and its determinants. We shall see that sexual processes are virtually ubiquitous in the animal kingdom. Sexual and evolutionary processes are so closely linked as to be almost inseparable. To understand evolution one must understand sex, and vice versa.

Finally, we are aware that many readers will begin, and perhaps finish, this book because they are required to; the book was conceived as a supplemental text for courses in animal behavior, comparative or physiological psychology, motivation, and the like. We have tried to keep these readers in mind and have striven to provide as much clarity and as little jargon as our talents, time, and space allowed. Even so, the book is not altogether an easy one to read, but we hope that students will be rewarded by the effort they put into it.

Biological
Bases
of Sexual
Behavior

Chapter 1
Understanding the Biological Psychology of Sex

The purposes of this chapter are to provide a brief review of some fundamentals of biological psychology, to introduce the basic concept of sex, and to show how the ramifications of this concept within biological psychology provide the material that makes up the rest of the book.

BIOLOGICAL PSYCHOLOGY

Biological psychology is a discipline concerned with the environmental, physiological, and historical determinants of the behavior of animals, including man. In order to clarify the general significance of each of these three classes of determinants we will discuss each briefly now.

ENVIRONMENT

An animal's environment is the set of physical and social stimuli and conditions that surround it. Some years ago the biologist Jacob von Uexküll (1921) cautioned against confusing the animal's environment as perceived by the animal and as perceived by us. Writing in German, von

Uexküll used the word *Umwelt* to label the phenomenal world of the animal. Although one may become immersed in murky philosophical waters by following this theme too far, there is utility in remembering that the sensory capacities and neural mechanisms for handling environmental information vary markedly among animal species. For example, honey bees see ultraviolet light that we do not; dogs, bats, and dolphins hear sounds that we do not; and many mammalian species smell odors that we do not. Among insects there is widespread use of chemical communication that is unavailable to the unaided human observer. When we study the environmental determinants of sexual behavior we must not overlook the particular sensory capabilities of the species under consideration.

There is another side of this coin. A good deal of social behavior, particularly of nonmammalian species, is regulated by the surprisingly limited portions of the environment available to the animal. The discovery of a number of these simple stimulus configurations (called *sign stimuli*) and the analysis of their evolutionary and functional significance were important early achievements of the European ethologists (von Uexküll, 1921; Tinbergen, 1951; Hinde, 1966). In Chapter 3 we provide several examples of the role of sign stimuli in regulating sexual behavior.

The response of an animal to its environment changes the environment and hence confronts the animal with a partially altered array of stimuli. This new array may elicit a further response that alters the environment again, and so on. The animal and its environment exert reciprocating influences: each regulates the other. This ecological principle works as effectively in the social environment as in the physical. As we shall see in a later chapter, it serves as the basis for the precise regulation of courtship displays and copulatory coordination.

The environment does more than elicit responses from an animal; it also selects and modifies responses through principles of positive and negative reinforcement. Animal species differ greatly in the details of their susceptibility to behavior modification through reinforcement. Earlier general theories of reinforcement, so dear to the hearts of several generations of American psychologists, are now being revised according to the demands of biological reality (Bolles, 1970; Rozin and Kalat, 1971; Seligman, 1970). Nevertheless, it is clear that the sexual behavior of some species, in particular nonhuman primates and man, is modifiable through the operation of environmental contingencies.

PHYSIOLOGY

When we speak of the physiological determinants of behavior we mean the demonstrable relations between changes in physiological functioning and changes in subsequent behavior. Critics of physiological psychology, particularly when it is applied to human sexual behavior, have maintained that this approach tends to "dehumanize" the behavior or "reduce" it to mere physiology (Farber, 1964; Kenniston, 1967). This is typically a mistaken criticism, for all a physiological approach to behavior does is to demonstrate functional relationships between events at one level of analysis and events at another. These relationships were just as operative before they were discovered as afterward; having knowledge about them does not in itself degrade the human significance of the behavior.

In the study of sexual behavior our interest centers on three physiological systems: the reproductive, endocrine, and nervous systems, which are interconnected in several ways. The reproductive and endocrine systems partially overlap anatomically (in vertebrates), for example, in the structure of the gonads. Within the testis are cells that produce male sex hormones (androgens) and cells responsible for the development of sperm; the adequate functioning of the latter depends on the adequate functioning of the former. Similarly, within the ovary there is an intimate relation between the tissues responsible for the production of the ovum and those responsible for the production of the ovarian hormones (estrogens and progesterone).

Gonadal hormones are secreted into the bloodstream and operate on many different tissues within the body. Of great importance is the action of these hormones on the nervous system, in particular the hypothalamus. This small but amazingly important system of nuclei and fiber tracts, located at the base of the brain behind the optic chiasma and just above the pituitary gland, contains cells that are sensitive to the levels of circulating gonadal hormones. These cells respond to changes in hormone level by secreting chemical factors that travel through a special blood supply (the portal vessel system) to the anterior lobe of the pituitary, where they produce changes in the amounts of some of the hormones secreted there. These hormones, called *gonadotropins*, travel through the circulatory system to the gonads, where they have several effects, including the regulation of gonadal hormone output. There is thus a closed loop of hormone function operating through the nervous system that serves in part to maintain the integrity of the reproductive system.

When the hypothalamus responds to changes in gonadal hormone levels it does more than simply pass that information along to the pituitary gland. It also operates on other brain areas in ways that may lead to large changes in the animal's behavior. The expression of sexual behavior is very closely related to the correlated activities of nervous and endocrine mechanisms.

Much of the material presented in later chapters expands on the basic physiological scheme just outlined. If this outline is totally unfamiliar to the reader, he may wish to consult a general text of physiological psychology or physiology in conjunction with the material in the later chapters.

HISTORY

In our context the word *historical* takes on three separate but related meanings. First, it refers to the development of a behavioral trait during the lifetime of an individual animal. For the animals with which we are most familiar—the vertebrates and particularly the mammals—an individual lifetime may be divided into several periods: prenatal, infancy, childhood or juvenile, adolescence, adulthood, and senescence or old age. Each of these may be subdivided according to several criteria. For our purposes it is important to note that the changes from juvenile to adolescent and from adulthood to senescence are determined in part by changes in reproductive abilities.

The second meaning of *historical* refers to the development of a behavioral trait across generations of an ongoing animal population. In particular, the concern is with changes in behavior that reflect changes in the frequencies of different genes represented in that population. Changes in relative gene frequencies arise from a number of sources, including mutations, genetic drift, and natural selection. Biological psychology is concerned primarily with changes produced by natural selection, that is, the evolution of the behavior of animal species. This interest leads naturally to a concern with the genetic determinants of behavior. For our purposes it is important to note the intimate relationship between sexual behavior and the transmission of genetic information across generations.

There is of course a third sense of history, particularly relevant to man, which is not captured in the first two; it refers to the course of human events as they have been recorded during the last 3000 years or so. This history records changes in human societies, cultures, and institutions; these changes reflect compounds of change in the behavior of individuals. But these developments of behavior are not necessarily accounted for in terms of either individual growth or physical evolution. Ideally, then, the behavioral repertoire of a human being should be described in terms of historical variables operating over three overlapping time bases: the evolutionary time of our species and the race to which the individual belongs (physical evolution), the time span of the individual's society or culture (social or cultural evolution), and the individual's own lifetime (individual development).

Biological psychology is not particularly well equipped to account for behavior in terms of social or cultural history. There are several reasons for this. First, this discipline is not concerned only with human behavior. Although processes like cultural evolution do operate in some nonhuman animal communities (for example in the troop of Japanese macaque monkeys that has taken to washing its sweet potatoes in the sea before eating them), their occurrence and importance seem now to be relatively minor. Biological psychology attempts to formulate principles that are applicable across species. Second, much of what we take to be evidence of cultural or social factors can be seen as a subclass of elements that operate on individuals during their lifetimes. Differences between cultures may be viewed as differences in the ways individuals are taught to behave. Biological psychology may be able to account for the forms in which teaching and learning occur, but it will not generally be able to account for the content of what is taught and learned. Finally, biological psychology is primarily an experimental discipline, and it is difficult to experiment meaningfully with the consequences of long-term social or cultural change. (Of course the same might be said for physical evolution, but in this case we have the experimental model of artificial selection to substitute for the slow process of natural selection. See Chapter 2.)

Like all experimental sciences, biological psychology deals primarily with repeatable events. Experimental results may be used to infer the processes that have operated in the unique lifetimes of nonexperimental individuals or populations, but these inferences are seldom if ever completely verifiable. Thus, this science is concerned with the development of behavior in individuals and the evolution of behavior in populations. Particularly with regard to man, biological psychology must be modest in the face of social and cultural histories if it is to avoid oversimplified conclusions about the causes of human behavior (e.g., Ardrey, 1966).

RELATIONS AMONG ENVIRONMENT, PHYSIOLOGY, AND HISTORY

These three classes of behavioral determinants are by no means completely separable. Some historical determinants of behavior may be operationally defined in terms of their contemporary environmental consequences. For example, the difference between incidences of homosexual intercourse occurring among boys growing up in a particular Melanesian culture and boys growing up in America (see Chapter 8) is produced by the different histories of the two societies. But this dissimilarity is also reflected in each generation by differing social environments, particularly regarding the public and private approvals and sanctions placed on this form of behavior.

There is a similar relation between historical and physiological determinants. For example, it is quite clear that certain kinds of experimental manipulations performed on prenatal or infantile animals may have profound consequences on their adult sexual behavior (see Chapters 3 and 7). In order to understand the behavior of the adult we need to know its childhood history. But obviously these historical variables have to be "contained" in the physiology of the animal from the time it is treated experimentally until it reaches adulthood. As our knowledge of the relevant physiology increases we ought to be able to specify exactly what has been changed and remained changed during subsequent development.

A traditionally confusing relation among these classes of determinants flies under the banner of such labels as "heredity vs. environment," "learning vs. instinct," or "the nature-nurture controversy." At first glance the question under consideration appears to be something like this: "Given a sample of behavior (anything from the way a mother rat brings her pups into the nest to the answers a child gives on an intelligence test), to what extent is that behavior due to the animal's physical heredity and to what extent is it due to experiences, training, or other factors acting on the animal after the time of conception?" The problem with this question is that, without a lot of further qualifications, it is simply unanswerable. But there are some simple concepts that help us understand how we must shape and focus the question to get at the important underlying issues.

One relatively simple way to resolve the initial question is to replace it by two related questions: "How best can we describe the processes that have resulted in this particular sample of behavior occurring under these conditions?" and "What can we do now, or what could we have done earlier, to increase, decrease, or in general modify the form or frequency of this behavior?"

It is clear that an answer to the first question will include a specification, in as much genetical detail as necessary, of the individual's hereditary material. Strictly speaking, all an animal inherits is DNA and a little cytoplasm. From the instant of fertilization our answer will be a description of an extraordinarily complex *developmental process* in which the original genetic material replicates and specializes again and again until the form of the behaving organism emerges. Throughout the process the organism's cellular mass and its environment are in continuous interaction. So long as we pay attention to the development of the single organism it makes no sense to question the relative importance of heredity and environment. As D. O. Hebb has pointed out, this question is as meaningless as asking whether the length or width of a rectangle contributes more to its area; take either without the other and the result is zero.

This view is correct so long as we attend to the behavior of a single animal. But it becomes incorrect when we enlarge our interest to include behavioral similarities and differences among groups of animals. In this case we can imagine doing experiments in which animals with identical physical heredities are, from conception onward, raised in different environments. Conversely, we can expose animals with different heredities to environments as identical as we can arrange. Most important, we can arrange experiments in which both heredity and environment are systematically manipulated as experimental variables. We can measure behavior of interest to us in all cases. These measurements will cover a certain range; that is, there will be variability among the responses of different animals to our testing situation. And most important, the techniques of quantitative genetics allow us to subdivide that total variability into separate classes, including genetic and environmental variabilities (Falconer, 1960). We need not be concerned with the details of these techniques here; they are essentially the analysis of variance procedures with which psychology and zoology students are familiar. The point is that our first naive question, when recast in terms of behavioral differences among groups or populations of animals, may receive a quantitative answer.

Our second question has to do with our ability to control or manipulate the behavior of the individual. In the case of experimental animals, we may make two major decisions about any individual: who its parents will be and what its life will be like after conception. We have as much control over both heredity and environment as we care to exercise. Our omnipotence over experimental animals makes resolution of properly asked nature-nurture questions a matter of experiment, not a matter of theoretical debate. In the case of human beings, however, our control is ethically and practically limited; we may not do many of the experiments required to obtain clear evidence. Thus the nature-nurture issue in human behavior typically revolves around indirect and *relatively* poorly controlled data collection procedures. This is particularly true with behaviors of social concern, for example alleged racial differences in intelligence. So long as strictly definitive data are lacking there will be room for argument. But this debate stems from unavoidable ignorance about specific cases. It should not be taken to mean that the relative importance of heredity and environment can never, in principle, be assessed.

THE BASIC CONCEPT OF SEX

There may be as many meanings of *sex* as there are persons who have used the word. At this point we want to stress a fundamental significance or meaning of this deceptively simple word.

Sex is separateness: a division of reproductive labor into specialized cells, organs, and organisms. The sexes of a species are the classes of reproductively incomplete individuals. In order for a sex member to contribute to the physical foundations of its species, to reproduce part of itself into the next generation, it must remedy its incompleteness. The remedy is found in the union of incomplete parts from complementary incomplete organisms: sperm and egg unite. The accomplishment of this union requires energy. The animals must locomote, gesture, posture, vocalize, display, or otherwise *behave* if union is to occur. Any behavior that increases the likelihood of gametic union may reasonably be called sexual behavior.

This is a simple theme. But its consequences ramify throughout animal life in complex, subtle ways. Sex is almost ubiquitous in the animal kingdom because of the genetic variability, hence evolutionary flexibility, provided by the separation of chromosome pairs into haploid cells and the union of two such cells from individuals with different histories. The genetic alternatives available to a population that practices this form of reproduction are greater than those available from any other form of reproduction. With this advantage comes the challenge of overcoming sexual separation. No two animal species meet this challenge in precisely the same way. As we shall see in Chapter 2, differences in sexual functioning among animal populations are a key feature in the definition of the species concept.

This theme stresses the relationship between sexual behavior and reproduction. Sexual behavior is a vital part of the process of sexual reproduction, but it does not follow that animals exhibit sexual behavior in order to reproduce. The phrase *in order to*, describing the reason why someone does something, carries an implication of intentionality that we cannot justifiably bring to bear in our discussion of nonhuman sexual behavior. Moreover, as we know very well, much sexual behavior among humans is performed for purposes other than reproduction. Hence a general account of the determinants of sexual behavior cannot rely on the relation of that behavior to reproduction, except in certain cases of human behavior.

Much if not all of the material in the rest of the book is an elaboration on the theme of the preceding three paragraphs within the context of biological psychology. In concluding this chapter we will provide a brief overview of the organization of subsequent chapters.

THE PLAN OF THE BOOK

Chapter 2 treats the general biological relations among sex, evolution, and reproduction. The so-called biological species concept is introduced in connection with the mechanisms that prevent species overlap (reproductive isolating mechanisms). Examples demonstrate the close relationship between the sexual behavior of a species and its genetic identity and integrity. This leads naturally to a discussion of the genetic determinants of sexual behavior. The chapter ends with a discussion of the distinction between sexual and nonsexual reproduction and includes treatments of behavioral features associated with parthenogenesis, gynogenesis, and functional hermaphroditism.

Chapters 3 through 7, the heart of the book, deal directly with the several classes of determinants of sexual behavior. Chapter 3 has to do with environmental determinants and includes discussions of reproductive cycles, the roles of stimuli in courtship and mating, the social control of mating opportunities and extended copulatory performances, and abnormal sexual response. The examples used in this chapter draw from a narrower range than those in Chapter 2; except for one or two descriptions of insect behavior, the material concerns vertebrates.

Chapter 4 treats neural determinants of sexual behavior. It begins with brief discussions of neurophysiological methodologies and the logic of neural organization. The key concepts in the chapter are inhibition and disinhibition of behavior following neurosurgical intervention. Differences between male and female animals in response to neural damage are described. The chapter concludes with a description of a simple model of the male ejaculatory mechanism. Species coverage becomes narrower in this chapter, dealing primarily with data collected from laboratory mammals. Discussion of neuroendocrine relationships is reserved for later chapters.

Chapter 5 presents the hormonal determinants of sexual behavior and reviews basic endocrinological concepts. Relationships between steroid hormones and sexual behavior are reviewed separately for males and females. The emphasis here is on experimental, as opposed to clinical, data. Hence species coverage remains restricted primarily to laboratory mammals.

Chapter 6 integrates the basic material presented in Chapters 3, 4, and 5 under the heading of Neuroendocrine Integration. The first section of the chapter attempts to show how changes in hormone levels operate on the nervous system to potentiate behavior changes. The second section discusses the effects of social stimuli on hormone production. Two important topics introduced in Chapter 3 are discussed in a more physiological context: social control of ovulation and pregnancy, and the concept of the pheromone, particularly as it applies to research with laboratory rodents.

Chapter 7 deals with the development of sexual behavior, particularly the differentiation of adult sexual behavior as a function of prenatal and neonatal hormonal events. Experiments are described in which male infants (rats primarily) are treated with female hormones, and vice versa. These and related results form the basis for a discussion of the physiological bases of sex differences in sexual behavior. Viewed broadly, the chapter deals with the importance of an individual's physiological history. The importance of environmental history is not dealt with directly here, but is considered under separate headings in the other chapters.

Chapter 8 is reserved for special consideration of four more or less scientific approaches to the study of human sexual behavior. Direct experimental results are presented along with clinical, anthropological, and psychoanalytic results. No attempt is made to provide a final evaluation of the worth of these several approaches; the reader is asked to judge for himself.

Chapter 2
Sex, Reproduction, and Evolution

In 1859 Charles Darwin revolutionized biology by publishing a massive amount of information in support of the theory that evolution proceeds primarily through the mechanism of *natural selection*: those animals survive and reproduce that best meet the demands of the environment. Animals whose characteristics are less well adapted to their environment are less likely to survive to reproductive age or to produce offspring. Thus the environment selects, and there is modification in the generations of a family line.

Natural selection depends on variation within animal populations. In Darwin's time the sources of variation in natural populations were not well understood. Genetics had not yet become a science. At approximately the time that Darwin was formulating and publishing his theory, Gregor Mendel was working quietly in the garden of his monastery, discovering fundamental genetic principles that were to remain in obscurity for thirty-five years. An immediate effect of the rediscovery of Mendel's laws in 1900 was a temporary discrediting of Darwin's views. This occurred because Darwinian theory dealt with the evolution of characteristics that vary more or less continuously (for example, body size), whereas Mendelian theory was based on the inheritance of traits that vary qualitatively or discretely (for example, eye color). The successful fusing of Mendelian and Darwinian views during the ensuing decades formed the basis for the synthetic theory of evolution (Stebbins, 1966). In more recent years the discovery of the basis for the genetic code within the DNA molecule (Watson and Crick, 1953; Watson, 1967) produced a tremendous surge in genetic research that now aids our understanding of the physical basis for evolution.

Appreciation of the modern theory of evolution depends in part on an understanding of the concept of the gene and its molecular basis; but equally important is a concept defined at a much higher level of biological organization: the concept of the *species*. Although not all biologists accord on the definition of the species concept, all will agree that the idea in one form or another has been central in evolutionary thinking (Mayr, 1963; Sokal and Sneath, 1963). Our concern will be with what has been called the *biological species concept* (Mayr, 1963).

REPRODUCTIVE ISOLATION

The natural selection of individuals occurs on the basis of their fitness for existence and reproduction in their home environment. Over the course of generations, a natural population of individuals will respond to changing environmental exigencies either by successfully changing its way of living, by leading a marginal existence with reduced numbers of weakened individuals, or by becoming extinct. In the course of evolutionary history, most of the species that have arisen have suffered extinction. Successful coping with the environment, from one generation to the next, *is* the evolutionary process. This means that the population is sorting out from the total genetic variation available to it those combinations of genes that, in interaction with the environment, produce individuals who can survive long enough to reproduce. If a particular environment remains relatively unchanged for a long time, we can expect the populations surviving in it to become highly adapted to it. Through a long series of biological improvements a population may achieve a genetic endowment (gene pool) that cannot undergo further major alteration without deleterious effects on the population's survival (Huxley, 1963). For example, the oyster has shown no appreciable evolution for many millions of years (Simpson, 1949). The introduction of new genetic material from outside the population is likely to have a negative effect. We can see the advantage of some form of *reproductive isolation* in this case.

Another advantage of reproductive isolation among animal populations is its effect on the distribution of available food supplies. It is naturally economical to utilize optimally the various food sources in the environment. This suggests that natural selection may lead to populations living side by side that are specially adapted to use different portions of the total range of available foods. Minimal overlapping of food preferences can prevent destructive competition between populations. To the extent that food preferences are genetically controlled, reproductive separation of populations will maintain the separation of food preferences, and thereby maintain each population's chances for continued survival (Klopfer, 1962).

These arguments lead to the conclusion that reproductive isolation between populations living in the same place (sympatric populations) at the same time (synchronous populations) is a phenomenon of substantial evolutionary significance. In fact, reproductive isolation is the critical feature in the modern definition of the biological species. *A species is a population of actually or potentially interbreeding individuals that is reproductively isolated from other populations under natural conditions* (Mayr, 1957, 1963).

If a species is defined in terms of the reproductive behavior of its members, then it is immediately clear that behavior plays a vital role in the maintenance of animal species. With some exceptions, an interbreeding population is one in which males and females join their gametes either within the body of the female or externally. We refer to the behavior immediately involved in the internal fertilization of ova as *copulation* and to that immediately involved in the external fertilization of ova as *spawning*. Both copulation and spawning typically are preceded by a more or less elaborate behavioral sequence called *courtship*. Courtship serves various functions in different animal species, and we shall have occasion to refer to courting behavior in various contexts throughout the book.

A species is a population of individuals that breed (copulate, spawn, etc.) together and, equally important, do not breed successfully with members of other populations under natural conditions. Conditions that in nature lead to reproductive isolation may be nonexistent or become broken down within the confines of a zoo or laboratory.

We have spoken of the advantages of reproductive isolation between species; our own species is the only one that is aware of these advantages. For example, we could not explain the maintenance of reproductive isolation between two sympatric species of toads by asserting that the toads know it will be to their evolutionary advantage not to mate with each other. Therefore, if we wish to understand how reproductive isolation is maintained between natural populations, we must search for the ways by which the anatomy, physiology, psychology, and ecology of these populations serve to prevent them from merging into a single, interbreeding population. We are looking for the *reproductive isolating mechanisms* (RIM). The classification of isolating mechanisms presented here is based on the work of Mayr (1963) and Spieth (1958).

Isolating mechanisms may be grouped into four classes, based on physiology, anatomy, habitat or season, and behavior. The latter three categories are *premating* mechanisms, that is to say, mechanisms that prevent gamete transfer. The physiological mechanisms, on the other hand, operate after mating has occurred. We shall describe them briefly now.

PHYSIOLOGICAL ISOLATION

There are four postmating RIM: gametic mortality, zygotic mortality, hybrid inferiority, and hybrid sterility. Gametic mortality refers to death before fertilization, zygotic mortality to death after fertilization but before birth. An example of gametic mortality is the death of heterospecific sperm in the vaginas of some species of fruitflies (*Drosophila*). Examples of zygotic mortality have been found for crosses between goats and sheep, rabbits and hares, and different genera of sea urchins (Dobzhansky, 1955). Hybrid inferiority refers to the failure of healthy, apparently fertile hybrids to produce offspring. The best-known example of hybrid sterility is the mule, the sterile offspring of a donkey and a horse.

ANATOMICAL (MECHANICAL) ISOLATION

Mechanical isolating mechanisms arise when differences in copulatory apparatus or related structures prevent insemination. At one time it was believed that the great diversity in the phallic structures of some groups of insects represented a "genital coding" that insured the prevention of successful cross-specific copulation. Although this situation exists among some fruit flies and tsetse flies, there are other insect groups in which an equally diverse array of genital structures does not preclude successful mating.

The domestic dog provides an exception to the rule that members of a species are anatomically compatible. For example, a Chihuahua and a Great Dane, both members of *Canis familiaris* as defined by most biological criteria, are mechanically isolated from interbreeding because of their relative sizes. This example points up the difficulties of applying the concepts of species and isolation to animal populations that have experienced selective breeding under human control.

HABITAT AND SEASONAL ISOLATION

Habitat isolation refers to species differences in preferred mating grounds. Examples for closely related species have been described in chipmunks, toads, fish, and dragonflies (Mayr, 1963).

Habitat isolation may also exist at the level of the subspecies, as is apparently the case in two subspecies of the deer mouse, *Peromyscus maniculatus gracilis* and *P. maniculatus bairdii*. Although individuals from these two groups will interbreed successfully in laboratory cages, hybrids are not found in their naturally overlapping range in the Great Lakes area. The explanation is apparently that *P. gracilis* lives in the forests and other covered areas, whereas *P. bairdii* occupies the open country. Scott (1958) has reported the results of an experiment by Harris (1952) in which members of the two subspecies, whose ancestors had been laboratory born and raised for several generations, were given the opportunity to choose between artificially contrived "forest" and "flatland" environments. *P. gracilis* chose the forest, and *P. bairdii* chose the flatland environment. It is important to note that these preferences existed in animals that had never lived outside the laboratory; moreover, their ancestors had not lived outside the laboratory for several generations. Thus the habitat preference successfully keeps the subspecies apart, and it is certainly possible that this habitat selection will eventually lead to the development of two distinct species.

Seasonal isolation refers to asynchronies in the reproductive cycles of closely related species. Differences in the breeding seasons have been reported for gulls, frogs, toads, and several species of insects (Mayr, 1963). The case of the gulls is particulary interesting. It involves two species that live sympatrically in northwestern Europe: the herring gull (*Larus argentatus*) and the lesser black-backed gull (*L. fuscus*). Hybridization is rare between the two populations. However, the two groups form the ends of an almost continuous ring of gull populations that circle the earth in the northern latitudes. Rings of this sort are also known for other species, and are given the German name *Rassenkreise*, or circles of races. The *L. fuscus* population is the westernmost of a series of Eurasian gull groups, and the *L. argentatus* population is the easternmost of a series of populations that expanded out of the North American continent. The population taken as a whole now shows gene exchange in all areas of contact except where *L. argentatus* and *L. fuscus* overlap. The isolation between these two populations is in part behaviorally maintained and in part seasonally maintained. The *L. argentatus* population reaches a peak of reproductive behavior during the latter part of April, two to three weeks before the *L. fuscus* population.

BEHAVIORAL ISOLATION

Behavioral reproductive isolating mechanisms are probably the most important type. Sympatric species are behaviorally isolated when some feature of the behavioral repertoire of one or both of the populations is effective in preventing the transfer of gametes between them. It will be appreciated immediately that one function of the courting behaviors of sympatric species might be to insure that the members of each population end up sharing gametes only with members of their own species.

One difficulty involved in demonstrating that a particular aspect of a species' courting behavior acts as an isolating mechanism is that it may work so well that it appears not to be working at all. For example, if there are several sympatric species that have similar or identical courtship and mating sequences, except that the masculine vocalizations are species-specific and clearly distinct, potentially breeding pairs of different species may never be formed because the females will not be attracted by the calls of foreign males. It is a difficult task to prove in the natural conditions of field research that it is in fact the vocalization that helps maintain separation; but some successful examples of this sort of research follow. In many cases, however, we simply infer that the existence of highly diversified courtship displays, vocalizations, and movements among closely related, sympatric species indicates that the behavioral differences are effective in maintaining species separation.

A convenient way to classify behavioral RIM is by reference to the sensory modality that is primarily affected by the particular mechanism. Some mechanisms involve visual displays, some involve auditory information, and others depend on chemical or tactual channels of communication. Examples from each of these categories will demonstrate the tremendous diversity of behavioral isolating mechanisms that have been evolved.

Cheliped-Waving Among Fiddler Crabs (Genus *Uca*)

Jocelyn Crane (1941, 1957) has provided an account of the species-characteristic courtship displays of many species of the crab genus *Uca*. These animals are called fiddler crabs because the males possess an oversized front claw (cheliped) that they wave vertically and laterally. Darwin published observations on *Uca* as early as 1871; he believed that the incongruously large chelipeds were evidence in favor of his theory of sexual selection. Later authors debated the validity of this and other interpretations of the functional significance of this unusual bodily development. Crane has now shown that the motions of the cheliped serve both to aid a female in recognizing a male of her own species and to warn other males away from the courting pair.

Figure 2.1 presents a frontal view of the display of three species of Indo-Pacific *Uca*. Crane has shown that the displays may differ along one or more of five dimensions: cheliped elevation, shell (carapace) elevation, cheliped lateral extension, speed of movement, and smoothness of movement. The diversities of displays among sympatric species all courting together on the beaches of the Panama Canal Zone allow the trained observer to make discriminations from a distance.

Once a male and female crab have made contact, the male must continue his courtship display for a period of time that may extend through most of the day. The female is at first very "timid"; that is, she will retreat from the male at the slightest environmental disturbance, for example the call of a bird passing overhead. In these animals, as in many others, courtship is a prolonged affair that may be broken off several times before it is consummated by copulation. There is ample opportunity for a mistake in species identification to be rectified. An interesting unsolved problem is if or how the males identify the females, which do not possess the species-specific behaviors and colorations of the males. It may be that the males are indiscriminate and the females alone "bear the burden" of maintaining species integrity. When at last the male's extended display has quieted the female and induced her receptivity, the pair retreat into a burrow in the sand, usually one occupied earlier by the male. Copulation takes place in the burrow.

In addition to their interesting semaphore signaling approach to courtship, the fiddler crabs provide a clear example of an important biological generality: when a species or higher taxonomic group is observed to possess a particularly unusual morphological structure, that structure is often associated with the courtship or copulatory behavior of the group.

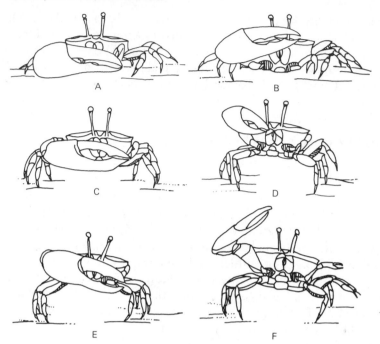

FIGURE 2.1 Three species of the fiddler crab, genus *Uca*, exhibiting distinct, species-specific cheliped motions during courtship. Rest positions between waves are shown on the left. The maximal elevation of the cheliped during the display is shown on the right. The species illustrated are: A, B, *U. rhizophorae;* C, D, *U. signata;* E, F, *U. zamboangana.* From Crane, J. (1957) Basic patterns of display in fiddler crabs (*Ocypodidae*, genus *Uca*). *Zoologica* 42:69–82. Drawing from New York Zoological Society.

"Flash-Answer" Patterns Among Fireflies (Genus *Photinus*)

Fireflies, sometimes also called lightning bugs, are in fact neither flies nor bugs but rather several genera of beetles (family Coleoptera) including two, *Photinus* and *Photurus*, that are observed in the eastern United States. There are 28 species of *Photinus* north of Mexico; Lloyd (1966) has done a careful analysis of the flashing behavior of 25 of them. His work provides several excellent examples of species-characteristic displays serving an isolating function.

Figure 2.2 is a diagrammatic representation of the typical flight elevations, paths, and flash patterns of nine *Photinus* species. The males fly and flash while the females observe from the grass. The flashing of a male stimulates a female to respond with a flash of her own, and the "flash-answer" sequence is begun. The male repeats his flash pattern, the female repeats hers, and so on until the male lands by the female and begins a tactile courtship that may be followed by copulation.

Figure 2.3 shows the temporal characteristics of male and female flashing patterns in 23 *Photinus* species. In a series of experiments Lloyd was able to demonstrate the following facts:

1. He could bring flying male fireflies to the ground by flashing a small flashlight at them with the pattern used by the females of their own species.

2. If, however, he flashed at the males with the flash pattern of another sympatric species, the males did not respond, except in one case.

3. If he placed females of one species in a transparent cage and exposed them to males of another species, the males were not attracted

FIGURE 2.2 Characteristic flash patterns and flight paths of males of several species (not all sympatric) of *Photinus* fireflies. Illustrated are: (1) *P. consimilis* (slow pulse); (2) *P. brimleyi*; (3) *P. consimilis* (fast pulse) and *P. carolinus*; (4) *P. collustrans*; (5) *P. marginellus*; (6) *P. consanguineus*; (7) *P. ignitus*; (8) *P. pyralis*; (9) *P. granulatus*. From Lloyd, J. E. (1966) Studies on the flash communication system in *Photinus* fireflies. *Mis. Publ. Mus. Zool.*, No. 130. Ann Arbor: University of Michigan.

to the cage except in two cases. In one of these cases (female *P. punctulatus*, male *P. curatus marginellus*—see Figure 2.3 for flash patterns) the males flew down to the cage. When they were permitted to enter, they crawled over the females but did not attempt copulation. Lloyd concluded that the isolation between these naturally sympatric species is based on chemical or tactile stimuli. In the other case (female *P. granulatus*, male *P. tenuicinctus*), the females attracted the males immediately and copulation followed shortly.

However, these two species are naturally allopatric (nonoverlapping in range). *P. tenuicinctus* occupies a tight range on the northern border of Oklahoma and Arkansas, whereas *P. granulatus* is located primarily in central Oklahoma and Kansas.

In summary, Lloyd demonstrated that among sympatric species of *Photinus* fireflies, the species-specific flash-answer visual displays serve exquisitely to maintain species integrity.

Mating Calls Among Narrow-Mouthed Frogs (Genus *Microhyla*)

Sound production and reception are important features in the courting behavior of frogs and toads. Males attract receptive females by croaking loudly while partially submerged in the pools in which breeding takes place. Blair (1955) has reported on the effectiveness of species-specific mating calls in minimizing hybridization between two species of narrow-mouthed frogs, *Microhyla carolinensis* and *M. olivacea*. These two species have a narrow zone of overlap in eastern Texas, with *M. olivacea* spreading to the west as far as Arizona and *M. carolinensis* continuing east into Florida.

The mating call of each species may be characterized roughly as a buzz. In analyzing the calls, Blair relied primarily on two of their physical dimensions: duration (measured in seconds), and midpoint frequency (measured in cycles per second). These measurements were facilitated by the use of a sound spectrograph, an instrument that allows the visual display of auditory signals. An example of a sound spectrogram is presented in Figure 2.4.

Blair collected samples of mating calls from males of both species in several parts of their respective home ranges, including the zone of overlap, the western extreme of the *M. olivacea* range in Arizona, and the eastern extreme of the *M. carolinensis* range in Florida.

Figure 2.5 displays the variation in mating calls observed in different

FIGURE 2.3 *(facing page)* Flash patterns and female response flashes in *Photinus* fireflies. Time zero for the female response is the beginning of the last pulse of the male flash pattern. Flash-pattern interval is indicated by r. Flash patterns are given only once; when several pulses are shown they are all part of one flash pattern. Flash patterns with variable pulse numbers are indicated by open (versus solid) pulses. From Lloyd, J. E. (1966) Studies on the flash communication system in *Photinus* fireflies. *Mis. Publ. Mus. Zool.,* No. 130. Ann Arbor: University of Michigan.

SPECIES	♂ FLASH PATTERN						♀ RESPONSE						
DIVISION I	1	2	3	4	5	6	1	2	3	4	5	6	7
			seconds						seconds				
Marginellus	▪		r(74°)			r	▪(70°)						
Curtatus x Marginellus	▪		r(73°)			r	▪(69°)						
Floridanus	▪			r(70°)		r	▪(75°)						
Sabulosus	▪			r(71°)			▪(69°)						
DIVISION II													
Pyralis	▬					r(73°)			▬(67°)				
Australis	▪			r(77°)			▬(77°)						
Scintillans	▪		r(70°)		r		▪(70°)						
Brimleyi	▪ r(74°)	r	r	r	r		▬▬(76°)						
Punctulatus	▪ r(76°)	r	r	r	r		▪(74°)						
Tenuicinctus	▪	r(70°)	r		r	r	▬(75°)						
Umbratus	▬				(72°)r			▬▬(70°)					
Collustrans	▪		r(75°)	r		r	▬(65°)						
Tanytoxus	▬		r(75°)		r		▬(75°)						
Granulatus	▪	r(72°)	r	r	r	r	▬(80°)						
Dimissus	▪ r(74°)	r	r	r	r		[?]						
Linellus	▪ ▫▫		r(74°)		r		[est](74°)						
Ignitus	▪					r(74°)				▬(77°)			
Consanguineue	▪ ▪				r(75°)		▬(70°)						
Macdermotti	▪	▪			r(71°)		▪(69°)						
Consimilis (fast-pulse)	▪ ▪ ▪ ▪ ▫ ▫ ▫ ▫ r(at 10.6/66°)								▪ ▫▫ ▫▫ ▫▫(70°)				
Consimilis (slow-pulse)	▬	▬	▫	r(at 14.7/67°)				?		[?] [?]			
Carolinus	▪ ▪ ▪ ▪ ▫ ▫ ▫ ▫ r(at 13.8/64°)									(61°)▬ ▫▫→			
Ardens	▬ ▬ ▪					r(62°)			▬ ▫▫ ▫▫(58°)				

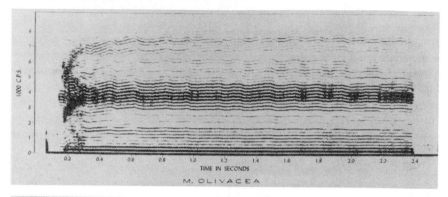

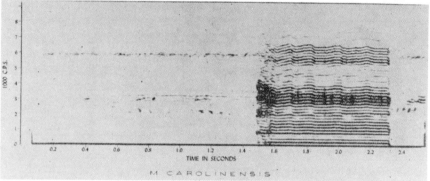

FIGURE 2.4 Representative sound spectrograms of *Microhyla* mating calls in eastern Texas. Time runs along the abcissa and sound frequency along the ordinate: the higher the frequency (pitch), the higher its position on the spectrogram. The intensity of the sound is represented by the darkness of the markings. Note that the emphasized frequency is at about 3000 hertz in *M. carolinensis* and at about 4000 hertz in *M. olivacea*. From Blair, W. F. (1955) Mating call and stage of speciation in the *Microhyla olivacea-M. carolinensis* complex. *Evolution* 9:469–480.

localities. Of particular interest are the differences between *M. olivacea's* call in Arizona (1 in graph) and in the zone of overlap (2 in graph). The calls of the westernmost *M. olivacea* are those displayed by a good portion of the *M. carolinensis* population, including that part of the population that resides in the zone of overlap. But the calls of the *M. olivacea* males that live in the zone of overlap are both higher and longer than those of their western conspecifics and of the *M. carolinensis* subpopulation with which they are sympatric. In this case the sexual isolation mechanism is in evidence only in that portion of the population in which it is required. Where the species are together, the behavior that advertises the location of breeding males has become relatively species-distinct.

The divergence of species characteristics in regions of overlap, as observed here for the mating calls of the *Microhyla* species, is called character displacement, and has been analyzed by evolutionary theorists (Brown and Wilson, 1956; Wilson, 1965).

Singing Stridulation Among Grasshoppers (Genus *Chorthippus*)

One of the most complete experimental studies of the factors isolating two species has been performed by Perdeck (1958) on two widely distributed species of grasshoppers, *Chorthippus brunneus* and *C. biguttulus*. These two populations form a pair of so-called sibling species, which means that they are morphologically extremely similar, yet they remain reproductively isolated in areas of sympatry (Mayr, 1963). The isolation of such pairs of species is of particular interest to the evolutionist and behaviorist.

As illustrated in Figure 2.6, acridid grasshoppers produce their acoustical signals by rubbing a file on the femur against a specialized portion of the forewing. This activity is called stridulation. Each raised segment of the file is called a stridulatory peg. One of the few anatomical characteristics that differentiates *C. brunneus* from *C. biguttulus* is the number of pegs on the file: for *C. brunneus* the number averages approximately 63; for *C. biguttulus* it averages approximately 103. The anatomical difference is probably in part responsible for the species-specific differences

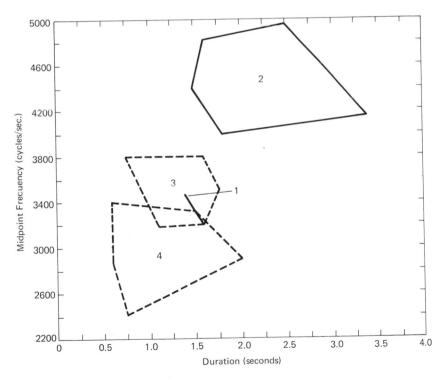

FIGURE 2.5 Duration and frequency of *Microhyla carolinensis* and *M. olivacea* mating calls in different geographical regions. (1) *M. olivacea* from Arizona; (2) *M. olivacea* from overlap zone in eastern Texas; (3) *M. carolinensis* from Florida; (4) *M. carolinensis* from overlap zone. Modified from Blair, W. F. (1955) Mating call and stage of speciation in the *Microhyla olivacea-M. carolinensis* complex. *Evolution* 9:469–480.

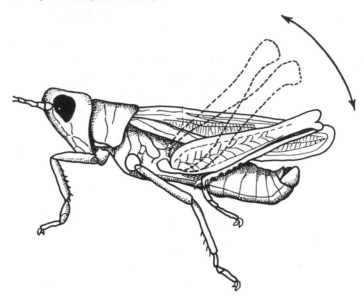

FIGURE 2.6 Movements comprising stridulation by a grasshopper. From Dumortier, B. (1963) The physical characteristics of sound emissions in Arthropoda. In R. G. Busnel (ed.), *Acoustic Behavior of Animals*, pp. 346–373. Amsterdam: Elsevier.

in courting songs. Oscillograms of sounds made by the two species are presented in Figure 2.7.

We have already mentioned that detailed experimental work is required to prove unequivocally that a particular behavioral difference between species is an effective isolating mechanism. Perdeck's experiments serve as an example of how the job can be done. The work is particularly interesting because it shows that differences in courting songs and in the behavioral responses to them are virtually the only effective reproductive isolating mechanism.

The method Perdeck employed can be described as the successive exclusion of possible RIM: starting with the observation that the species are in fact well isolated (as judged from the very small number of hybrids found in nature), he showed that several types of RIM that might be operative were in fact ineffective. To begin with, he showed that hybrids could be produced readily in the laboratory; this finding ruled out the operation of gametic and zygotic mortality and mechanical isolation. The hybrids were fertile and apparently not at an obvious disadvantage to the parent species; these findings ruled out a large-scale effectiveness of hybrid sterility or inferiority. Observations in the field indicated that the members of both species lived side by side in areas of overlap, and that there were no large differences in breeding times either on a daily or an annual basis. These observations ruled out the operation of habitat and seasonal RIM. Hence species separation must be maintained by behavioral mechanisms.

Having established the necessity of behavioral RIM, Perdeck made a close analysis of the various modes of communication by which species distinctiveness might be transmitted: visual, olfactory, tactile, and auditory. For each of these modes he investigated the several segments of the behavioral sequence that begins with the male encountering the female and ends with copulation. These included the original orientation of the male to the female on the basis of initial contact with her (not necessarily visual), the male's locomotion to the female, his copulation attempt, and the female's response to the copulation attempt.

The major result of the work can be stated simply: the animals determined species identity almost entirely on the basis of auditory stimulation provided by stridulation. In addition, *C. biguttulus* females made some discrimination between males on the basis of olfactory cues. But all other differences between the species were either nonexistent or irrelevant.

Perdeck summarized the role of the song in species separation by listing five behavioral differences that he discerned between homogamic (same species) and heterogamic (different species) courtship-copulation sequences: in the homogamic sequence there were more locomotor behaviors by the male, more effective orientation by the male to the female, more effective orientation by the female to the male, a larger number of copulation attempts by the male, and an increased tendency for the

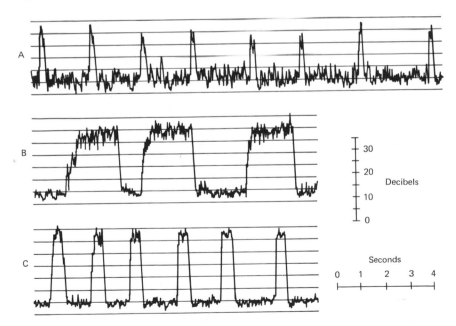

FIGURE 2.7 Oscillograms of courting songs by males of two species of acridid grasshoppers, *C. brunneus* (A) and *C. biguttulus* (B), and the hybrid between them (C). From Perdeck, A. C. (1958) The isolating value of specific song patterns in two sibling species of grasshoppers (*Chorthippus brunneus* Thunb. and *C. biguttulus* L.). *Behaviour* 12:1—75.

female to accept the male's copulation attempt without moving away. Finally, Perdeck was able to show that the viable, fertile hybrids that could be produced under artificial conditions in the laboratory were partially behaviorally isolated from the parent species. This was particularly true for the males, whose song, being intermediate to those of the parents, was relatively ineffective in attracting females.

Chemical Signals Among Fruit Flies (Genus Drosophila)

Few if any genera of animals have been as important as the fruit fly, genus *Drosophila*, in contributing to biological knowledge. Initial scientific interest in these small flies was due in part to the very large chromosomes found in the salivary glands of the larvae. These giant chromosomes have been intensively studied by geneticists interested in the physical localization of genes (Srb, Owen, and Edgar, 1965). Additionally, the flies have a rapid generation time and may be bred very easily in the laboratory. These features combine to make *Drosophila* an ideal experimental animal for studies of genetics and artificial selection (Tinbergen, 1965). Consequently it is not surprising that the courtship and copulatory behavior of fruit flies has been studied in substantial detail. The classic paper on the subject describes the behavior as it occurs in eight subgenera within the genus (Spieth, 1952). For our example of chemical isolation we will discuss the behavior of the females of two sibling species within the subgenus *Sophophora*, *D. pseudoobscura* and *D. persimilis* (Mayr, 1950). The species are sympatric along the west coast of the United States.

Experimental verification of the importance of chemical stimuli in isolation was accomplished by surgically removing olfactory receptors from the females of both species; these receptors are located on the third segment of the antennae. The females were then placed with male *D. pseudoobscura*. The logic of this experiment was straightforward: if olfactory differences between male *D. persimilis* and *D. pseudoobscura* are utilized by females for species recognition, then *D. persimilis* females without their antennae should be more likely to accept *D. pseudoobscura* males than they would be if their antennae were intact.

When this experiment was conducted under ideal environmental conditions, Mayr (1950) was able to show that antennaeless *D. persimilis* females were as likely to accept *D. pseudoobscura* males as *D. pseudoobscura* females were. However, only about 30 percent of each group of females was inseminated; with intact antennae, 86 percent of the *D. pseudoobscura* females and 2.5 percent of the *D. persimilis* females were inseminated by the *D. pseudoobscura* males. These results meant that olfactory discrimination was a relevant isolating mechanism, but also that the removal of the antennae resulted in a reduced rate of successful copulation within the same species. In another set of experiments, using *D. melanogaster*, Mayr demonstrated that the reduction in intraspecific inseminations resulting from antennae removal was not necessarily produced by generalized surgical trauma. Females whose probosci had been removed were more likely to be inseminated than antennaeless females, yet the proboscis operation is more debilitating and traumatic than the antennae operation.

SUMMARY OF RIM

Two sympatric populations are called separate species if they are reproductively isolated from one another. Reproductive isolation is often maintained by behavioral differences between the two populations. In particular, the behavioral differences frequently relate to the courtship and/or mating sequences of the populations. To put it slightly differently, the sexual behavior in one population can prevent an exchange of genetic material between that population and a sympatric one; the behavior is a cause of isolation. In addition, the differences in sexual behavior between the two species may also be an effect of the isolation. That is, the members of populations differ in their sexual behavior because of their partially nonoverlapping genotypes.

GENETIC CORRELATES OF SEXUAL
BEHAVIOR—EXPERIMENTAL ANALYSES

Up to this point we have been able to use terms and phrases such as *genotype* and *genetic constitution* without specifying their referents in terms of chromosomes or genes. We will consider now some examples of experiments that have attempted to make explicit the relations between specified genetic conditions and the expression of sexual behavior. Some of these experiments have been motivated by an interest in the genetic bases of the reproductive isolating mechanism operating between two closely related (e.g., sibling) species, whereas others have assessed the consequences for sexual behavior of certain genetic conditions within the members of a single species. In either case, the experiments can be classified under the rubric of the genetic correlates of behavior or, as the field has been recently labeled, behavior-genetic analysis (Hirsch, 1967).[1]

We may distinguish two general experimental approaches in this area, each of which can be elaborated in several ways. We will first give an overview of these approaches, then return to consider some examples.

The first approach compares the sexual behaviors of animals with different genetic characteristics. Several sorts of genetic characteristics have been considered: single-gene differences, differences in relatively large

[1]During the last decade or so, there has been a substantial increase in the number of experiments directed toward assessing the relation between genotype and behavior; only some of these have been concerned with sexual behavior. There are several adequate source works (Fuller and Thompson, 1960; Hirsch, 1967; Parsons, 1967; Wilcock, 1969) to which the reader is referred for the history and current status of the field.

chromosome segments, and differences in chromosome (particularly sex chromosome) complement or number. In the same vein, comparisons have been made among several highly inbred strains of the same species. The primary genetic characteristic of inbred strains is a high degree of homozygosity (i.e., identical genes existing at corresponding loci of chromosome pairs). Comparisons of highly inbred strains are comparisons of different, highly homozygous populations. Hence it is often useful to make additional comparisons between inbred strains and their hybrid offspring. If each of the parent strains is completely homozygous, then each of the hybrid progeny will be genetically identical and heterozygous at loci where the parents possessed different genes.

The second major experimental approach consists in selectively breeding animals that exhibit particular behavioral characteristics. During each generation, the only animals allowed to reproduce are those that show extreme values on the behavioral scale that is being studied. By restricting the genetic input to successive generations and by measuring behavioral changes across generations, one can estimate the heritability of the particular behavioral measure. Because of the similarity of this procedure and natural selection, it is sometimes called artificial selection.[2]

DIFFERENCES IN A SINGLE GENE: WILD-TYPE AND YELLOW FRUIT FLIES (DROSOPHILA MELANOGASTER)

On the X chromosome of D. melanogaster there is a locus that may be occupied by the recessive, mutant gene "yellow." The gene receives its name because of the yellow body color associated with its presence. It has been known for more than 50 years that yellow males are not as successful as normal males in courting and mating females. Bastock (1956) has made a thorough comparative analysis of the courtship sequences of normal and yellow males in an attempt to specify the nature of the mutant gene's effect on behavior. We begin our discussion of this work with a general description of the courtship and copulation sequence in this species; this description is based on one provided by Bastock (1967).

The precopulatory courtship sequence of the male has three major constituents: orienting, wing-vibrating, and licking (see Figure 2.8). Licking is usually accompanied by a copulation attempt. These components are preceded by tapping, in which the male taps the female with a foreleg. Discrimination of chemical information picked up by receptors on the leg serves as a sign of species identity, and hence prevents cross-specific courtship. Additionally, the function of the male's wing vibrations seems to relate not to visual stimulation (for courtship proceeds as rapidly in the dark as in the light), but rather to maximizing the chemical stimulation at the female's antennae; the male's odor reaches the female's antennae more efficiently if he beats his wings. Thus for D. persimilis as well

[2]The concepts of inbreeding, selection, and heritability all have rigorous definitions within the framework of quantitative genetics. See McClearn (1967) and Roberts (1967) for fuller discussion of these terms.

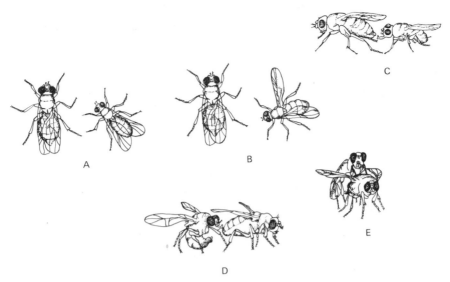

FIGURE 2.8   Courtship sequence of *Drosophila melanogaster*. (A) Orientation. The male faces the female, stands close to her, and will follow if she moves. (B) Wing-vibrating. While orienting, the male holds the wing that is closest to the female's head at right angles to his body and vibrates it vertically. (C) Licking. The male circles behind the female, quickly extends his proboscis, and "licks" her genital region. (D) Attempted copulation. (E) Copulation. The receptive female spreads her wings and genital plates. After Manning, A. (1965) *Drosophila* and the evolution of behavior. In *Viewpoints in Biology*, Vol. 4, pp. 125–169. London: Butterworth.

as *D melanogaster* as described earlier, the integrity of the female's antennae is very important for successful courtship.

As the first step in her analysis of the effects of the yellow gene, Bastock (1956) prepared two groups of flies that were genetically very similar except for the presence or absence of the yellow gene; then she measured the percentage of normal and yellow males that mated successfully as a function of duration of exposure to females. After 15 minutes, 60 percent of the normal and 30 percent of the yellow males had copulated, and after 60 minutes, 75 percent of the normal and 48 percent of the yellow males had copulated. Yellow males were slower than normals to begin courtship (about 10 vs. 5 minutes), and slower to progress from the beginning of courtship to copulation (10.5 vs. 6 minutes).

Bastock next analyzed the courting sequences of the males into their three major components. She found several reliable differences between normal and yellow males, of which the most important were that the yellow males spent more time orienting and less time vibrating than the normals did. The average duration of yellow male wing vibration episodes was less and the interval between episodes was greater than the corresponding values shown by normal males.

With this analysis Bastock had provided a quantitative description of the courting difference between normal and yellow males. However, with only this result she could not conclude that the single-gene difference had produced the behavioral difference directly. There were some alternative interpretations that needed to be investigated experimentally. For example, the differences in male behavior might have been attributable directly to preferential treatment of the normal males by the females, which themselves were all normals. In this case the genetic difference between the males would have related to their behavioral differences in a very indirect fashion. Bastock tested this possibility in several separate experiments. These included direct measurements of the number of rejection movements made by the females to both kinds of males and of the effects of removing the females' antennae. In all cases the hypothesis of feminine preference could be rejected.

Bastock advanced two alternate interpretations of her results. In the first, which she favored, she argued that the yellow mutation resulted in lower "sexual motivation." By this term she apparently meant some (at present unknown) neural, endocrine, or peripheral physiological process that regulates or is otherwise involved in the production of the relevant courtship responses. In this argument, whatever the physiological details should turn out to be, the yellow gene will be found to be directly associated with deficiencies in the relevant physiological system.

The alternative interpretation is that the yellow mutation produces a generalized decrement in mobility or responsiveness. Consequently, yellow males cannot keep up with the females as they move around the mating chamber; they must spend more time orienting to the female and hence less time vibrating. In this argument the question of sexual motivation becomes irrelevant. This interpretation of Bastock's results has been readvanced by Wilcock (1969) in his general review of the experimental literature on single-gene effects on behavior. At present it appears that more experiments will be required to decide the issue for the yellow gene in *Drosophila*. However, the general point that Wilcock makes needs to be developed briefly here because it bears as well on the other examples that follow and in general on studies in the genetic correlates of behavior.

Wilcock is concerned with the concept of *pleiotropism*, which refers to the fact that a single gene may affect two or more characters or attributes of an organism. We must understand that pleiotropism is the rule and not the exception. As Mayr (1963) put it, "one can safely state that nonpleiotropic genes must be rare, if they exist at all" (p. 265). In considering studies of gene action on behavior, Wilcock argues that most, if not all, of the experiments have dealt with a kind of "trivial pleiotropism" that is of little psychological interest. That is, the available data do not show interesting relations between genes and behaviors; rather, at best, they demonstrate a causal relation between obvious anatomical-physiological features and behavioral capacities. His argument about the reduced mobility of yellow *D. melanogaster* is a case in point.

In general, the issue of pleiotropism in behavioral genetics must be appreciated with the following considerations in mind: behavior, per se, is not inherited by individuals. Genes, as biochemical units of action, do not act on behavior, but, as logical units of action they may be manipulated in ways that lead to behavioral changes in populations across generations. These changes will always be correlated with other changes that are describable (at least in theory) in physiological or biochemical terms. Hence any behavioral unit that can be shown to vary quantitatively as a function of genetic manipulations is inevitably related to changes in some physiological variable or variables. Whether a particular relation between behavior and physiology or structure is trivial or significant depends not so much on the facts as on their usefulness for the plans and theories of the scientists who are studying them. From the viewpoint of the general biological psychologist, study in behavioral genetics is a promissory note on an eventual physiological-biochemical analysis of mechanism. The strength of the promise will depend on both the genetical and behavioral sophistication that have been brought to bear on the experiment. Bastock's (1956) study represents one of the best of its kind; we will now move on to some additional examples.

THIRD-CHROMOSOME PAIR ARRANGEMENTS AND MATING SPEEDS IN DROSOPHILA PERSIMILIS

Spiess and his colleagues (Spiess and Langer, 1961, 1964; Spiess and Spiess, 1967) have performed a series of experiments to determine the relation between different structural arrangements (inversions) of the third-chromosome pair and mating speed differences in Drosophila persimilis. Here we describe some findings from one of these studies, in which the behaviors of animals carrying the Whitney (WT) and Klamath (KL) chromosome arrangements were compared (Spiess and Langer, 1964). Both groups of flies descended from stock originally captured high in the Sierra Nevada Mountains of California.

As a first step in the experiment, Spiess and Langer measured the mating speeds of pairs in which each animal was a homokaryotype.[3] Four combinations of males and females were tested: WT/WT male and female, WT/WT male and KL/KL female, KL/KL male and WT/WT female, and KL/KL male and female. There were large differences in rates at which pairs in these four groups completed the mating sequence. Table 2.1 shows the percentages of completed mating after 30 and 60 minutes.

Pairs containing WT/WT females showed more rapid mating than pairs containing KL/KL females. WT/WT males were also more effective with both kinds of females than KL/KL males, but this difference

[3]Karyotype refers to the chromosome complement, considered structurally. In the current example, homokaryotype means that both members of the third-chromosome pair had either the WT arrangement (symbolized WT/WT) or the KL arrangement (KL/KL). The converse of homokaryotype is heterokaryotype.

TABLE 2.1   Percentage of Completed Matings:
WT/WT and KL/KL Homokaryotypes

Karyotype		Time	
F	M	30 min.	60 min.
WT/WT	WT/WT	77	84
WT/WT	KL/KL	48	63
KL/KL	WT/WT	25	30
KL/KL	KL/KL	11	18

(From Spiess and Langer, 1964.)

was relatively small. Hence, Spiess and Langer concluded that most of
the observed difference could be attributed to differences between the
females. In detail, they noted that WT/WT females accepted males for
copulation (i.e., spread their genital plates) after little or even no courtship, whereas the KL/KL females were likely to reject courtship over-
tures by extruding their ovipositors or moving away abruptly. Also, WT/
WT males were more persistent in courtship than KL/KL males, who
tended to intersperse brief periods of courtship display with general
locomotion not oriented to the female.

The next step in the analysis was to compare the mating rates of heterokaryotypes (one chromosome WT and one KL) with the rates of the
homokaryotypes. Male and female heterokaryotypes were mated with
each other and with homokaryotypes.[4]

The mating rate of heterokaryotype pairs was intermediate between
the fast WT/WT and the slow KL/KL animals. The data are shown in
Table 2.2. However, when the heterokaryotypes, either males or females,
mated with homokaryotypes, the mating speeds were determined by the
homokaryotype animals. Thus when WT/KL flies mated with WT/WT
flies, mating proceeded rapidly, but when they mated with KL/KL flies,
mating proceeded slowly. This was a particularly interesting result because

TABLE 2.2.   Percentage of Completed Matings:
Homogamic Matings of Homo- and Heterokaryotypes

Karyotype	Time	
	30 min.	60 min.
WT/WT	75	81
WT/KL	49	57
KL/KL	14	20

(From Spiess and Langer, 1964.)

[4]The mating of animals with similar genotypes, in this case homokaryotype ×
homokaryotype or heterokaryotype × heterokaryotype, is called *homogamic* mating.
The mating of different genotypes (e.g., homokaryotype × heterokaryotype) is called
heterogamic mating. Thus, in Table 2.1, WT/WT × WT/WT and KL/KL × KL/
KL are homogamic matings, and the others are heterogamic.

it demonstrated a behavioral plasticity associated with the heterokaryotypic condition. Further behavioral studies using these genetic combinations should lead to deeper insights into the relation between chromosome complement and mating speed.

DIFFERENCE IN SEX CHROMOSOME COMPLEMENT:
NORMAL MALE (XY) AND SUPERMALE (YY) JAPANESE MEDAKA FISH

Normal male Japanese medaka fish (*Oryzias latipes*) have an XY sex chromosome complement, and normal females have an XX complement. However, males with YY chromosome complement may be produced through a series of special procedures. Males with two Y chromosomes are called supermales.[5] Hamilton et al. (1969) have reported the results of an experiment in which pairs consisting of one XY and one YY male were placed with one female and allowed to court and mate (spawn) with her. The object of the experiment was to determine whether YY males were more or less successful than XY males in inducing spawning when placed in direct competition.

Fourteen pairs of males were observed in 265 tests; in 155 of these (60 percent) the female was induced to spawn by one male or the other. In these positive tests the YY males induced spawning in 137 cases (88 percent) and the XY males in 18 cases (12 percent). In 7 out of the 14 pairs of males the YY male was responsible for induction in 100 percent of the cases, and in the remaining pairs the YY male was responsible for at least 60 percent of the inductions. A more detailed analysis of the several behavioral components of the courtship sequence revealed that the YY males were more active than the XY males in each component. Moreover, the YY males occasionally circled, chased, and bit the XY males, but the reverse never occurred.

These results show, for a situation involving direct competition between the two kinds of males, that the YY males were more effective in courting the female and inducing spawning. At present it is impossible to make any statements concerning the series of biochemical mechanisms that link the different chromosome complements with these behavioral outcomes. Moreover, the possibility that this result is explicable in terms of obvious structural differences between males cannot be discounted. However, it is interesting to note that *reproductively capable females with a YY chromosome complement* may be produced by treating YY fry with estrogen (Hamilton et al., 1969).[6] A behavioral analysis of these YY-bearing females, particularly in regard to their tendency to court XX females in the male manner, should add substantially to our understand-

---

[5]The concepts of supermale and superfemale arose originally in the study of mechanisms of sex determination in *Drosophila*. A supermale is one in which the ratio of X chromosomes in the sex chromosome complement to the number of autosomes in a homologous group is less than normal. For example, a normal diploid male has a ratio of one X chromosome to two autosomes (1/2). The YY male, with a ratio of 0, is thus a supermale. See Dobzhansky (1955) for further discussion.

[6]See Chapter 7 for a discussion of the hormonal basis of sex differentiation.

ing of the relation between this chromosome complement and sexual behavior.

COPULATORY BEHAVIOR OF MALE MICE:
STRAIN DIFFERENCES AND THE EFFECTS OF HYBRIDIZATION

During the last 60 years there has been a substantial effort to produce and maintain large stocks of inbred strains of the common mouse, *Mus musculus*. An inbred strain is one that has been produced through brother-sister matings for at least 20 generations (Staats, 1966b). The consequence of strict inbreeding is a high degree of homozygosity; for a single locus, the probability of heterozygosity after twenty generations of brother-sister matings is only 0.012 (Green, 1966). At the present time more than 200 inbred strains are being maintained in laboratories throughout the world (Staats, 1966a). McGill and his colleagues have done a series of experiments on the relations between genotype and copulatory behavior utilizing several of these inbred strains; we will describe some of their results (McGill, 1962; McGill and Blight, 1963; McGill and Ransom, 1968).

We begin with a brief description of the copulatory sequence in the mouse. The male mouse mounts the female from the rear and while grasping and palpating her flanks with his forelegs begins a series of rapid, shallow, pelvic thrusting movements. These thrusts typically result in the initial insertion of the penis into the vagina (*intromission*). Following the initial intromission the rate of the male's thrusting decreases; at this stage each thrust takes approximately 0.7 second (McGill, 1962). The male keeps one hindfoot on the floor and puts the other high on the female's hindquarters. During the multiple thrusts of each intromission the fully receptive female stands quite still with her hindquarters raised and her tail deflected, a posture termed *lordosis*.[7] Following the series of deep thrusts, the male dismounts and moves away from the female. He has not yet ejaculated. Typically ejaculation is achieved only after numerous intromissions, each of which includes a number of thrusts. During the final intromission the male's rate of thrusting increases. Strong thrusting is terminated when the male quivers, lifts his remaining hindfoot off the floor, and grasps the female with all four legs. From this position he falls off to the side; the female sometimes falls with him. In this way, the copulatory sequence is completed. The duration of the male's postejaculatory refractory period, from ejaculation until he attempts to mount again, is at least 1 hour, and may last in certain strains as long as 24 hours.

In order to study this complicated behavioral sequence systematically, one must break it down into separate components that can be described quantitatively. McGill (1962) has defined 16 such measures. The diagram in Figure 2.9 provides a description of the most important of these.

---

[7]See Chapter 5 for a consideration of the hormonal conditions that produce female receptivity.

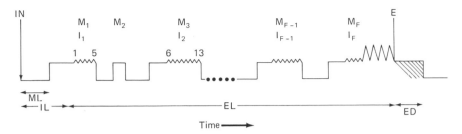

FIGURE 2.9   A schematic representation of the copulatory sequence of a male mouse. Time progresses from left to right. The entrance of the female (IN) into the male's cage begins the sequence. The time from the female's entrance to the male's first mount is called the mount latency (ML). The male's shallow thrusting results in intromission. After the initial penetration the male engages in several deep thrusts before he dismounts; as shown in the figure, each thrust may be counted (e.g., 5 thrusts in intromission #1, 8 in #2, etc.). The time from the female's entrance to the first intromission is called the intromission latency (IL). Not every mount results in an intromission; in the illustration, the second mount does not result in intromission. The numbers of mounts with and without intromission can be counted; in the illustration M_F stands for the final mount and I_F for the final intromission, and M_{F-1} stands for the penultimate mount and I_{F-1} for the penultimate intromission. The time from the beginning of the first intromission to ejaculation (E) is called the ejaculation latency (EL). Just prior to ejaculation the male's thrusting becomes more vigorous. Ejaculation duration (ED) is the time from the initial ejaculatory response to the final separation of the animals. The broken segment in the middle of the line signifies that only the beginning and ending segments of the sequence are pictured. Diagram based on descriptions by McGill, T. E. (1962) Sexual behavior in three inbred strains of mice. *Behaviour* 19:341–350.

With this material as background, we can consider some of McGill's experimental results. To begin with, there are large differences between strains on many of the fundamental measures of the copulatory sequence. For example, each of the three strains BALB/c, C57BL, and DBA/2 (see Staats, 1966a, for an explanation of this nomenclature) is significantly different from the other two on 5 of the 16 measures, including IL, EL, and number of intromissions (McGill, 1962). Thus on the average, BALB/c males take about 60 minutes to achieve ejaculation; DBA/2 males take about 30 minutes; and C57BL males take about 21 minutes. During these times the numbers of intromissions achieved by the three strains are 36 (BALB/c), 7 (DBA/2), and 23 (C57BL). Note that there is not a monotonic relation between ejaculation latency and number of intromissions. This means that the strains are also differentiated in terms of the amount of time they spend away from the female between intromissions. The C57BL males show the shortest intervals between intromissions (33 seconds), the BALB/c males the next shortest (56 seconds), and the DBA/2 males the longest (215 seconds).

These results make it clear that different inbred strains of mice show quantitatively different patterns of copulatory response. The next step in the genetic analysis of these behavioral sequences is to compare the behavior of the inbred strains with the first-generation hybrid progeny (F_1 generation). McGill and Blight (1963) performed this experiment with the C57BL/6J and DBA/2J strains and their F_1, and McGill and Ransom (1968) checked the generality of the findings by doing a similar experiment with DBA/2J and AKR/J and their F_1. There were three ways in which a single behavioral characteristic of the F_1 might be related to the same characteristics of the parents: *dominance, intermediacy,* or *heterosis.* Dominance describes the cases in which F_1 performance remains identical to that of one parental strain; that strain is the dominant one. Intermediacy describes the cases in which F_1 performance falls quantitatively between the performances of the parents. Heterosis describes the cases in which F_1 performance is significantly greater than that of either parental strain. McGill and Blight were able to assign 14 of their behavioral measurements into one or the other of these three "modes of inheritance." However, McGill and Ransom were not able to substantiate these findings. Predictions from one set of results to the other were unsuccessful for 10 of the 14 behavioral measures. On the basis of this result, McGill and Ransom cautioned against oversimplified conclusions regarding the mode of inheritance of quantitatively defined behavioral characteristics.

ARTIFICIAL SELECTION FOR MATING SPEED IN *DROSOPHILA MELANOGASTER*
As a final example in this section we will consider the work of Manning (1961, 1963) on artificial selection for fast and slow mating speeds in *Drosophila melanogaster.* A general description of the behavior has been provided in earlier examples. In the first experiment (1961) Manning began with two groups of flies from a genetically diverse population; each group contained 50 males and 50 females. Manning allowed each group to copulate en masse and picked from each the 10 fastest and the 10 slowest pairs. This resulted in four groups of 10 pairs each, two fast and two slow. These four groups were the parental stocks out of which subsequent generations were produced. At each generation the fastest 10 pairs from each of the fast lines and the slowest 10 pairs from each of the slow lines were used to produce the subsequent generations. The experiment was terminated after the twenty-fifth generation.[8] Table 2.3

TABLE 2.3.   Average Mating Times (Minutes) Before and During Selection

Line	Parental Stock	Generation			
		6	12	18	24
Fast	6	5	4	3	3
Slow	6	60	15	25	45

(Approximated from Manning, 1961, Figure 1.)

[8]In order to put these procedures into some time perspective, it should be noted that the flies were bred four to six days after hatching.

presents the average mating times of the parental stock and of the fast and slow lines during generations 6, 12, 18, and 24. For simplicity of presentation we have averaged the results within the two fast and two slow lines.

The response of mating speed to this selection pressure was evident within the first several generations. Differences between fast and slow groups were as large in generation 7 as they ever became. Quantitatively, the selection effect was more pronounced in the slow lines than in the fast. This suggests that the mating speed of the parental stock already approximated the physiological maximum.

During the course of the selection procedure Manning performed several additional experiments that provided more detailed information about the genetic changes that were taking place under this regimen. Fast and slow animals of generation 17 were crossbred and the mating speeds of the F_1 animals were measured. On the average, the hybrids were intermediate between the fast and slow lines. F_1 animals were then bred back to members of the fast and slow lines and mating speeds of progeny from these backcrosses were measured. The backcrossed flies tended to mate with speeds intermediate between those shown by their parents. These results suggested that many genes were involved in the determination of mating speed and that their effects were additive.

Manning also conducted several studies on the selected flies in order to get a clearer picture of the behavioral changes that contributed to the large differences in mating speed. First he demonstrated that both the males and the females had been affected by selection; in this case, one sex was not dominant in the determination of mating speed. Next he showed that both male and female fast-mating flies, when tested individually for general locomotor activity, were less than half as active as slow-mating flies or the general population of unselected flies. This finding ruled out the possibility that the long mating times of the slow flies were an obvious consequence of their lowered mobility or general activity. In fact, the converse was true: the long mating times of slow flies were due in part to the long "lag times" they exhibited between the time they entered the mating chamber and the time they began the courtship sequence; this lag time was taken up by general exploration of the mating chamber. This was not the only behavioral feature that distinguished the groups, however. In a final experiment it was shown that the slow males exhibited significantly fewer licking responses during courtship than did fast or unselected flies. Because licking frequency is highly correlated with the frequencies of other courtship responses, Manning concluded that the courting vigor of the slow flies had been reduced.

In summary, the behavior of the fast-mating flies was characterized by low general activity and high rates of courtship display, while the opposite was true of the slow-mating flies. Normal (unselected) flies showed both relatively high levels of general activity (like the slow flies) and courtship activity (like the fast flies). Hence Manning concluded that the behavioral effect of selection in both fast and slow lines was to raise performance thresholds. In the fast line the general activity threshold was raised, whereas in the slow line the threshold for exhibiting courting and copulation responses was increased. Although Manning does not use the phrase *sexual motivation* in his 1961 paper, it is clear that the concept of *raised threshold for sexual performance* is very close in meaning to *lowered sexual motivation.*

Manning (1963) extended his analysis of selection effects by using a procedure in which selection pressure was placed either on males or females, but not, as in the earlier study, on both simultaneously. Generations of males were selected either for fast or slow mating, and females were selected for slow mating; technical difficulties prevented an attempt to select for fast females.

Selection pressure exerted on one sex was effective in producing slow males but ineffective in producing fast males or slow females. Manning accounted for the failure to produce fast males by noting that the parent stock was already mating at near the maximum rate. Selection pressure limited to just the male, and hence to just half the genotype of the subsequent generation, was simply insufficient to push behavior beyond this already high level of performance. Reasons for the failure to produce slow females are not so clear. It may be, in the relatively early stages of selection, that the behavioral effects associated with the changing genotypes become manifested primarily or altogether in the males. An accumulation of relevant genes beyond a certain minimum number will be required before the effect on mating speed will be seen in females. If females are not bred with males that have accumulated sufficient numbers of relevant genes, their female progeny will not exhibit the changed behavior. Some evidence in favor of this idea was found in the behavior of the female progeny that were produced in the line in which males were being selected for slow mating. In generation 11 these females mated with unselected males at normal speeds. By generation 21, however, the mating speeds of these females had slowed significantly. Again, the testing was conducted with unselected males.

Finally, Manning noted that the males selected for slow mating speeds in this experiment showed correlated decrements in their levels of general activity. This relation was opposite to the one found in the earlier experiment, in which the general activity of the slow maters remained at approximately control levels. Hence, Manning concluded that rates of sexual activity could be manipulated independently of rates of general activity, at least as these were measured in his experiments.

SEXUAL AND NONSEXUAL REPRODUCTION

We have completed our main discussion of reproductive isolation and genetic determinants of sexual behavior. We will conclude the chapter by discussing some differences between sexual and nonsexual reproduction. The material in this section will seem more difficult or technical to many readers, and perhaps will appear removed from the main line of interest. But we believe that it is important for students, particularly students of psychology who might not otherwise encounter this information, to appreciate obligatory sexual reproduction and its attendant behavior as one of several strategies for species survival. We attempt to foster that appreciation by describing some alternative strategies and by showing how some species combine sexual and nonsexual reproductive modes.

When we speak of reproductive isolation between two animal species we usually have in mind that each of the species is composed of two sexes, and that premating isolation involves the separation of the males of each species from the females of the other. In general, the concept of sexual behavior is closely related to the typical relation between sex and reproduction: specialized cells (sperm and ova) within the two sexes possess half of the total genetic material required to produce a new individual. The behavior immediately required to get the two halves together so that sperm can fertilize egg is one clear form of sexual behavior; the beginning of the reproductive process is the result of that behavior. However, there is no logical or biological necessity that reproduction always begin with the joining of sperm and egg. In nature there are many examples of reproduction occurring without sexual behavior. We will consider some examples.

BINARY FISSION, CONJUGATION, AND AUTOGAMY IN PARAMECIA

The paramecium is a genus of ciliated protozoa, that is, a group of single-celled animals that move by the coordinated beating of hairs (cilia) that line the body surface in orderly rows. These animals are of interest to us because of the diversity of reproductive forms they exhibit.

In the paramecium one individual becomes two through the process of *binary fission*: the cell elongates and splits in half. The macronucleus (see Figure 2.10) participates in the elongation and splitting process, and the micronuclei (one or more depending upon the species) undergo mitosis. Half of the resulting micronuclei migrate to the posterior portion. Each of the two new cells (daughter cells) is identical to the single parent, and only one parent is required for the production of the two new individuals. Binary fission is, therefore, an unambiguous example of nonsexual reproduction (Moment, 1958; Srb, Owen, and Edgar, 1965). It is called *asexual* reproduction to distinguish it from *parthenogenesis*, the development of an unfertilized ovum.

If binary fission were the entire story of reproduction in the paramecium, then each individual and its asexual progeny (known as a *clone*) would, by our definition, constitute a species, because they would be reproductively isolated from all other individuals. Moreover, descendants could differ from parents only insofar as there had been genic mutations

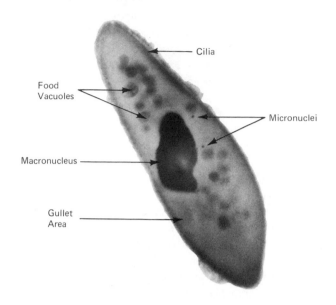

Cilia

Food
Vacuoles

Micronuclei

Macronucleus

Gullet
Area

FIGURE 2.10 *Paramecium multimicronucleatum* at a magnification of about 800. This species has four micronuclei although only two are shown here due to limitations of the optical plane. Photograph courtesy of Professor T. M. Sonneborn.

or chromosome aberrations at the time of mitosis. As it turns out, paramecia are not limited in this way. They have the means to exchange and modify combinations of genetic material.

Two paramecia may exchange genetic material through the process of *conjugation*. Conjugation begins when two animals of the proper mating types (see below) join together at their oral surfaces. The micronuclei in each cell undergo meiotic divisions; in the case of *Paramecium aurelia*, for example, the two micronuclei in each animal divide meiotically into eight daughter nuclei, seven of which disintegrate, and the eighth daughter nucleus divides mitotically to produce two identical nuclei. One of these nuclei from each animal migrates across the bridge of cytoplasm that has formed between the animals and joins the nucleus that has remained stationary in the other animal. This newly combined nucleus then undergoes a series of divisions while the two paramecia (now called *exconjugants*) separate. After separation each exconjugant undergoes binary fission, so that now there are four animals where once there were two, and each of the four contains genetic material from each of the original two (Walker, 1967).

Paramecia do not conjugate with other members of the population at random. Instead, there are systems of *mating groups* within each species; each mating group contains two or more *mating types*. In general, an animal is capable of conjugating with another of a different mating type but in the same mating groups. In this sense mating types in a paramecium species correspond roughly to the sexes of higher animals, with the important difference that in some mating groups there may be as many as eight interbreeding mating types.

Conjugation allows for genetic recombination that cannot be achieved in binary fission. However, it does not necessarily lead to genetic uniformity on the paired chromosomes within each animal (homozygosity). The presence of a certain proportion of homozygous individuals within a population is advantageous in that it permits genetically recessive traits to be tested against environmental demands. This genetic characteristic is achieved in *P. aurelia* by a process called *autogamy*.

In autogamy the micronucleus in a single animal undergoes the reduction divisions of meiosis while the macronucleus disintegrates. Three of the four haploid products of meiosis also disintegrate, and the remaining one duplicates itself by a mitotic division. These two haploid nuclei fuse to form a new diploid micronucleus. This structure now behaves like the micronuclei in animals that have recently separated after conjugation; it provides material for the new macronucleus. The major differences between autogamy and conjugation, then, are that autogamy takes place within the individual animal and that it results inevitably in a homozygous chromosome condition.

Let us now summarize briefly the relations among binary fission, conjugation, and autogamy, with an eye toward providing sharp criteria for the distinction between sexual and nonsexual reproduction. Binary fission involves neither genetic exchange between individuals nor systematic genetic recombination within a single individual; it is a very clear form of asexual reproduction. Conjugation involves genetic exchange and thereby provides for genetic recombination. Experts in this field generally agree that conjugation is a form of sexual reproduction (Caspari, 1965; Dougherty, 1955; Moment, 1958; Srb, Owen, and Edgar, 1965). But what about autogamy? Here is a process that provides for genetic recombination but does not include genetic exchange. If genetic recombination is considered a sufficient condition for sexual reproduction (Mayr, 1963), then we ought to include autogamy in the category of sexual reproduction. But if transfer of genetic material from one organism to another is taken to be a necessary condition for sexual reproduction (Dougherty, 1955), then autogamy cannot be included in that category.

REPRODUCTION IN HYDRA

The hydra is a genus of coelenterate that ranges in length from 3 mm to 5 cm and is found in bodies of fresh water in virtually all parts of the world. It receives its name from the mythical nine-headed monster slain by Hercules. Although each hydra has only one head, numerous long tentacles extend from it. The tentacles are used by the animal to capture prey. Once the prey is entwined, highly specialized harpoon-like structures (nematocysts) are ejected from the tentacles into it. A paralyzing poison is then secreted from the nematocyst. This unusual method of prey capture has been the subject of intensive experimental study (Brien, 1960; Lentz, 1966).

The reproductive adaptations of the hydra are also of substantial scientific interest. Figure 2.11 shows a hydra in the final phase of its process of nonsexual reproduction. New hydra are formed out of buds on an area of the body called the *budding region*, which is located between the long cylindrical stomach and the peduncle. The peduncle is connected to the base of the animal, which makes contact with the substrate. As illustrated in Figure 2.11, several new hydra may be formed at one time from cells in the budding region.

Budding is the typical form of reproduction in hydra, and under constant laboratory conditions colonies of animals may be maintained for a year or more without any appearance of deviation from this asexual process (Burnett and Diehl, 1964). Under certain circumstances, however, budding declines or stops, and each hydra begins to develop gonadal cells. In some species an individual develops both ovaries and testes; in other species the sexes are separated. But in either case, the

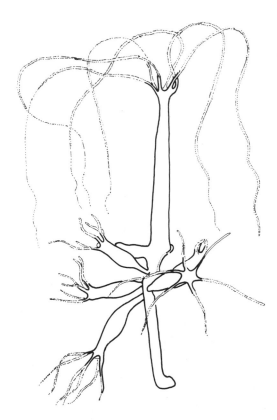

FIGURE 2.11 *Hydra fusca* with several buds. From Brien, P. (1960) The freshwater hydra. *American Scientist* 48:461–475.

gonadal tissue develops along the outer (epidermal) wall of the animal. When an ovum is mature the epidermis immediately above it ruptures, leaving the egg exposed to the water but secure in a shallow cup. At about this time sperm erupt from the testes into the water and eventually reach the egg and fertilize it. As the fertilized egg develops, it becomes surrounded by a protective shell that eventually separates it completely from its parent. The encapsulated embryo then remains dormant for a variable period of time. When it finally emerges it has become a fully mature hydra (Lentz, 1966; Loomis, 1959).

We are now in a position to consider two important questions about the hydra's reproductive processes. First, what are the stimuli and mechanisms that trigger and effect the shift from asexual to sexual reproduction? Second, what is the general evolutionary significance of the changes in reproductive mode and of what value is it to hydra to shift its sexual gears?

Our understanding of the mechanisms that prompt the change to sexual reproduction was forwarded substantially by the ingenious experiments of Loomis (1959). To begin with, he showed that the shift was caused by changes in the animal's environment. Sexual reproduction in this species is a temporary adaptation to changing external conditions. Moreover, he believed he had isolated a single, simple stimulus that was necessary and sufficient for the change: an increase in the pressure of the carbon dioxide surrounding the animal. The elegant simplicity of this result led Loomis to call carbon dioxide the "sex gas" of hydra.

One of the nice features of this result was that it could be interpreted in evolutionary terms. One of the determinants of the quantity of carbon dioxide in the environment is the number of individuals existing in that environment, because each animal gives off carbon dioxide as a product of respiration. Also, if the volume of water within which a stable population lives decreases, then the concentration of carbon dioxide increases. In either case the level of carbon dioxide reflects the degree of crowding of hydras in the environment. As pointed out, sexually produced hydras can exist in an encysted form for extended periods; in this way they might survive adverse environmental conditions (overcrowding, diminished living space) that were heralded by the gradually increasing pressure of carbon dioxide.

Subsequent research has suggested that increased carbon dioxide pressure may not be the single necessary stimulus for sexual reproduction. Burnett and Diehl (1964) have suggested that sexual reproduction occurs when normal growth processes are inhibited. There are several kinds of environmental changes that might inhibit normal growth, for example a lack of adequate food. The details of the mechanisms by which sexual reproduction is initiated remain unclear. Nevertheless, it is obvious that in hydra the shift from asexual to sexual reproduction is a behavioral adaptation, under the control of the environment, which enhances the survival chances of the species.

FORMS AND FUNCTIONS OF PARTHENOGENESIS

Parthenogenesis (Greek *parthenos* [virgin], *genesis* [origin or generation]) refers to the development of an offspring from an unfertilized ovum. There are several detailed reviews of parthenogenesis in the literature (Beatty, 1967; Suomalainen, 1950; White, 1954), to which the reader is referred for a more complete treatment than we shall provide here.

It is simple enough to understand the fundamental definition of parthenogenesis, but it is much more difficult to appreciate the myriad forms that the basic process takes in different animal species. There is a rather large list of technical terms that has come to be used in discussing various aspects of the topic. We will present these terms in this discussion and hope that the reader will neglect them should they get in the way of his understanding the basic ideas.

Arrhenotoky

To begin with, there is one form of parthenogenesis (*arrhenotoky*) in which the unfertilized egg always develops into a male animal. This is the case, for example, in the insect order Hymenoptera, which includes the honey bee. Thus the queen bee possesses ova, all of which have the haploid (unpaired) chromosome number. During her mating flight the queen copulates with a number of male (drone) bees, which deposit sperm cells (also haploid) into her spermatheca. When the queen subsequently returns to the hive and lays her eggs into the cells built by the worker bees, she allows some of the eggs to be fertilized before deposit and others to pass into the cell unfertilized. The fertilized eggs develop into the new generation of worker (genetically female) bees, and the unfertilized eggs develop into drones. A number of environmental and physiological factors determine the relative numbers of fertilized and unfertilized eggs laid (Eckert and Shaw, 1960). The parthenogenesis of the males is called arrhenotoky from the Greek *arrheno* (male) and *tokos* (offspring).

There are four points about this form of parthenogenesis that need to be emphasized. First, the egg out of which the male animal develops initially has unpaired chromosomes; hence there is a sense in which arrhenotoky could be called *haploid* parthenogenesis. However, this would be an oversimplification if taken literally because the chromosome number may be increased by several factors (become polyploid) in the somatic cells of adult males that were produced through arrhenotoky Thus, arrhenotoky is haploid parthenogenesis only to the extent that the arrhenotokously produced male began development as an ovum with the haploid chromosome number.

Second, arrhenotoky is available in each generation of the species. Thus every generation of queen honey bees produces the next generation of males by depositing unfertilized eggs and the next generation of females by depositing fertilized eggs. In this sense parthenogenesis is not a *cyclic* phenomenon across generations of this species, nor is it an *obligatory* means of reproduction within a generation. We can say then that across generations the parthenogenesis is *complete*,[9] and within a generation it is optional or *facultative*.

Third, it is clear from the foregoing discussion that arrhenotoky is a part of the general process of sex determination in the species in which it occurs. To know that a honey bee is a male is to know that he was produced parthenogenetically, and vice versa. This relationship is not so tight in other forms of parthenogenesis.

Finally, arrhenotoky is a relatively rare occurrence within the animal kingdom; it has arisen only seven times in the course of evolution. However, lest the phrase *relatively rare* be misunderstood, we need to point out that arrhenotoky apparently does occur in all hymenopteran species, which number more than 100,000.

Thelytoky

Far more common than arrhenotoky are the cases in which the unfertilized egg always develops into a female (*thelytoky*) or either a male or a female, depending on other conditions (*deuterotoky*). We will limit our treatment to a consideration of thelytoky. Moreover, in order fully to grasp the details of thelytokous parthenogenesis, it is necessary to appreciate its development. A presentation of these details is beyond the scope of this book; hence the following presentation is merely the barest outline of an incredibly rich and complex set of data.

We begin by noting that the ovum that will develop into a female originally has paired chromosomes (the diploid number).[10] This contrasts with the case of arrhenotoky. Given this as a starting point, there are two different ways in which the ovum may begin parthenogenetic development. In one case (*apomictic* or *ameiotic* parthenogenesis) the chromosomes within the nucleus of the ovum do not undergo a reduction in number; that is, meiosis does not occur. The changes that take place in the cell resemble the mitotic divisions that occur in regular somatic cells. The important thing to understand about this process is that it leaves no room for genetic recombination, so that an offspring is likely to be genetically identical to its mother.

In the second case (*automictic* or *meiotic* parthenogenesis) there is a reduction in chromosome number, but this reduction is followed by a refusion of chromosomes early in the development to restore the diploid chromosome number in the cells that are subsequently proliferated. There is a wide variety of different mechanisms that produce refusion in

[9]This use of the phrase *complete parthenogenesis* is slightly different from that of some other authors.
[10]In some cases the egg may be polyploid.

different species. For our present purposes, however, we need note only that these processes allow for genetic recombination; hence they stand in contrast to the ameiotic case.

It should be clear that if ameiotic parthenogenesis were both complete and obligatory for an animal population, that is, if it occurred in every generation and were the only means of reproduction available within a generation, then that population would consist of more or less identical individuals that would have no way of exchanging genetic material. The population would exist in a state of great genetic stability, but it would be relatively inflexible in the face of rapidly changing environmental demands. Such populations do in fact exist (e.g., the "walking stick" insect *Saga pedo*). However, this condition is never the only means of reproduction in large groups of related species.

We have pointed out that the genetic consequences of meiotic parthenogenesis are different from those of the ameiotic variety. Naturally occurring meiotic parthenogenesis is known to take place in a vertebrate, the lizard *Lacerta saxicola armeniaca*. Another vertebrate example can be found in the Amazon Molly fish, *Mollienisia formosa*. We will discuss this case in detail later. For now we want to emphasize that both meiotic and ameiotic parthenogenesis are often cyclic characteristics of reproduction in a species; that is, some generations reproduce parthenogenetically and others bisexually. The best example of cyclic parthenogenesis occurs in well-known insects, the aphids.

During the spring and summer aphids reproduce through obligatory thelytoky: parthenogenesis is the only means of reproduction available, and all the offspring are female. However, the structure, behavior, and habitat of the several thelytokous generations are not identical. In *Tetraneura ulmi*, for example, some of the generations are winged and some wingless; some live on (and damage) elm trees, and others live on grass. Nevertheless each individual is a reproductively capable female that produces only other females. Then in one generation there is a change in the pattern of reproduction. Obligatory thelytoky is replaced by obligatory deuterotoky: both males and females are produced parthenogenetically. In some species a female can produce both males and females, whereas in other species some females produce only males and others produce only females. But in either case pairs of individuals from this new, bisexual generation copulate. The fertilized eggs are laid in the winter. The offspring are always female, and they form the reproductive foundation of the new series of parthenogenetic generations.

Several authors have pointed out that cyclic parthenogenesis provides an evolutionarily advantageous compromise between obligatory bisexual reproduction and obligatory thelytoky. Obligatory thelytoky, whether meiotic or ameiotic, allows every individual to be a complete reproductive unit; the number of individuals in the population can grow very rapidly in a short time. On the other hand, the population lacks the flexibility of a population in which the genotypes of the next generation are produced by combining half of the genotypes of two individuals. Cyclic parthenogenesis combines both advantages during the course of a year.

Gynogenesis (Pseudogamy)

The term *gynogenesis*, or *pseudogamy*, refers to a form of parthenogenesis in which a sperm cell activates an ovum into development but does not fertilize it. It is a prevalent phenomenon among several groups of worms, particularly those in which individuals are hermaphroditic. It also occurs in a very interesting fashion in a vertebrate population, the Amazon Molly (*Mollienisia formosa*), a small fish from the waters of southern Texas and the east coast of Mexico. One feature of this group of animals, which it shares with many parthenogenetic species,[11] is that it is composed virtually entirely of females. The question then immediately arises: if a species composed entirely of females needs sperm cells to initiate the developmental process in its eggs, where does it get the sperm?

The original work of Hubbs and Hubbs (1932) suggested an answer. First, *M. formosa* is a population that originated as a hybrid between two related species, *M. sphenops* and *M. latipinna*. Second, *M. formosa* always lives sympatrically with one of these "parent" species. Hubbs and Hubbs suggested that the *M. formosa* females "borrow" the sperm from the males of the sympatric species. Within the framework of our earlier discussion this means that the reproductive isolation between *M. formosa* and the other species is certainly not of a premating variety; if it were, *M. formosa* would immediately become extinct. Hubbs and Hubbs were able to demonstrate experimentally that mating occurs readily between *M. formosa* and males of the other two species.

Further experimentation by Hubbs and other investigators (Meyer, 1938; Kallman, 1962) seems to confirm this interpretation of the Amazon Molly's mode of reproduction. The work of Kallman was particularly interesting because it utilized a test for histocompatibility: transplants of tissue were made between animals in such a way that predictions could be made about which grafts would succeed and which would fail if the gynogenesis hypothesis were correct. The results of these experiments confirmed the hypothesis.

Summary—Parthenogenesis in Vertebrates

We have referred to two vertebrate species that are normally parthenogenetic: the lizard *Lacerta saxicola armeniaca* and the Amazon Molly. Although apparently parthenogenetic offspring have also been reported

---

[11]The reader may question the use of the word *species* in the case of a noninterbreeding population. White (1954) has argued that complete parthenogenesis is a relatively recent development in the history of populations that currently utilize it; these populations were probably originally bisexual and still possess morphological characteristics that make them distinct. Hence, the term *species* is still appropriate.

for some other species (amphibia and reptiles), there are no known examples of naturally occurring parthenogenesis in species of birds or mammals. However, there has been considerable research on the experimental stimulation of parthenogenetic development in all the vertebrate classes (Beatty, 1967). Viable parthenogenetic offspring have been produced experimentally in at least two species of frog. But perhaps the best-known example is the Beltsville Small White turkey. By a program of selective breeding, Olsen and his colleagues (e.g., Olsen and Marsden, 1954; Olsen, 1962) have been able to increase substantially the proportion of eggs that will show at least some parthenogenetic development. Some viable adults have been produced in this manner; all have been males.

There has been no experimental success in bringing a parthenogenetically developing mammalian ovum to term. Beatty (1967) provides a discussion of the problems involved in evaluating the relevant data for humans as well as for other mammals.

HERMAPHRODITISM

The word *hermaphroditism* comes from the Greek *Hermaphroditus*, the name of the legendary son of Hermes and Aphrodite who had both male and female sexual capacities. We will limit our use of the term for the time being to the description of animals that function completely both as males (sperm bearers) and females (egg bearers) during their lifetimes.

Functional hermaphroditism is almost always the anatomical condition in several animal phyla: Ctenophora ("comb bearers," also called comb jellies, a marine phylum), Bryozoa ("moss-like animals," a minor freshwater and marine phylum) and Platyhelminthes (flatworms, including the well-known planarians, the flukes, and the tapeworms). In the phylum Annelida there are two major hermaphroditic groups: earthworms and leeches. In the phylum Mollusca some of the land and water snails (opisthobranchs and pulmonates) exhibit hermaphroditism. The largest animal phylum, Arthropoda (insects, spiders, crustacea, centipedes, etc.), contains very few hermaphroditic groups. And finally, in our own subphylum Vertebrata, functional hermaphrodites are found only among some fish and amphibians. Thus, no reptilian, avian, or mammalian species is normally hermaphroditic (Armstrong and Marshall, 1964; Hyman, 1967; Mayr, 1963; Moment, 1958; White, 1954).

From this brief overview we conclude that functional hermaphroditism exists primarily in the simpler or phylogenetically older animal groups. In the more complex or phylogenetically younger groups sperm and eggs are typically separated into different individuals. The technical term for the separation of sex cells into different individuals is *gonochorism*.

Some hermaphroditic individuals are capable of fertilizing their own eggs and some are not. Whether self-fertilization occurs in a species depends in part on the means the species employs to bring sperm and egg together. In the comb jellies, for example, self-fertilization is possible because the gametes are shed together through a series of canals into the main gastrovascular cavity of the animal and thence out of the animal's mouth into the sea. This process also permits the gametes of separate individuals to contact each other; thus cross-fertilization is possible as well (Borradaile and Potts, 1935; Moment, 1958).

The parasitic tapeworms (class Cestoda of the flatworm phylum) offer an interesting example of self-fertilization. The individual worm grows by budding off new segments, called *proglottids*, which remain attached to one another as in a chain. Each proglottid is a complete reproductive unit containing ovaries and testes. Sperm and eggs are shed from the gonads through separate channels (sperm duct and vagina) into a genital pore on the edge of the proglottid and thence into the intestine of the host. However, an egg cannot be fertilized after it has left the proglottid because it has been encased in a shell during its trip from the ovary. A sperm must fertilize the egg by traveling along the vagina to the oviduct. Sperm from the same or a different proglottid may make this trip. Normally it is sperm from a different proglottid that penetrate the vagina because the worm has partially doubled back on itself, and the genital pores of several proglottids lie pairwise against each other. Additionally, cross-fertilization may occur if two worms have attached themselves adjacently on the intestinal wall of the same host. In any case, the fertilized egg, protected by its shell, is carried out of the intestinal cavity of the host in the fecal material; one stage of the parasite's life cycle has been completed (Buchsbaum, 1938).

Self-fertilization may also occur in vertebrate hermaphrodites. The sperm and eggs of individual fish from the sea bass species *Serranellus subligarius*, which lives off the Florida coast in the Gulf of Mexico, can unite to produce normal offspring. Normally, however, these fish shed their gametes into the water after pairs or small groups of them have engaged in a display of courtship behavior. During courtship one fish takes the male role and emits sperm while the other fish sheds eggs. The synchronization of spawning in this way increases the likelihood that eggs will be cross-fertilized (Clark, 1959). Internal self-fertilization has also been described in the cyprinodont fish *Rivulus marmoratus* (Harrington, 1961).

As a rule, self-fertilization does not occur in hermaphroditic species that practice copulation, for example the planarian flatworms, the earthworms, and the snails (for exceptions see Hyman, 1967, p. 599). In some of these species copulation results in *reciprocal* cross-fertilization: sperm is transferred from each animal to the other. Some very unusual and interesting examples of copulatory behavior are to be found among these hermaphroditic species; we will describe the behavior of earthworms and great slugs.

The Copulation of Earthworms (*Lumbricus terrestris*)

An obvious morphological feature of the common earthworm is the collarlike thickening of the body that begins at approximately the thirtieth segment and extends back for five or six segments. This ring of glandular substance is called the *clitellum*, and it figures importantly in the copulatory behavior of the species. When two worms come to copulate they appose their ventral surfaces, with their heads pointing in opposite directions, so that the clitellum of each worm lies over the sperm receptacles of the other. The openings of the receptacles are between segments nine and eleven. Once situated in this way, the worms secrete a tunnel of mucus that encapsulates them from the ninth segment to the end of the clitellum. Each worm then emits sperm from an opening on its fifteenth segment. Muscular contractions of the worms cause the sperm to migrate backward along grooves until they reach the sperm receptacles of the other worm, which they enter. The worms separate after the sperm have been transferred.

The next stage of the reproductive process occurs when the clitellum begins to secrete a cylinder of mucus that moves forward toward the worm's head. As this tube passes over segment 14 it collects eggs from the opening of the oviduct; similarly, the tube receives the sperm that were deposited in the receptacles between the ninth and eleventh segments. Fertilization takes place inside the ring of mucus, which continues to move forward until the worm shrugs it off over his head. The ring seals at both edges to become a suitably closed environment for the development of the fertilized eggs (Buchsbaum, 1938; Moment, 1958).

The Copulation of Great Slugs (Limax maximus)

The mating behavior of the 6-inch-long great slug, a native British species also found in North America, offers one of the more dramatic examples of sperm transfer in the animal kingdom. During the autumn breeding season a pair of slugs meet on the underside of a tree branch. After the animals have traded caresses with their tentacles, they form a head-to-tail ring and circle each other while laying down a thick blanket of mucus. The circling behavior may last for as long as 90 minutes; during this time each animal becomes covered by the mucus strands secreted by the other. Then, rather suddenly, the slugs release their grips on the branch and fall off, while at the same time they wrap their bodies around each other. Their secreted mucus follows them like a rope, and because it is firmly attached at the other end to the tree branch it stops their fall before they hit the ground. Thus suspended as much as 18 inches below the branch, the slugs begin to copulate.

As in the earthworms, copulation consists of an exchange of sperm. The transfer takes place when each slug extends his copulatory organ downward and entwines it around the organ of his partner. When fully extended the organ is approximately 4 inches long (i.e., two-thirds the length of the slug). Once entwined the two organs undergo a remarkable transformation of shape (see Figure 2.12). The details of this configuration are apparently important for successful sperm exchange.

After sperm have been exchanged the slugs unwrap their copulatory organs and retract them into their bodies. Then one at a time they ascend the rope of mucus, move along the branch of the tree to the trunk, and descend the trunk. Once on the ground each animal finds an appropriate spot in which to lay its fertilized eggs (Chace, 1952).

Normal or functional hermaphroditism is not limited to the *simultaneous* hermaphroditism that we have discussed so far. Successive herma-

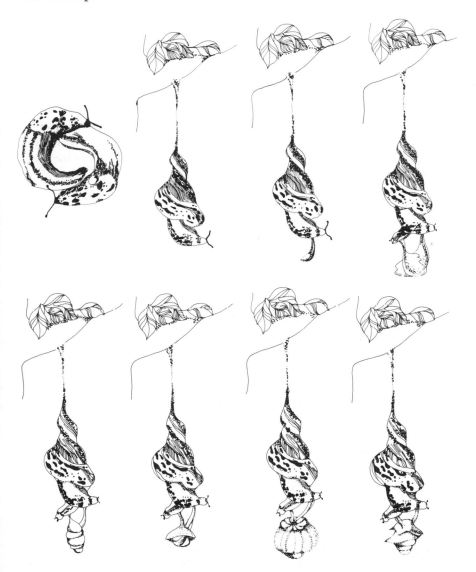

FIGURE 2.12 Aerial copulatory sequence of the great slug, *Limax maximus*. Once suspended, the slugs extend their cylindrical copulatory organs. The organs become club-shaped at the tip, then frilled, and finally twist together in a tight spiral. The spiraled organs assume an umbrella shape with sperm transfer effected by the free edges. Redrawn from Chace, L. (1952) The aerial mating of the Great Slug. *Discovery* 13:356–359.

phroditism also occurs; in this case the animal functions first as one sex and later in life as the other. If the animal is first a male, the condition is called *protandry*; if it is first a female, the condition is called *protogyny*. Successive hermaphroditism is particularly common among species of hermaphroditic fish. In one fish family (*Sparidae*, the porgies and breams) protandrous, protogynous, and simultaneous hermaphroditism are all represented (Atz, 1964).

A hermaphroditic species is not necessarily limited to self- or cross-fertilization as the means of reproduction. For example, planarians may reproduce either by fragmentation or by hermaphroditic cross-fertilization, and gynogenesis may occur in other hermaphroditic worms. In many instances the mode of reproduction is determined by the condition of the animal's environment at the time. We will discuss the environmental regulation of sexual behavior and reproduction in Chapter 3.

Finally, we need to point out that *hermaphroditism* has a broader scientific usage than we have so far accorded it. In addition to its meaning of *functional bisexuality* (simultaneous or successive) as a typical or natural condition of single individuals within a species, the word is also often used to refer to nonfunctional or pathological conditions of gonadal or genital development. And in the chapter on human sexual behavior we define hermaphroditism very broadly in order to allow convenient classification of the various pathological conditions that may arise during the development of human gonadal and genital structures.

RECAPITULATION—NONSEXUAL AND SEXUAL REPRODUCTION

We offer the following definition of sexual reproduction: it is any form of reproduction that allows for genetic segregation and/or recombination. In terms of the examples discussed, we include conjugation, autogamy, meiotic thelytoky, arrhenotoky, and self- and cross-fertilizing hermaphroditism.

Nonsexual reproduction is any form of reproduction that does not include these genetic processes. In addition to the examples of binary fission, ameiotic thelytoky, and budding presented, the category includes sporulation and fragmentation.

From our definition of sexual reproduction it follows that the most obvious forms of sexual behavior, copulation and spawning, are neither necessary nor sufficient conditions for sexual reproduction. The examples of meiotic thelytoky and self-fertilizing hermaphroditism show the lack of necessity, and ameiotic parthenogenesis triggered by pseudogametic mating (as in some nematode species) shows the lack of sufficiency. However, these examples are relative rarities in the animal kingdom. In general, the success of sexual reproduction in a species usually depends on a form of social behavior that is involved directly with the transfer of genetic material, that is, obvious sexual behavior.

Chapter 3
Environmental Determinants of Sexual Behavior

Imagine that you are sitting in front of an aquarium in which fish are swimming among the plants. You have at your disposal a miraculous magnifying device that permits you to see the details of the activities in the aquarium without losing sight of the edges or borders. As you turn up the gain on the magnifier you see each animal and plant represented as a collection of separate units—first tissues, then cells, then molecules, and so on. At every twist of the magnification knob a new collection of units is seen to constitute the earlier smallest visible unit. The final resolving power of the magnifier will not show things as ordinarily perceived, but rather energy relationships in particular configurations. At this level the aquarium is an energy system in which the concepts of tissue, cell, and so on are no longer explicitly defined. These concepts become explicit again as the resolution of the magnifier is gradually turned down. Where one pauses for a closer look, to describe, classify, or attempt to manipulate what one sees, is a matter of taste as well as utility. There is just one aquarium and, in some sense, just one thing going on.

Repeated observations at different levels of magnification make it clear that a useful distinction can be made in terms of events that occur inside and outside sets of "energy fences" that correspond to what we generally think of as *skin*. Individual plants and animals are located and delimited by the configuration of their skins, that is, boundaries making reasonably clear alterations of energy relationships. We call events occurring inside the boundaries properties of the *organism* and events outside properties of the *environment* of that organism. If we concentrate our attention on the influence that events outside a boundary have on events inside, then we are studying the influence of environment on an organism. If we pay particular attention to those environmental influences that cause the organism in turn to alter the environment, then we are dealing with environmental influences on the organism's *behavior*. It is crucially important to understand that the behavior of an organism can modify the environment in ways that will in turn modify subsequent behavior. As a bird builds a nest in the pattern typical of its species, for example, the changing appearance of the nest produces changes in the bird's behavior that further modify the shape of the nest, and so on. Organisms and environments continually adjust to each other. Our task in this chapter is to consider environmental influences relevant to sexual behavior.

VARIETIES OF ENVIRONMENTAL EVENTS

We need to make distinctions among the kinds of environmental events that influence sexual behavior. To begin with, some events are relatively brief and localizable, for example, a light turned on overhead for 5 seconds. Such relatively discrete events are called *stimuli*. Other environmental events occur relatively slowly and surround the organism, for example, the increase in average ambient temperature from winter to spring. In these cases the gradually changing *conditions* may result in behavioral changes, even though we cannot point to a discrete stimulus.

Some environmental events, whether stimuli or conditions, may be importantly related to behavior that does not appear for a relatively long time after the event. Some trace or memory of the event must remain within the organism. In such cases we speak of the effects of particular *experiences* of the organism as being influential in determining its behavior.

Finally, from the viewpoint of the individual organism, we can distinguish environmental events that are directly related to other organisms from those that are not. In the first case we refer to the *social* environment and in the second to the *physical* environment. In this chapter we will discuss the importance for sexual behavior of stimuli conditions and experiences, derived from both physical and social environments.

VARIETIES OF EFFECTS ON SEXUAL BEHAVIOR

Just as there are several classes of environmental events, there are also classes of effects these events have on sexual behavior. First, an environmental change may alter the *mode of reproduction* in some species from nonsexual to sexual or vice versa. The sexual behavior associated with gamete transfer may thus be present or absent as a function of changing environmental events. We provided an example of this effect in Chapter 2, when we described the change to sexual reproduction in *hydra* due to increased carbon dioxide tension in the surrounding water.

Second, the *reproductive cycles* of sexually reproducing species are often related to seasonal or daily fluctuations in environmental conditions: day length, temperature, rainfall, or other variables.

Third, environmental events serve to bring reproductively capable individuals together, often from substantial distances; they serve an *assembling function.* Stimuli involved in assembling may also serve as species indentifiers; examples of such stimuli were discussed in Chapter 2 in the context of reproductive isolation. Stimuli regulate the interaction of male and female during courtship and mating. The complex reciprocal courtship displays of some species are examples of highly elaborated communication systems that synchronize the activities of each animal so that the probability of fertilization is maximized.

Fourth, environmental events can affect quantitative measures of *sexual opportunity* and *performance.* The social structure of some populations puts definite constraints on random mating. And in several species the number of ejaculations achieved by males depends greatly on how many females are available and the sequencing of their availability.

Fifth and finally, *sexual behavior pathology* may result from certain cnvironmcntal events, particularly experiences encountered relatively early in life. The expression of biologically functional sexual behavior in the adult depends on adequate environmental support during earlier stages of development.

PLAN OF THE CHAPTER

The rest of this chapter is organized under five major headings, corresponding to five of the aspects of sexual functioning we have just listed: reproductive cycles, courtship and mating sequences, control of mating opportunities, control of extended copulatory performance, and abnormal sexual responding. In the section on reproductive cycles we describe the various forms of reproductive cyclicity and their dependence on different kinds of environmental events. The section on courtship and mating sequences describes the roles of stimuli in eliciting, regulating, and terminating sexual behavior and gives special attention to the problem of the regulation of aggressive tendencies that arise in the sexual context. Our discussion of the regulation of mating opportunities is limited to a discussion of the sexual consequences of social organization among different species of African baboons. Under the heading of extended copulatory performance we present data from experiments with some domesticated species and offer a speculation concerning the significance of these results for other species. And finally, in the section on abnormal sexual responding, we provide a discussion of what *natural* means in the study of animal behavior, introduce the concept of imprinting, then describe the results of experiments involving the social isolation of immature rodents and monkeys.

REPRODUCTIVE CYCLES

Reproduction takes time and energy. Animals are not continuously involved in the making of progeny. The phrase *reproductive cycle* refers most generally to the periodic fluctuations that occur in the reproductive processes of individuals and species. A considerable amount of information has been accumulated concerning the environmental determinants of these fluctuations in vertebrates. See Amoroso and Marshall (1960), Asdell (1964), and Bullough (1961) for further information and bibliography. We will concentrate primarily on birds and mammals.

ANNUAL REPRODUCTIVE CYCLES

The most obvious reproductive cycle is the annual breeding season of wild birds living in temperate zones. Almost all such species of birds mate and lay their eggs during the spring and early summer. The distinctiveness of the four seasons increases with the distance away from the equator; hence the breeding seasons also tend to be shorter and more compactly distributed. This is illustrated in Figure 3.1, which shows the relative likelihoods that eggs will be laid in particular months of the year in particular latitudes. The figure represents an agglomerate picture for many species. Although equatorial birds in general lay eggs at all times of the year, individual species may have more restricted breeding seasons.

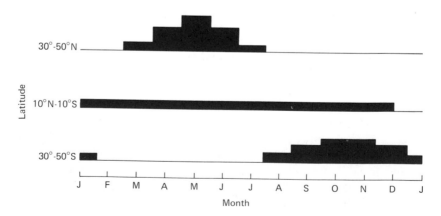

FIGURE 3.1 Changing distributions of egg-laying times as a function of latitude. Modified from Bullough, W. (1961) *Vertebrate Reproductive Cycles*, 2d ed. London: Methuen.

The breeding seasons of many mammalian species are also closely related to seasonal changes. Some species of hooved animals have breeding seasons in the fall, for example, the elk (*Cervus canadensis*), the red deer (*Cervus elaphus*), and the bighorn sheep (*Ovis canadensis*). And several species of wild canines breed in mid-winter (Asdell, 1964).

Why do these annually breeding species mate at different times of the year? This question has no single, simple answer, but we can gain some insight into the interaction of sexual behavior and physiology with environmental conditions by considering the relations between the time of the breeding season and the length of the gestation period in different groups of mammals. For example, Table 3.1 presents the relevant data for several species of deer (family Cervidae), cats (Felidae), dogs (Canidae), and some seals, sea lions, and walruses (suborder Pinnipedia, families Otariidae and Odobenidae). Most of the deer mate in the fall, the cats in the spring, the dogs in the winter and spring, and the pinnipeds in the summer. The cats and dogs have relatively short gestation periods, the deer are intermediate, and the pinnipeds take the longest. The net result of these differences is that the young of all these species are born during the spring and summer, that is, in the mildest weather. There are many additional examples (Asdell, 1964).

There are some exceptions to the rule that annually breeding mammals living in temperate climates give birth in the spring and summer. A striking example is some of the hibernating bears, which gave birth in mid-

TABLE 3.1. Breeding Seasons, Gestation Periods, and Birth Seasons for Several Mammalian Species

Species	Breeding Season	Gestation Period	Births
Fallow deer	October–November	8 months	June–July
Wapiti (elk)	September–October	8 months	May–June
Hanglu	October	6 months	April
Japanese deer	September–November	8 months	May–June
Mule deer	November–December	6 months	May–June
Moose	September–October	8 months	May–June
Reindeer	September–October (Siberia)	8 months	May–June
Roe deer	July–August	10 months	May–June
European wildcat	March–May	2+ months	May–July
Bobcat	February–April	1.5–2 months	May–June
Puma	March–June	3 months	June–September
Northern coyote	February–April	2 months	April–June
Wolf	Late January–March	2 months	Late March–May
Arctic fox	April–May	2 months	June–July
Alaska fur seal	July–August	12 months	July–August
California sea lion	June 15–July 15	11.5 months	June
Steller's sea lion	May–June	12 months	May–June
Walrus	June	11 months	May

(Data collected from Asdell, 1964.)

winter. Birth occurs in the hibernation cave after a gestation period of five to seven months. The mother and young emerge from hibernation in the spring (Burton, 1967). Exceptions do not seriously weaken the importance of the principle that for many species there is a selective advantage to giving birth in periods of clement weather and adequate food supply.

Next we need to consider the nature of the environmental events that are related to the onset of the reproductive process. Speaking casually, we might ask how animals know that their proper breeding season is upon them. More technically, we ask what conditions in the environment synchronize or trigger the physiological events required for the initiation of courtship and mating behavior. For the time being we will continue to limit our attention to the case of annual breeding seasons.

Two obvious physical signs of the changing seasons are day length and temperature fluctuations. Members of a given species could be sensitive to neither, either, or both. Because day length and mean temperature are highly correlated, it is difficult to separate their effects except through controlled experimentation.

Some of the earliest experiments of this sort were performed by Rowan, who worked originally with the Canadian junco, *Junco hyemalis* (Rowan, 1938). Normally this bird species migrates south from Canada in the fall, winters in the United States, and returns to Canada in the spring to breed. The testicles of the males regress and become nonfunctional in the fall and grow again in the spring. Rowan captured juncoes in Canada on their southward migration, caged them outdoors, and then exposed them to gradually increasing day lengths by providing artificial light after the natural sunset. Thus these birds were living in spring light conditions but fall temperature conditions.

The results were clear. After six weeks of increasing "daylight," the testicles had begun to grow. At the end of three months, approximately January 1, the testes weights were equivalent to those of males arriving from the south in the spring. Birds in control cages, in which there was no additional light, showed no testicle growth. Hence the changes in the experimental animals could not be attributed to an internal, automatic cycle of testicle growth. Moreover, decreasing temperatures (down to $-51°$ F) clearly did not inhibit testicular growth in animals provided with increasing light; temperature could be discounted as a relevant variable in this case.

Similar results have been reported for a number of avian and mammalian species (Amoroso and Marshall, 1960; Aronson, 1965; Bullough, 1961; Hafez, 1968; Rowan, 1938; Thorpe, 1967). Changing day length can affect the reproductive and neural physiology of both males and females. Aronson (1965) summarized the effects of changing conditions of illumination into four categories: increased anterior pituitary gland activity that leads to gonadal development (see Chapter 5); increased anterior pituitary gland activity that leads to ovulation from mature ovaries; induction of egg placement and sperm emission (in spawning species); and direct effects on the neural machinery associated with courtship and mating. In all cases we need to emphasize that the effects of changing conditions of illumination must be understood as a *signal interpreted by members of the species.* In some species increases in light lead to reproductive capability and sexual behavior; in others decreases in light produce the same effects. The "interpretation" of the signal is the set of neurological processes that transform information about the changing conditions of illumination and thus mediate its effects on reproductive processes.

Clear examples of natural temperature changes leading to the onset of the annual breeding season are found among the cold-blooded (ectothermic) vertebrates (Aronson, 1965). Among warm-blooded (endothermic) animals, the male ground-squirrel shows a cycle of testicular activity that is related to changes in temperature and independent of changes in illumination (Amoroso and Marshall, 1960). Moreover, it is well known that extreme temperatures or temperature changes can adversely affect reproductive function and behavior (Hafez, 1968; Schein, 1965).

Another significant environmental variable is rainfall. The breeding behavior of several species follows the first rains of the season. However, as Aronson (1965) has pointed out, and as we suggest in an example to follow, it is not necessarily rainfall per se that triggers courtship and mating. The addition of large amounts of water to the landscape alters the environment in many ways, for example, changing the chemical balance in rivers and ponds. Some of these subtler effects may be involved in the control of breeding seasons.

Annual Seasons in Primates

During the last decade we have become increasingly aware of the existence of annual breeding seasons in several species of nonhuman primates. Evidence collected in the 1930s, based primarily on monkeys living in zoos, indicated that nonhuman primates mated continuously throughout the year. This idea was probably accepted without serious criticism for two reasons: almost all primates live in the tropics (Napier and Napier, 1967), where breeding seasons are generally not so closely tied to the seasons; and man himself appears not to be a seasonal breeder. But we now have good evidence that several monkey species do have seasonal cycles, and in a small way, so do we ourselves.

We will take as an example the rhesus monkeys (*Macaca mulatta*) that have been brought to and now live freely on the islands of La Cueva and Guayacan, near Puerto Rico (Vandenbergh and Vessey, 1968). These islands, at latitude 18° N, experience only slight day length and temperature changes throughout the year. However, there is a significant annual cycle of rainfall (the rainy season lasts from mid-July to mid-November), and there is a cycle of leaf growth associated with the rainfall.

Vandenbergh and Vessey found that the frequency of copulation among the monkeys was closely tuned to the occurrence of the rainfall. Mating rates showed sharp peaks toward the end of the rainy season and were very low before and after. The gestation period of the rhesus is approximately 5.5 months. Hence most of the births occurred between May and July. This happens to be the time at which the amount of fresh vegetation is greatest; thus, the young are born at a time when the food supply is maximal.

Vandenbergh and Vessey compared their data with those collected at another location in Puerto Rico and in northern India. In all cases a strong correlation was found between periods of heavy rainfall and mating. From this they concluded that the annual cycle in rhesus was not an artifact of the particular environment in which their study group lived. A more substantial account of annual reproductive cycles in nonhuman primates has been presented by Lancaster and Lee (1965).

Cowgill (1966) has reviewed the monthly changes in human birth rates that occur in various parts of the world. Her data indicate that there is usually one major peak, and sometimes a minor peak, during the 12-month period. However, the timing of the peaks is not uniform throughout the world. For example, over a wide range of European countries the peak occurs in late winter and early spring; in the United States the major peak occurs in late summer and early fall. Although it is real, the peak in the United States is a relatively minor affair, as can be seen from Table 3.2. The table is based on approximately 4.25 million births recorded in 1959 by the National Office of Vital Statistics. The numbers represent relative changes in births from month to month; that is, a month that contained exactly one-twelfth of the annual total would be assigned a score of 100.0.

If we think in terms of conception dates instead of birth dates, we find that European conceptions are highest in late spring and early summer, whereas American conceptions peak in the winter.

How are we to consider these data for our own species? First, given the general tendency of vertebrate species to show marked seasonal variations in reproduction, we might ask why there is relatively little variation in the human case. Here we may note that our closest living relatives, the chimpanzees and gorillas, can conceive the year around; however, this does not mean necessarily that there is no seasonal peak

TABLE 3.2. Relative Birth Frequencies, United States, 1959

Month	Relative Frequency	Month	Relative Frequency	Month	Relative Frequency
January	97.4	May	93.3	September	107.8
February	99.6	June	98.0	October	102.5
March	99.6	July	103.6	November	99.3
April	95.3	August	105.1	December	99.8

(From Guttmacher, 1971.)

of births under natural conditions (Schaller, 1965; Van Lawick-Goodall, 1968). Further, man must be considered the most domesticated of animals, and a major effect of domestication is a spreading or blurring of annual reproductive cycles (Hafez, 1965). Improvements in shelter and nutrition relax the selection pressures of inclement weather and inadequate food sources on infant survival.

Second, how are we to account for the small but reliable differences in seasonal rates that exist within our own country? We can provide no satisfactory answer to this question. It seems unlikely that the relatively simple environmental variables we have heretofore considered for other species are sufficient to explain the observations.

Third, how are we to account for the different patterns of seasonal reproductive rates that characterize different geographic or cultural areas? Here again we cannot shed much light on the matter, although in at least a few cases there are well-known cultural factors, related to environmental variables, that can account for some of the observed data. For example, Cowgill (1966) points out that there are regions in Mexico where high birth rates occur approximately nine months after the time of annual corn planting. The time of planting is seasonally controlled, and there are ritual ceremonies associated with the planting that are conducive to sexual intercourse.

ESTRUS CYCLES

When a female mammal is sexually receptive she is said to be "in heat" or "in estrus" (from the Greek and Latin *oestrus*, meaning gadfly, sting, frenzy). The duration of the *estrus cycle* is the length of time between the onset of one period of receptivity and the next, when the female is not given the opportunity to copulate. Copulation typically disrupts the estrus cycle by producing pregnancy or pseudopregnancy.[1] An exception to this rule occurs in some marsupials, in which the gestation period is shorter than the estrus cycle (Asdell, 1964). But in general, the duration of the estrus cycle is the natural minimum time between periods of receptivity.

Estrus cycles relate to annual cycles in the following way. In some species (e.g., laboratory rodents), the females have estrus cycles throughout the year; they are *totally polyestrous*. Other species (e.g., horses, sheep, goats) have several estrus cycles during a restricted breeding season; they are *seasonally polyestrous*. And other species (e.g., bears, deer in temperate zones) come into heat just once during a restricted season. If the female is not inseminated at that time, there will not be another opportunity for a year; these females are *monestrous*.

A full understanding of estrus cycles depends on an appreciation of the physiological events that produce them. We describe these events in

---

[1]Pseudopregnancy refers to changes in hormone balance and uterine condition that are similar to those of pregnancy but that are not accompanied by the implantation of a fertilized egg and the development of a placenta. See Chapter 6.

detail in Chapter 5, but it will be helpful to introduce a few fundamentals now. To begin with, sexual receptivity in most female mammals is closely tied to the occurrence of ovulation: females mate when the likelihood of fertilization is greatest. In some species the occurrence of ovulation is relatively independent of the female's behavior during estrus (spontaneous ovulation); examples include the laboratory rodents and their wild relatives, cattle, sheep, goats, horses, and dogs. In other species, by contrast, ovulation does not occur unless the female copulates (coitus-induced or reflex ovulation); the best known examples are the rabbit and the cat.

The time between the onset of estrus and spontaneous ovulation has been determined for several species of laboratory rodents. It is 2 to 3 hours in the mouse, 6.5 to 10 hours in the rat, 8 to 9 hours in the hamster, and 10 hours in the guinea pig (Harvey, Yanigimachi, and Chang, 1961). As a point of general biology, one can see that there is a substantial advantage in having sexual receptivity precede the occurrence of ovulation. Note also that the induction of ovulation by copulation is an efficient mechanism for preventing gamete waste. The stimulation received from the male is a clear example of an environmental determinant of the reproductive process.

Table 3.3 represents the average durations of the estrus cycles of several species of totally and seasonally polyestrous species. Note that the cycle duration of the guinea pig is substantially longer than those of the other common laboratory rodents. This difference is related to a significant difference in the endocrinology of "long-cycle" and "short-cycle" rodents that we will discuss in Chapter 4. For now we point out only that for the initiation of a successful pregnancy, female rats and mice are dependent on the sensory stimulation accompanying copulation (Wilson, Adler, and LeBoeuf, 1965; Whitten, 1966). The physiological processes normally triggered by copulatory stimulation are also sensitive to olfactory stimulation; they are involved in the olfactory control of estrus cycles, which we discuss later in this chapter.

The definitive criterion of estrus is that the female accept the male for copulation. Yet as we have mentioned, copulation may disrupt estrus cycling by producing pregnancy or pseudopregnancy. Therefore it is not always possible to measure the duration of the cycle by exposing females

TABLE 3.3.   Estrus Cycle Durations (days)

Seasonally polyestrous	
Domestic sheep (coarse-wooled breeds)	16–17
Domestic goat	20–21
Domestic horse	22

Totally polyestrous	
Laboratory rat	4–5
Guinea pig	16–17
Golden hamster	4
Laboratory mouse	4–8

to active males. Fortunately there are both physiological and behavioral correlates of receptivity that can be measured easily. The females of some rat strains, for example, may be tested for receptivity by stroking their sides and genital areas by hand. If they are in heat they will elevate their hindquarters and deflect their tails to the side; this is *lordosis*, the posture that the receptive female assumes when the male mounts her.

The most intensively studied behavioral correlate of the estrus cycle in the rat is changes in running activity (e.g., Finger, 1969; Wang, 1923). In the standard experimental situation the female lives in a small cage from which she has free access to a vertically mounted running wheel ("squirrel cage"). The wheel revolves as she runs in it. Each revolution is counted automatically, and daily records are kept of the number of revolutions run. In another procedure the female lives in a stationary cage, and her locomotor activity is monitored by light beams and photocells. Figure 3.2 shows the relation between locomotion and receptivity for a single female (Finger, 1969). Maximal activity and sexual receptivity occurred together, every fourth day.

Kuehn and Beach (1963) measured the duration of receptivity during a single cycle by briefly exposing female rats to active males once every hour. Average duration of estrus was 19 hours. The degree of receptivity was calculated by a *lordosis quotient*: for each exposure to a male, the lordosis quotient was the number of times the female exhibited lordosis when mounted relative to the total number of times she was mounted. Thus if the female was totally receptive, the index was 1.0; if she was totally unreceptive, the index was 0.0. During this experiment, the female

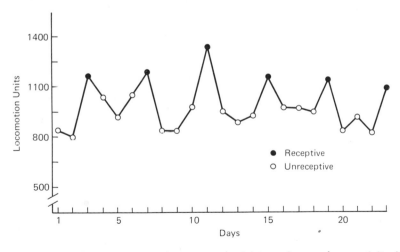

FIGURE 3.2 Relation between locomotor activity and sexual receptivity in a female rat. Receptivity was measured by the occurrence of lordosis in response to manual palpation. Modified from Finger, F. (1969) Estrus and general activity in the rat. *Journal of Comparative Physiological Psychology* 68:461–466. Copyright 1969 by the American Psychological Association, and reproduced by permission.

reached an average maximum quotient of 0.8, which usually occurred 4 to 7.5 hours after initial receptivity. Note that this is approximately the time of ovulation.

Determinants of Estrus Cycle Characteristics

The timing of rodent estrus cycles is dependent on lighting conditions. Under natural lighting, receptivity begins in the early evening. However, if an artificial light cycle is imposed, with lights on at night and darkness maintained during the day, the females will gradually accommodate to it by becoming receptive during the new "nighttime." Under conditions of continuous darkness the females will continue regular cycles, but in continuous light the cycles become irregular or fail completely (Alleva et al., 1968; Browman, 1937). It is interesting to note that male laboratory rodents, which do not exhibit 4-day cycles of sexual activity congruent with female estrus cycles, are nevertheless responsive to daily changes in illumination. For example, the copulatory behavior of male rats is more vigorous in darkness than in light (Larsson, 1956). Dewsbury (1968) found that males copulated more vigorously after 8 hours of darkness than after a half hour. Thus males are most vigorous at approximately the same time that females are most receptive.

The social environment can be extremely important in the regulation of estrus cycles. Very clear examples of social control mediated by olfactory stimulation have been found in several strains of laboratory mice. For example, van der Lee and Boot (1955, 1956) discovered that female mice caged in groups of four tend to become pseudopregnant spontaneously; this seldom occurs when the females are caged alone. Moreover, if the females are made anosmic by excising their olfactory bulbs, and are then placed in groups, they cycle normally. This consequence of anosmia suggests that olfactory stimulation is critical in the disruption of normal cycling.

The effects of grouping become more pronounced if the number of females per cage is increased from 4 to 30. In this case the females tend to become completely unreceptive (anestrous). Tactile contact between the females is not responsible for the effect because the anestrous condition persists if the females live in small individual cages formed by partitioning the larger enclosure. This is further evidence that estrus is suppressed by olfactory stimulation generated among the females (Whitten, 1966).

Quite a different result occurs if stimuli from male mice are presented to the females. For example, if a male is allowed to live in a wire basket inside the female's cage, the duration of the female's estrous cycle will be shorter than if she is living alone. The data in Table 3.4 demonstrate this effect. When the male was present, over 90 percent of the cycles were 4 or 5 days long; when he was absent, only 41 percent of the cycles were this brief.

TABLE 3.4. Effect of Male Presence on the Distribution of Estrus Cycle Durations: Percentages of All Observed Cycles

Duration	Male Present	Male Absent
4 days	55.1	3.8
5 days	37.2	37.2
>5 days	7.7	59.0
	100.0	100.0

(Modified from Whitten, 1966.)

In addition to shortening the cycle, stimuli from the male can synchronize the estrus cycles of females living in groups. Females tend to come into heat after approximately 48 hours of exposure to the male. They need not be able to touch, see, or hear him; what is essential is the presence of olfactory stimuli arising from a volatile fraction of the male's urine (Marsden and Bronson, 1964; Whitten, Bronson, and Greenstein, 1968).

Although the chemical structure of the "synchronizing compound" has not yet been determined, some relevant facts about it are known. First, not all male mice produce it; males bearing the mutant gene "yellow," for example, do not synchronize estrus when placed with females (Bartke and Wolff, 1966). And second, the production of the compound is related to the presence of male sex hormone, androgen. Urine from castrated males does not synchronize estrus, whereas urine from ovariectomized, androgen-injected females does (Bronson and Whitten, 1968).

The spontaneous pseudopregnancy observed in groups of females is commonly called the *Lee-Boot effect*; the shortening and synchronizing of cycles produced by the male urinary factor is called the *Whitten effect*. A third, related effect that has been named after its discoverer is the *Bruce effect*, which refers to the prevention of successful pregnancy by male odors. Specifically, if a female mates with one male (the stud) and then is exposed to vapor from the urine of another (the blocking male), she is less likely to become pregnant or pseudopregnant than if she remains unexposed. The likelihood that exposure will block pregnancy depends on several factors. First, the block is more likely if the stud and blocking males belong to different strains. Parkes and Bruce (1961) reported that the percentage of blocks ranged from 16 to 48 percent when the males were from the same strain and from 52 to 88 percent when they were from different strains. The strain of the female did not matter. Second, the exposure must occur within 5 days of mating and be maintained for 12 to 48 hours. There are clear physiological reasons for the 5-day limit of effectiveness, which we will discuss in Chapter 6. Third, the blocking male must have his gonads intact; castrated males are ineffective blockers (Bruce, 1967). The pregnancy-blocking factor in the urine is similar in this way to the synchronizing factor. And finally, blocking does not occur in females whose olfactory bulbs have been removed. But removal of other sensory modalities does not prevent blocking. These findings demonstrate that olfactory stimuli are crucial for production of the effect in the mouse.

Pregnancy blocking has also been observed in another "short-cycle" rodent, the deer mouse (*Peromyscus maniculatus bairdii*). But in this species exposure to the odor of a blocking male is only one of several powerful environmental blocking agents. Females living in relatively large cages are more likely to exhibit blocking than are females living in smaller cages; moreover, merely moving a female from one cage to another may produce blocking (Eleftheriou, Bronson, and Zarrow, 1962). Like the laboratory mouse, the deer mouse responds to these procedures only during the first several days after insemination. As we will see in Chapter 6, the pregnancy-block phenomenon does not represent an abortion but rather the failure of the fertilized egg to become implanted in the wall of the uterus.

We must not overlook the fascinating implications of the Bruce effect for our understanding of olfactory communication between animals. The data show that female mice can discern the identity of individual males on the basis of odor alone. Each male apparently has a unique identifying scent. It may be that the "identifier" substance is chemically related to the "blocking" substance, and that both of these are related to the "synchronizing" substance of the Whitten effect. At present we do not know (Bronson, 1968). But in any case, these volatile substances carry specific messages between animals. The technical name for such a substance is *pheromone* (Karlson and Butenandt, 1959). The mouse pheromones are some of the best examples we have for mammals, and we will have occasion to refer to some other examples later in this chapter and in Chapter 6.

PRIMATE MENSTRUAL CYCLES

The word *menstruation* stems from the Latin word for month; physiologists used the word originally to refer to the approximately monthly discharge of blood from the woman's uterus. Subsequent observations that many species of apes and monkeys showed similarly spaced discharges led to an extension of the term to cover these nonhuman cycles.

How do menstrual cycles relate to estrus cycles? To begin with, the two cycles are defined in terms of different criteria. Estrus cycles are defined behaviorally, whereas menstrual cycles refer to changes in the walls of the uterus. Monkeys, apes, and humans are the only animals to show this form of periodic bleeding.[2] Hence we need to know whether female primates have estrus cycles, and if so, how they are related to their menstrual cycles. Of course there are almost 200 species of extant primates; this means that there will be exceptions to the generalizations presented here.

We noted that estrus in nonprimate mammals is closely related to the occurrence of ovulation. In primates ovulation typically occurs at the

[2]The bleeding of female dogs just prior to estrus has a different physiological basis.

midpoint of the menstrual cycle, that is, 13 to 15 days after the onset of bleeding. Thus we might expect female primates to be sexually receptive at that time and unreceptive at other times. How does this expectation relate to the known facts?

In general, female primates do not exhibit the close relationship between ovulation and receptivity that characterizes other mammals. This generalization is particularly valid for humans (see Chapter 8) and for Old World monkeys and apes living in seminatural, zoo, or laboratory environments. We will consider as an example the behavior of rhesus monkeys living in outdoor pens in England (Rowell, 1963). Under these conditions the females would copulate at any time during the cycle. There were no peaks of copulation around the time of ovulation; however, copulation was least often observed during and just before menstruation. On the basis of these findings, Rowell suggested that we consider the female rhesus monkey to be more or less continuously receptive except for the several days just before and during menstruation. However, as we point out in a different context in Chapter 8, recent studies of tree-living Old World monkeys have discovered a somewhat tighter relationship between position in the menstrual cycle and behavioral receptivity.

In some primate species, for example the baboons, there is a marked swelling and reddening of the female's hindquarters that begins after menstruation and reaches a peak at about the time of ovulation. Female baboons are sexually most active when they are most swollen. It may be that the engorged, colorful hindquarters serve as a sexually exciting stimulus to the males. The obvious biological advantage of this stimulation would lie in its making copulation most likely at the time of ovulation (Kummer, 1968; Rowell, 1967; Wickler, 1967).

Finally, we need to emphasize that the sexual gestures and postures of monkeys and apes are woven tightly into the larger fabric of their systems of communication. This means that the analysis of their sexual behavior depends very heavily on an understanding of the details of their social organizations. In many species, for example, the "sexual invitation" of the female consists in her standing directly in front of a male with her hindquarters displayed fully (the *presenting response*). But the presenting response is not tied to sexual receptivity in the same direct way that lordosis, for example, is tied to estrus in female rodents. As Marler (1968) has pointed out, the presenting response is used in a variety of social contexts by both males and females. It serves generally as a signal that reduces the distance between individuals within a single social group.

THE ROLES OF STIMULI IN COURTSHIP AND MATING

In Chapter 2 we gave some examples of the roles that social stimuli play in reproductive isolation. We observed that such stimuli may serve more than a single function, for in addition to maintaining species separation they also lead to the assembly of males and females in breeding condition. Social stimuli also play other roles in sexual behavior: they can elicit species-typical courtship and copulatory responses, regulate the time course of copulation, and signal the presence of conflicting behavioral tendencies, for example sexual and aggressive tendencies, to potential mates. We will consider examples in each of these categories.

STIMULI THAT ELICIT COURTSHIP AND COPULATORY BEHAVIOR

The assembly and/or early courtship of the males of many insect species are produced by the action of volatile substances (pheromones) secreted by the females from specialized glands. Insects as diverse as roaches, bees, and butterflies are all known to rely on pheromones (Butler, 1967; Roelofs and Comeau, 1969). In the case of the honey bee, the queen's pheromone secretion during her single mating flight serves as a clue to her whereabouts for the drones who are chasing her. This would be a formidable task, without the "trail" of pheromone to follow, for Gary (1961) estimates that the drones have to find the queen within a flight space that contains at least 50 million cubic yards. Once the drones locate the queen they mount her and copulate while remaining in the air. The presence of the pheromone is required if the drones are to attempt copulation (Butler, 1967).

Under what conditions do females release their sexually attractive pheromones? As a rule, the production and secretion of the substance begins shortly after the female has emerged as an adult; several days may be required for the production of enough substance to attract males. In many female insects, apparently, pheromone secretion stops after mating. The queen bee is an exception, for she continues to secrete the relevant substance (9-oxodec-2-enoic acid) after her mating flight, when she has returned to the hive and is laying eggs. However, the substance now serves other biological functions: it is the "queen substance" that inhibits ovarian development and queen-cell-building behavior in the worker bees (Butler, 1967).

An interesting instance of environmental control of sex pheromone secretion has been reported for the polyphemus moth, *Antheraea polyphemus*. Males mate with females only after their antennae have been stimulated by the appropriate pheromone. But the females of this species will release the pheromone only if they have been stimulated by a vaporous product of oak leaves (Riddiford and Williams, 1967). Oak leaves are one of the preferred foods of the larval stage of this species, the polyphemus silkworm. Volatile products from other food sources used by the larvae (elm, maple, birch, etc.) do not cause the adult female to secrete the pheromone. Riddiford (1967) has characterized the critical plant substance as trans-2-hexenal, which is very common in green leaves. Apparently the combination of this substance with other substances in leaves other than oak prevent it from stimulating the antennae of the female.

The males of some insect species also secrete sex pheromones. With some exceptions, male sex pheromones are not used to attract other animals from a distance; rather they are used during courtship to induce the female to copulate. They are true chemical aphrodisiacs, at least for the species in question (Butler, 1967). The chemical stimulus of the pheromone is linked with visual and tactile stimuli in the regulation of the courtship and copulation sequences. One of the best-studied cases is the Queen butterfly, *Danaus gilippus berenice*.

Brower, Brower and Cranston (1965) have summarized the sequence of courting and initial copulatory behaviors of a pair of Queens as shown in Figure 3.3. The male is attracted by the sight of the flying female and follows her. He overtakes her and while flying above her lowers his abdomen above her antennae. Protruding down from each side of his abdomen is a specialized organ, called a hairpencil, which is about 4 mm long and 0.75 mm in diameter. At the end of the hairpencil is a bunch of

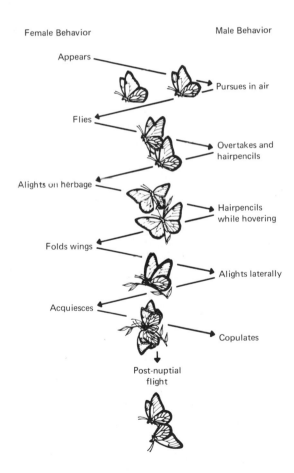

FIGURE 3.3 Courtship and copulatory sequence of the Queen butterfly, *Danaus gilippus berenice* (Cramer). From Brower, L. P., Brower, J, and Cranston, F. (1965) Courtship behavior of the Queen butterfly, *Danaus gilippus berenice (Cramer). Zoologica* 50:1—54. Photo from New York Zoological Society.

tufted hairs. The male's sex pheromone is located on these hairs. It is a "dust" composed of two compounds; one is biologically active in stimulating the female's antennae, and the other is a viscous substance that allows the active substance to stick to the hairpencils and the antennae (Meinwald, Meinwald, and Mazzocchi, 1969; Pliske and Eisner, 1969; Schneider and Seibt, 1969).

The male flies in front of and over the female in this fashion (called aerial hairpenciling) until she slows and lands on a stem or branch. This response of the female is apparently due directly to the chemical influence of the pheromone, which is mediated through receptors on her antennae. When the female has landed the male continues to hover and bob just above her, keeping his hairpencils over and on her antennae (ground hairpenciling). If the female is completely receptive she folds her wings and awaits the male's next maneuver. If she does not do this, the male is likely to drop repeatedly onto her back, until she flies again (hovering and striking). The male then repeats his aerial hairpenciling maneuver. Repetitions of these maneuvers increase the likelihood that successful copulation will occur, indicating that continued exposure to the pheromone increases the female's receptivity (Brower et al., 1965). Indeed, receptivity in this case seems to consist primarily in the female remaining immobile.

When the female has folded her wings, the male retracts his hairpencils and drops down beside her in a lateral position. He grasps her with his legs and attempts to appose his genitalia to hers by vigorous movements of the tip of his abdomen. It appears that the male continues to move his abdomen under the female until he contacts her genitalia; this contact stimulates him to attach his claspers. At this point he is in the copulatory position. But he does not consummate the relationship immediately. Rather, after a snap of his wings, the male flies off with the female dangling head down from the tip of his abdomen. He alights again under the cover of a bush or tree, and only then does he inseminate her and terminate the copulation.

As Figure 3.3 suggests, this sequence of courtship and copulation maneuvers can be understood as a passage of interlocked stimuli and responses. The stimuli are in three sensory modalities: visual, tactile, and olfactory. The "post nuptial flight," which apparently is released by tactile feedback from the male's achievement of proper copulatory posture, places the animals in a position relatively free from predators. The exigencies of achieving the copulatory posture and making the flight seem to be lessened by the passivity of the female, which has been induced by the hairpencil aphrodisiac.

Copulation in Domestic Turkeys

Schein and Hale (1965) have provided a thorough experimental analysis of the role of stimuli in regulating the normal copulatory sequence in domestic turkeys. The completion of virtually every step in the sequence depends on the presence of a particular stimulus. In this sequence, as in many others, the naive observer's expectation that the animals somehow know what they are doing is weakened by the observation that biologically inappropriate objects may be used to elicit and guide "normal" behavior if they possess the proper stimulus characteristics. In the case of the turkey, we shall see that the hand of the experimenter and a turkey's head impaled on a stick can elicit the same responses from a turkey cock as would a normal and receptive turkey hen. Thus, experiments of this sort, in addition to providing detailed analysis of how the behavioral sequence is regulated, also instruct us against building more "mental machinery" concerning the animal's "intentions" or "purposes" than is minimally required to account for the observed behavior.

Figure 3.4, taken from the work of Schein and Hale (1965), describes the sexual interaction of turkeys as a chain of discrete events. The receptive female responds to the initial sexual display of the male by crouching and drawing in her neck. The male approaches the crouched female, and she extends her neck partially. The neck extension stimulates the male to move behind the female and begin his mount. As the male puts pressure on the crouched female's back, she extends her neck to its maximal extent and remains quite still. The male can now strengthen his grip by grasping the base of her wings with his claws. Stimulation of this area (usually by the male, but in experiments by the hand of the experimenter) causes the female's head to lower and her tail to be deflected upward. If this tail deflection occurs too soon or too late in the sequence, the male cannot achieve cloacal contact. The timing of this response is also important because it produces a partial eversion (pouching out) of the oviduct opening in the cloaca. As the male brings his cloaca down to cover the female's, the stimulation his body provides to the base of her tail causes complete oviduct eversion. This has two consequences. First, it allows the male's genital organs to be stimulated, which produces ejaculation. Second, it terminates the female's receptive posture very rapidly. She will shake the male off her back just a few seconds after her oviduct has become totally everted. Hence the proper stimulus response relationships must obtain within rather narrow time limits if insemination is to occur. The female's response is not to ejaculation per se, for regardless of the outcome of the first mating, she will not crouch in response to further displays by the male if her oviduct is everted.

Some receptive turkey hens will crouch in the presence of objects other than displaying turkey cocks; Schein and Hale (1965) suggest that large size is the fundamental characteristic to which the females are responding. These experimenters have also demonstrated that it is the sight of the female's head that is primarily responsible for eliciting the precopulatory responses of the male. They found that both experienced males and androgen-injected four-week old males would court and attempt to mount a stand on which a female's head was impaled; less often,

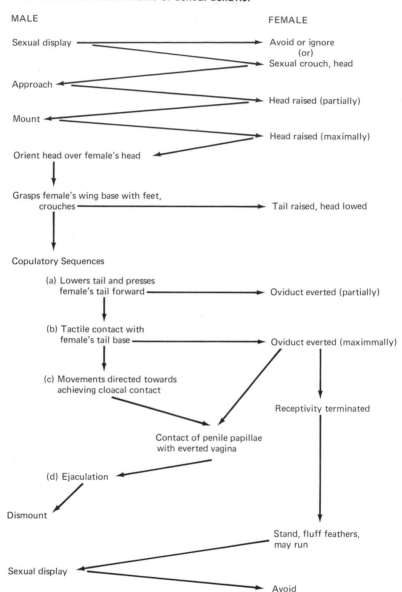

MALE   FEMALE

Sexual display ──────────────────→ Avoid or ignore
                                                  (or)
                                     ──→ Sexual crouch, head

Approach ◄──

                                     ──→ Head raised (partially)

Mount ◄──

                                     ──→ Head raised (maximally)

Orient head over female's head ◄──

Grasps female's wing base with feet,
        crouches ──────────────────→ Tail raised, head lowed

Copulatory Sequences

    (a) Lowers tail and presses
        female's tail forward ──────────→ Oviduct everted (partially)

    (b) Tactile contact with
        female's tail base ──────────────→ Oviduct everted (maximmally)

    (c) Movements directed towards
        achieving cloacal contact

                                          Receptivity terminated

            Contact of penile papillae
            with everted vagina

    (d) Ejaculation ◄──

Dismount

                                          Stand, fluff feathers,
                                          may run

Sexual display ◄──

                                     ──→ Avoid

FIGURE 3.4 Courtship and copulatory behavior of the domestic turkey. From Schein, M. W., and Hale, E. (1965) Stimuli eliciting sexual behavior. In F. Beach (ed.), *Sex and Behavior*, pp. 440–482. New York: Wiley.

the males would also respond to the head of another male or to wooden carvings in the shape of a turkey head. Schein and Hale speculate that the head has acquired tremendous importance in the release of turkey courtship because it is the only part of the female the male sees during preejaculatory mounting. Except for minimal feedback through his feet, this is the only contact he has with her.

THE ROLE OF STIMULI IN REDUCING FIGHTING AND FLEEING
RESPONSES DURING COURTSHIP

One secure generalization about animal behavior is that most intraspe-cific fighting occurs between males during the breeding season. The physiological foundation for increased hostile tendencies is hormonal: the changes in hormone levels, particularly increases in androgen, that make males more likely to copulate or spawn also make them more likely to threaten and fight. The environmental circumstances that promote the release of threats or fighting vary from species to species. In some cases fighting becomes more likely because males gather together in close groups only during the breeding season; this is often the case in large hooved animals, such as the buffalo, for example. In other species males prepare for reproduction by establishing territories, areas that they de-fend against intruders by threatening, chasing, or fighting. Territory for-mation is particularly prevalent in migrating birds and several groups of fish. In both of these cases and others, the initial response of a male to any other individual of appropriate size and shape will be aggressive.

One can see possible biological advantages in this relationship between sexual readiness and an increased tendency to fight. To the extent that skill and success in fighting are positively related to other survival skills and are genetically determined, males who successfully compete against other males for the possession of a female will contribute their superior genes to the next generation. The general importance of this process must be regarded cautiously, however, because in many if not most in-stances of aggressive interactions between males the exchange is termi-nated by threatening gestures and postures rather than by overt combat. A second function of aggression is the spacing out of individuals over the available area; this function is obviously served by the formation of territories. Spacing out, in turn, can serve several adaptive functions, including optimal utilization of the available food supply and protection from mass damage by a single predator. Third, aggressive behavior can serve to maintain a previously developed social order during the gener-ally chaotic times when a number of females are in heat; later we will consider an example of this, in baboons (Lorenz, 1964; Tinbergen, 1960).

But with these advantages comes an obvious problem: males must not continuously and indiscriminately attack receptive females, and females must not flee permanently from an initially hostile male. How are the hostile tendencies of the males to other animals to be controlled so that mating can take place?

Different species have evolved various solutions to this problem, which involve the presentation of social stimuli to the male by the female. In some cases the relevant stimulus is a particular characteristic of the fe-male's shape or coloration. In other cases, where the males and females look very much alike, the female may behave in a distinctively different way from a male in similar circumstances. We will now consider one example in each of these categories, using three-spined sticklebacks and herring gulls. Then we will present a third example, the Japanese quail, in which the role of aggression in courtship remains obscure.

Sticklebacks

The three-spined stickleback, *Gasterosteus aculeatus*, is unquestionably the best-studied fish in the world. It may fairly be called the white rat of European ethology. Its reproductive behavior has been studied especially closely; for references see Hinde (1966) and Tinbergen (1951). Here we will describe only a small portion of the reproductive behavior cycle.

Under natural conditions, in the spring sticklebacks migrate upstream to shallow water. Upon arrival at the breeding grounds, the males establish territories and build nests. As the male works on his nest his coloration changes from a rather uniform grayish green to a highly accented pattern with a conspicuous red belly. Eventually a female will deposit her eggs in the nest and the male will fertilize and care for them there. However, females do not develop a red underside; rather, as their eggs develop, their bellies become swollen.

The male becomes increasingly pugnacious with time. He will approach and threaten another male that enters his territory. In the typical threat posture, at the edge of his territory, the male goes vertical in the water with his mouth just above the ground, his side turned toward his opponent, and his spines erect. If the intruder enters farther into the territory, the "host" fish will attack him and usually drive him out.

Tinbergen (1951) has shown experimentally that the threat and fighting responses of the male (in breeding condition) are released by the characteristic red underside of other males. Males were much more vigorous in attacking crude wooden models of approximately stickleback size and painted red on the bottom than they were in attacking quite accurate and detailed models of male sticklebacks that lacked only the red belly. Thus the characteristic red belly is a critical stimulus (sign stimulus) for aggressive behavior in this species. However, it is not the only factor that influences the strength of the response, for the male will respond more vigorously to a dummy presented in the vertical threat position than to one presented horizontally.

The female stickleback lacks a red belly, and hence the male's initial response to her is relatively temperate. Additionally, the male's sexual responsiveness is heightened by the sight of the female's swollen belly. In an experiment using models, Tinbergen showed that a crude wooden dummy with a swollen belly elicited more intense courtship from the male than did an accurate model of a female without a swollen belly.

The male's red underside is important in releasing appropriate courtship behavior in the female. This means that the same stimulus releases quite different responses in the reproductively active males and females of this species. Once the initial contact between male and female has

FIGURE 3.5 Zig-zag courtship of the three-spined stickleback, *Gasterosteus aculeatus*. From Tinbergen, N. (1951) *The Study of Instinct*. Oxford: Clarendon.

been established, the male exhibits a zig-zag courtship dance that eventually leads the female into the nest he has prepared. After she is in the nest the male prods the base of her tail with his snout, which causes her to release her eggs. She then leaves the nest, and the male enters it and fertilizes the eggs. The chain of events from the appearance of the female to fertilization is diagrammed in Figure 3.5.

Herring Gulls

Tinbergen (1960) has provided an excellent description of the behavior and social organization of the herring gull (*Larus argentatus*) during its spring and summer breeding and nesting seasons on the sandy beaches of Holland. His treatment can serve as a model for the conduct of field work in animal behavior and for the shrewd combination of fact and theory to provide a compelling account of the adaptive value and immediate causation of species-specific behavior.

In contrast to the stickleback, unmated male herring gulls do not set up territories immediately upon arriving at the spring breeding grounds. Rather, they and unmated females join together in loosely knit aggregations that Tinbergen has called "clubs." It is on the grounds of the club that the ceremonies leading to pair formation begin.

Males on the grounds of the club are likely to respond to an intruder, or another animal that approaches too closely, with a characteristic behavior called the upright threat posture. As the name implies, the male draws himself up to full height, extends his neck up and out, and points his bill toward the ground. He also lifts his wings slightly up and away from his body. While in this position he walks in a jerky, stiff manner. The position of his wings and bill prefigure the typical fighting maneuvers of pecking, grabbing, and batting with the wings. Tinbergen argues that this posture, like many threat and courtship postures in animals, results from the male's ambivalence regarding fighting and fleeing. A discussion of the merits of this argument is beyond the scope of this book; the reader is referred to Hinde (1966, Chapters 16 and 17) for a fuller discussion. It is sufficient here to point out that two males that adopt this posture in each other's presence and refuse to give ground will most likely end up in a full-scale fight.

A female who approached a male on the grounds of the club would most likely be driven away were she not to adopt special signalling procedures; everything else being equal, the male will behave aggressively, and he takes no initiative in approaching females. Moreover, male and female herring gulls are not distinguished by sexually dimorphic plumage. There are no obvious morphological sign stimuli relating to courtship, such as the swollen belly of the female stickleback. Thus when a female approaches a male in the club she relies on behavioral signals to lessen the likelihood that he will threaten her or drive her off.

The female's courting posture is in some ways the opposite of the upright threat posture. She keeps her neck well tucked in and stretches her body out horizontally. As she approaches the male she moves her head back and forth in a characteristic way that is labeled "head-tossing." And she vocalizes in a particular way that Tinbergen has transcribed as "kiloo." A male's response to the female's courting behavior varies with a number of factors, including the presence or absence of other males nearby and the number of times the female has already approached him. Initially the male may respond with an upward stretching of his body, along the lines of a partial upright threat posture. If there are other males nearby, the courted male may exhibit the full threat posture and then chase them. If the other males also are being courted, a fight may break out. However, the male does not attack the courting female.

Another masculine response to the female consists of adopting a horizontal posture and walking with her for a few yards to a place where both animals engage in what is called "aggressive choking." The response is called choking because of the characteristic vocalization that accompanies the posture; it is called aggressive because a paired male and female engage in it when defending their joint territory against intruders. A mated pair also "choke" during their nest building. Thus this same posture can be seen to function in different ways in different parts of the reproductive behavior cycle. The significance of the signal is interpretable only within the context of the immediate situation.

Finally, a male may respond to a courting female with a characteristic turning and twisting of his head. This response stimulates the female to increase the rate of her head-tossing. Then the breast and neck of the male begin to heave until, after what appears to be considerable effort, the male opens his beak and regurgitates a mass of food. As soon as he opens his beak the female approaches and puts her beak in his and pecks and eats the food that he has regurgitated.

This last sequence of events is more likely than the other two to be followed by copulation. But before the male mounts the female he exhibits head-tossing in much the same way that the female has and, additionally, vocalizes with the "begging call," so named because of its similarity to the vocalization of immature birds begging for food. The birds circle each other until the male finally reaches a position behind the female and at an angle to her body. Then his vocal behavior changes to the characteristic "copulation call" that immediately precedes his mounting attempt. He jumps on the female's back and maneuvers to bring his cloaca next to hers. All this time the female continues her head-tossing movements. After cloaca contact and seminal expulsion the male dismounts and both birds preen. Copulation may be repeated several times during a single copulatory bout.

Copulation usually marks the formation of a permanent pair-bond between herring gulls. As far as Tinbergen could tell, the pairs remain monogamous for at least one season and probably for more. Mated pairs that return to the breeding grounds do not go through the extended "rating and dating" procedures of unmated animals in the club.

It is important to note that the horizontal posture of the courting female is very similar if not identical to the "submissive" posture shown by subordinate individuals during aggressive encounters. By exhibiting this posture in the initiation of courtship, the female effectively prevents the male from aggressing against her. But the deflection of his aggression is not sufficient to generate his active participation in courtship and copulation. Typically the female must repeat the courtship maneuvers several times before the male is prepared to drop his aggressive postures and respond with head-turning or feeding.

Japanese Quail

The Japanese quail (*Coturnix coturnix japonica*) is a small brownish-yellow bird that was introduced from the Orient into the United States during the 1950s as a potential game bird. Although its use as game has been limited, its small size, ease of maintenance, and short generation time have made it a valuable subject for research in avian husbandry and genetics (Woodard, Abplanalp, and Wilson, 1965). The bird has not been studied extensively under field conditions; the available data suggest that it is migratory, territorial, and seasonally monogamous (Wetherbee, 1961).

At least as it is observed in the laboratory, the courtship and mating sequence of the Japanese quail is much simpler than the behaviors of the stickleback or herring gull. Before he mounts the female, the male may exhibit a characteristic posture in which he extends his neck out parallel to the floor and walks around the female stiff-legged. There is some evidence that the sight of the female releases the execution of this movement, for Farris (1967) found that it could be classically conditioned to the sound of a tone when the sight of a female was used as the unconditional stimulus. However, the posture does not invariably precede copulation; Wilson and Bermant (1970) found that when pairs were tested in relatively small cages, males assumed the extended-neck posture before mounting in only 5 percent of the cases.

Copulation itself is brief and rather violent. The male grabs the female by the feathers on her head or neck, then moves behind her and mounts. As he mounts he spreads his wings, apparently for balance, and moves his hindquarters down to appose his cloaca to hers. The entire sequence may be described in abbreviation as HG-M-WS-CCM, for head grab-mount-wingspread-cloacal contact movement. The male relies on his grip to maintain his position on the female; sometimes he pulls her over on top of himself. An active male paired with a completely receptive female may achieve several complete copulation sequences during a 5-minute period (Beach and Inman, 1965; Wilson and Bermant, 1972).

Beach and Inman (1965) discovered that the size of the enclosure in which a pair of quail was tested affected the amount of aggressive behavior that females directed against males. Females were more likely to repel the attempts of males if the cage were less than 4 feet square. In smaller cages the initially receptive females would eventually turn against the males and peck at them whenever they attempted to mate. This result showed that the conditions in the animals' physical environment could alter the relationship between sexual and aggressive behavior.

Selinger and Bermant (1967) investigated the behavior of pairs of males placed together for 5 minutes a day in small cages. Under these circumstances the males exhibited all the components of the complete sexual sequence, including the horizontal-neck extension posture and the HG-M-WS-CCM sequence. In each pair there was always one male who became dominant in the execution of the copulation-like behaviors. However, the relatively submissive males seldom permitted the sequence to reach the stage of cloacal contact. Both the heterosexual mountings observed by Beach and Inman and the homosexual mountings observed by Selinger and Bermant were dependent on the presence of gonadal androgen, for the behavior dropped out within 4 days after castration. Replacement of the hormone by injection reinstated the behavior. Moreover when the noncastrated submissive males in Selinger and Bermant's experiment were given additional androgen, they markedly increased their own "sexual-aggressive" activities, and in some cases actually reversed the dominance relationship that had been established.

Wilson and Bermant (1972) made a direct comparison of the behaviors occurring in male-male and male-female pairs, and found that the sexual components of the behavior were virtually identical in form and frequency except at the transition from wingspread to cloacal contact movement. In the heterosexual pairs the occurrence of WS was more likely to lead to CCM than in the homosexual pairs. The reason for this was clear: in homosexual encounters the mounted male resisted the mounting male more than the female did.

How are we to account for the occurrence of homosexual mounting in the quail? One hypothesis is that males mount other males when they are deprived of sexual contact with females. This was ruled out by Wilson and Bermant when they found that when males were exposed to females and other males on alternate days they continued to mount both. Second, one might argue that the homosexual behavior was a pathological outcome of the physical conditions in which the animals were tested. In the absence of data collected under more natural conditions, we cannot rule this out. In the small testing cages neither males nor females have much space in which to act out ceremonies of threat or submission that play an impoitant role in courtship and fighting. However, Wilson and Bermant did observe homosexual head grab and mounting attempts when males were paired with a single female in a large indoor enclosure; hence it is not the small cage alone that produces the behavior.

Third, one could speculate that the head grab through cloacal contact movement sequence serves two separate functions in the quail's behavioral economy. In addition to the role in the execution of copulation it serves to establish dominance among males. Again, this interpretation will remain insecure until the birds have been observed in ecologically valid surroundings. The activities of these quail under laboratory conditions points up the difficulties of interpreting behavioral data without a baseline generated under natural conditions in the physical environment.

SHORT-TERM SEXUAL REGULATORY STIMULI
As described, when the turkey hen's cloaca becomes fully everted she becomes sexually unreceptive. This is a clear example of a specific stimulus (cutaneous feedback from the everted cloaca) regulating sexual receptivity. Moreover, it is obvious among humans, nonhuman primates, and at least some other mammals, that direct mild stimulation of the genitalia, whether by a member of the same species or another species, or even stimulation received by rubbing against an inanimate object, often leads to erection, mounting and/or pelvic thrusting, orgasm, and ejaculation. In these cases genital stimulation turns sexual responsiveness on, whereas in the turkey hen it turns responsiveness off. Specific stimulation of the genitalia may also act during intermediate points of the copulatory sequence to regulate its detailed form and timing. The roles of these regulatory stimuli have been demonstrated experimentally for male cats and male and female rats.

When a male cat mounts a receptive female he grips the fur on her neck between his teeth and begins a series of shallow pelvic thrusting movements. After one or more of these thrusts the male's penis enters the vagina and he can thrust more deeply and ejaculate. Aronson and Cooper (1968, 1969) have shown that the transition from shallow to deep thrusting depends on penile stimulation; the function of the shallow thrusting is to locate the vaginal opening. When the major sensory nerve (the dorsal nerve) of the penis was surgically severed, the males were still capable of erection and mounting but became incapable of intromission. Some of the males eventually recovered the ability to achieve intromission, and subsequent histological investigation suggested that the dorsal nerve had regenerated. The other animals showed continuing intromission failure and demonstrated a general lack of balance while mounting. Eventually they mounted the female only rarely, which suggested that the capacity for sexual arousal had been diminished by the repeated failures of intromission.

Male rats also require feedback from the penis in order to achieve intromission. Each of the rat's intromissions is preceded by a number of shallow pelvic thrusting movements that serve to bring the penis into line with the vaginal opening. Rats whose penises have been anesthetized continue to mount and thrust against the female but do not achieve intromission. Thus, for both cats and rats the sequence of events from mounting to intromission has two components, the shallow and deep thrusting stages, which are successfully linked only when proper feedback is received from the penile receptors (Adler and Bermant, 1966).

A different kind of regulatory stimulation arises from the characteristics of the vagina, in particular its size and texture. This variable is probably of greater importance for species in which the male needs a relatively large amount of penile stimulation in order to reach ejaculation: primates, carnivores, and rodents, for example. The very rapid ejaculation of many hooved animals suggests that this variable is of small importance for them.

The surface characteristics, and in some species, the natural size of the vaginal barrel, vary with the estrus or menstrual cycle. In fact a common means of detecting behavioral receptivity in female rodents is by examining a smear of cells from the vagina; around the time of ovulation the vaginal walls are lined with large cornified epithelial cells that are eventually sloughed off and replaced by a large invasion of leucocytes (see Chapter 5 and Nalbandov, 1964, for further explanation).

Among human females, as we describe more completely in Chapter 8, the surface characteristics of the vagina normally change rapidly at the onset of erotic stimulation. The vaginal walls secrete a fluid that serves two important functions: it decreases the acidity of the vaginal environment and thus enhances the life expectancy of any deposited sperm, and it lubricates the vagina and thus makes penetration easier. Without adequate lubrication, vaginal penetration is difficult or painful for the man and painful for the woman (Masters and Johnson, 1966).

There appears to be no evidence that in nonprimate species sexual stimulation per se produces lubrication of the vagina and hence facilitates active participation in copulation. However, there is good evidence that nonhuman females can voluntarily regulate their participation in copulation in response to the intensity of genital stimulation just received. The details of this regulatory behavior have been worked out in some detail for the female rat.

Copulation in rats follows a pattern in which the male makes 8 to 15 brief intromissions, each lasting about a quarter of a second and separated by intervals of about a minute before he finally mounts and ejaculates. After ejaculation he waits approximately five minutes before initiating the second series of intromissions (Bermant, 1967). Under normal testing conditions the female appears to exercise relatively little control over the timing of the intromission responses. Although she may approach and nuzzle a male who has recently completed an intromission or dart around him in the characteristic gait of the receptive female, it is not clear that these maneuvers are effective in pacing his behavior. In order to test the hypothesis that females can and will voluntarily regulate their participation in copulation, it was necessary to give the female direct control over the male's access to her and then to determine if she would take advantage of her opportunity to make available a copulating male, and if so, how she would regulate the timing of his appearances.

Two procedures were used to test the hypothesis. In one (Pierce and Nuttall, 1961), a group of sexually active males were placed in an enclosure above which was suspended a balcony. A receptive female was placed on the balcony, from which she could jump down to the males and return. The experimenters prevented males from jumping up to the balcony by rapping them on the nose with a long stick. The experimenters measured the time the female remained on the balcony as a function of what happened to her when last she jumped down to the males, that is, whether she received a mount without intromission, an intromission, or an ejaculation.

In the second procedure, females were trained to press a lever to produce a sexually active male who was allowed to remain until he had achieved a mount, an intromission, or an ejaculation (Bermant, 1961a, b). Again, the time between sexual event and the female's next voluntary response was measured. Average times for both procedures are presented in Table 3.5.

The two procedures generated very similar results: females came quickly to regulate their participation in copulation, and the quantitative characteristics of their behavior were closely related to the degree of stimulation provided by the male. They returned to the copulatory situation most rapidly after a mount without penetration (a third to a fourth

TABLE 3.5. Time (Seconds) Between Three Classes of
Sexual Event and Females' Next Sex-Producing Response

Sexual Event	Balcony Method	Lever-Press Method
Mount	13	20
Intromission	60	62
Ejaculation	170	123

(From Pierce and Nuttall, 1961; Bermant, 1961b.)

of a minute), somewhat more slowly after an intromission (about one minute), and most slowly after an ejaculation (two to three minutes).

If the female's tendency to engage in further copulation is partially determined by the amount of stimulation she has just received, then the experimental manipulation of her sensitivity to stimulation should affect the pace of her regulatory behavior. Bermant and Westbrook (1966) tested this hypothesis in three experiments using the lever-pressing method. In the first they treated the males with a drug (guanethedine sulfate) that prevented the expulsion of seminal fluid, which usually hardens in the vagina to form a plug. The male thrusts more vigorously and more often and grips the female more tightly during the ejaculatory response than during an intromission, and the drug leaves this behavior intact. Thus in this experiment only the stimulation from the vaginal plug was absent.

Figure 3.6 shows that females pressed the lever again sooner after an "ejaculation" when no semen was emitted than after normal ejaculations. The difference between plug-present and plug-absent ejaculations increased during the latter part of the experimental session, suggesting that under normal conditions the effects of successive ejaculations accumulate to make the female progressively slower in responding. Also, there was no appreciable change in the rates at which females paced themselves after intromissions or mounts; the effect of eliminating the vaginal plug was restricted to the first response after ejaculation.

In a second experiment the sensitivity of females' vaginas was decreased by application of a topical anesthetic. Because the general lubrication produced by placing a paste in the vagina might itself affect the female's behavior, a control condition was used in which the vagina was swabbed with an inert paste of the same consistency as the anesthetic. As shown in Figure 3.7, females responded generally more rapidly when their vaginas were anesthetized. The largest differences followed ejaculations and the smallest followed mounts.

In the third experiment sexual stimulation was reduced to a minimum by anesthetizing the males' penises so that they could not achieve intromission. Every time the female pressed the lever the male was placed with her until he mounted and thrusted against her, and then he was removed. Each session lasted 1 hour. Under these conditions females continued to respond with relatively brief delays throughout the session.

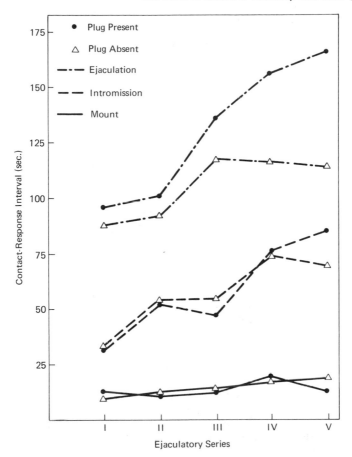

FIGURE 3.6 Average time between sexual contact with a male and the subsequent lever-press response across five ejaculatory series, for female rats in which a vaginal plug was either present or absent. From Bermant, G., and Westbrook, W. (1966) Peripheral factors in the regulation of sexual contact by female rats. *Journal of Comparative Physiological Psychology* 61:244–250. Copyright 1966 by the American Psychological Association, and reproduced by permission.

Average delays ranged from about 10 seconds during the first quarter of the session to about 30 seconds during the fourth quarter.

These results demonstrated conclusively that the rate at which females voluntarily participate in copulation is influenced by the degree of stimulation they have just received. Their "preferred" intervals between sexual events parallel the rates at which males return to copulation after each event. When a male has mounted without intromission he approaches the female again sooner than if he has achieved intromission. In absolute terms, the intervals following these events are approximately the same for males and females. After an ejaculation, however, females are prepared to return to copulation more rapidly than males. Males sel-

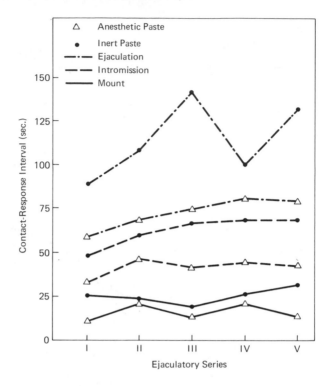

FIGURE 3.7   Average time between contact with male and response across five ejaculatory series for females whose vaginas were either swabbed with an inert paste or anesthetized. From Bermant, G., and Westbrook, W. (1966) Peripheral factors in the regulation of sexual contact by female rats. *Journal of Comparative Physiological Psychology* 61:244–250. Copyright 1966 by the American Psychological Association, and reproduced by permission.

dom begin their second series of copulations in less than 4 minutes after their first ejaculation, and the duration of their subsequent postejaculatory intervals increase rapidly (e.g., Larsson, 1956).

SOCIAL CONTROL OF MATING OPPORTUNITIES

One of the most obvious facts about human sexual life is that numerous social forces operate to prevent sexual intercourse between certain groups of people. Differences in age, race, and socioeconomic status are potent nonbiological inhibitors of sexual behavior; and monogamy, which is by far the most prevalent form of marriage, markedly reduces the number of available sexual partners. Obviously these forces are not perfectly effective, nor are they necessarily always desirable. The details of the rules prescribing sexual partners vary from culture to culture; we discuss this further in Chapter 8, but for now we want to emphasize that no human group is without a set of rules about who has sex with whom, and when, and how. Having and breaking some rules about sex seems to be an invariant feature of organized social life.

Many people internalize the sex rules of their group and obey them so well that the group as a whole views the behavior of these members as natural. The internalization of the rules during childhood and adolescence reduces the need for obvious physical constraints on inappropriate matings in adulthood. For the most part, sexual control is exercised without direct physical containment or threat.

Nonhuman species possessing organized social forms also exhibit sexual constraints. Although many more data are needed before the generalization can be completely secure, it appears that random mating among all reproductively capable members of groups of mammals, particularly primates, does not occur. In fact, the sexual organization of a group (who copulates with whom and when) is one of the very salient features of the entire social structure. We can illustrate these points by reference to the behavior and social organization of groups of baboons.

Although the exact taxonomic status of African baboons is a question of some debate, we may distinguish here three species of the genus *Papio*: *P. ursinus*, the chacma baboon of South Africa; *P. anubis*, the olive baboon of Kenya; and *P. hamadrayas*, the sacred or hamadrayas baboon of Ethiopia (DeVore and Hall, 1965; Kummer, 1968). The sexual behavior and social organization of all three species have been studied intensively since the late 1950s. The chacma and olive species, which together represent the baboons of the African savanna, are behaviorally rather similar when contrasted with the hamadrayas, which lives in desert country. We will consider the savanna baboons first.

The unit of baboon social organization is the *troop*. A single troop may contain as few as 8 to 10 or as many as 185 animals, but on the average, for both chacma and olive troops, it contains 30 to 50 individuals. Troops contain adult, subadult, and immature animals of both sexes, but not in equal proportions, for there are always fewer adult males than adult females. For every 10 females, there are approximately 3 males in a chacma troop and 4 males in an olive troop. When the number of females per male falls much below these values, as it occasionally does, marked changes in social behavior occur (Hall and DeVore, 1965). In troops of both species there are usually two immature animals for every adult female. In general, once a troop has been formed it remains a relatively closed social system, that is, there is little movement of animals between troops.

It is the social relations among males that constitute the backbone of the social organization of the entire troop. Because the troop as a whole moves about its home range foraging for food, going to water, and nesting in trees or on rocks at night, individual males do not possess spatially delimited areas that they defend against other males (territories). Rather there is a definite social ranking or ordering of males based on several behavioral dimensions; the name given to this social ordering is the *dom-*

inance hierarchy. As used by Hall and DeVore (1965), the dominance hierarchy of male baboons in a troop is the consequence both of a linear dominance ranking (analogous but not identical to a peck-order in chickens, and based on power relations among individual males) and of a ranking with regard to membership in what is known as the *central hierarchy* in the troop. This latter concept refers to the operation of two or three males as a group in threatening or fighting other males. Thus, if a male is a member of the central hierarchy, he will be aided in some of his hostile encounters with nonmember males by other members of the central hierarchy. This cooperation gives him an advantage over animals that would not otherwise defer to him. The central hierarchy creates a complexity in the dominance hierarchy that is not found in the hierarchies of nonprimate groups.

The linear dominance rank component of the dominance hierarchy is influenced substantially by physical size: the largest male is likely to be the most dominant. This position in the hierarchy is characterized in several ways, of which the most important for our discussion is the access of the dominant male to completely estrous females. It is characteristic of the dominant male that he copulates with females at the time of maximal perineal swelling, which is correlated with the time of ovulation. We pointed out earlier that during the first half of the menstrual cycle of female baboons there is a marked swelling and reddening of the perineum; this is produced by the gradually increasing levels of circulating estrogen. The swelling is maximal at midcycle (ovulation), and the female subsequently "deflates" just before and during menstruation. Hall and DeVore (1965) have shown that the dominant male (or males) of a troop are much more likely to copulate with a fully swollen female than with one who is inflating or deflating and that subadult or submissive adult males are relatively less likely to copulate with these females. The biological significance of this relationship seems clear: most pregnancies in the troop will result from matings with dominant males; thus the heritable characters of these males will have greater frequencies in the subsequent generation than those of the submissive animals.

Saayman (1970) has provided a detailed analysis of mating behavior in the chacma baboon that partially confirms the earlier observations of Hall and Devore on both chacmas and olives. He observed a troop of 77 animals that included 3 adult, 15 subadult, and 8 juvenile males. He found that all classes of males mounted and thrust against females, but that all of the adult male ejaculations occurred when the females were fully swollen. He observed 63 ejaculations, of which 37 were attained by the 3 adults, and the remaining 26 were distributed among the 15 subadults. All 37 adult ejaculations occurred with fully swollen females, as contrasted with fewer than 10 of the subadult ejaculations. Because adult males are dominant over subadults, the generalization made by Hall and DeVore is strengthened. Moreover, Saayman reports that the oldest male, who had lost his teeth and in general was not the most dominant, occasionally lost the fully swollen female with whom he was then consorting to a more dominant male. In general, however, there was little fighting over females among the adults or large subadults. It appears that the degree of social disruption attendant on pair formation with a fully estrous female depends on the female-male ratio. When the ratio is relatively large there is little or no harassment of a consorting pair. Nevertheless, the opportunity to mate with a fully swollen, ovulatory female is restricted to fully adult males, with the more dominant of these having priority when the number of females is relatively low.

One must not conclude from this description that the social-sexual organization of savanna baboons is representative of all nonhuman primates, or even of all baboons. For example, the closely related hamadrayas baboon of Ethiopia has evolved a different system of sorting males and females; Kummer (1968) has provided a detailed description of this system. Here we will describe only two major contrasts between the hamadrayas and savanna baboons.

First is the contrast in overall social organization. The savanna baboon troop is a single, closed group that operates as a foraging and breeding unit without breaking into subgroups for specialized activity, whereas the hamadrayas troop is a looser collection of individuals of all ages and both sexes that may come together only in the evening to share a single sleeping place. Kummer found the average size of these troops to be about 120 individuals. Within each troop he distinguished several subgroupings. These included *bands* of several males with attendant females, subadults, and juveniles (corresponding in size and age-sex composition to the savanna baboon troop), which often formed during intratroop fights and were involved in troop disintegration; *teams* of two males with females, which were formed during periods of travel; and, most important for our purposes, *one-male units*, which, in their fullest form, contain one adult male and two to five females who are adult or nearly so, and the juveniles belonging to these females.

The copulatory behavior of these baboons is organized within the context of the one-male unit: adult males rarely if ever ejaculate with females outside their own unit. The initiative in maintaining the polygynous relationship seems to be taken by the male. Adult estrus females often stray away from their male, and at a certain point the male will run to the female and threaten her or bite her on the back or neck. The effect of this aggressive behavior is to bring the female back into the immediate area of the rest of the male's unit. As Kummer points out, this consequence of aggression (viz., bringing an adult animal closer to the animal attacking it) is relatively rare among animals. It appears to be the essential component in maintaining the integrity of the one-male unit.

The second major contrast with the savanna species is thus that the one-male unit is *the* reproductive unit of the hamadrayas baboon. One might imagine that the male's opportunities for copulation would be the motivating factor controlling the development and stability of the unit.

However, the way that the single-male unit is initially formed is not compatible with this interpretation.

The unit begins its existence when a young adult male begins to herd a single juvenile female (12 to 18 months of age) with great regularity. This will not have been the first time the male has attended to a young female, for as a subadult he probably "kidnapped" even younger males and females for brief periods. Now, however, the physical distance between the male and female remains short and relatively constant, and the male utilizes the facial and gestural cues used by fully adult males in herding their groups of adult females. There are two additional features of the developing social relationship that are of particular importance: first, the pair does not engage in any copulatory behavior. Thus there is no sense in which the male's herding behavior is directly rewarded or reinforced by sexual contact. And second, the male exhibits many of the postures, gestures, and complex responses that adult female baboons show with very young animals; in short, the male mothers the juvenile female. Copulation in a new unit probably does not occur for about a year after initial formation. During this time, therefore, when the male shows progressively less interest in associating with other males and may acquire a second female into his unit, the control of his behavior is unrelated to opportunity for copulation. Moreover, Kummer observed a unit in which one female had begun menstrual cycling while the other had not. During this transitional stage the male copulated regularly with the cycling female during estrus but continued to exhibit the typical care-taking behavior to the younger female.

In conclusion, it is clear that the development of the reproductive unit of hamadrayas baboons does not depend on immediate sexual reinforcement. In its fully developed form, when all the females are adult, males do watch their estrous females very closely when they are near other males and will drive the other males away or, if necessary, fight them. But again, the unit does not disband when females are not sexually receptive, for example during pregnancy. Hence the role of primary sexual reinforcement as a controlling factor in the maintenance of the group must be, at best, a partial one.

SOCIAL CONTROL OF EXTENDED COPULATORY PERFORMANCE

We have already seen some examples of the roles played by physical and social stimuli in the initiation and maintenance of mating. These stimuli act on animals that are already physiologically predisposed to mate (e.g., the females are in physiological estrus or the males have adequate levels of circulating androgen). Thus the production of the mating sequence in a pair of animals is the consequence of an interaction of physiological and environmental determinants.

Moreover, mating itself produces alterations in the animals and their local environments that lead eventually to its cessation. An obvious example for females was the case of the turkey hen. A similar occurrence seems obvious in males of many species. The achievement of a single ejaculation in males of some species leads to relatively long periods of sexual inactivity; male guinea pigs, for example, seldom achieve more than one ejaculation in a 24-hour period. Males of some other species normally ejaculate more than once during a single sexual encounter with one female, but even in these cases the intervals between successive ejaculations increase. Male rats, for example, who may achieve five to eight ejaculations with a single female during a 90-minute period, wait about 5 minutes after their first ejaculation before mounting again but wait 15 minutes or more after their fifth. Male sheep also show regularly increasing postejaculatory intervals with successive ejaculations with a single female.

Taken alone, these data might suggest that the occurrence of ejaculation results in the gradual fatiguing of neural mechanisms associated with erection and ejaculation, and that this fatigue is independent of the characteristics of the social environment in which it develops. However, this interpretation would be incorrect; in some species, at least, ejaculatory frequency can be substantially enhanced by changes in the environment, in particular by the successive introduction of different females.

Domestic ungulates such as cattle and sheep provide excellent examples of the enhancing effect of female novelty.[3] An average bull, for example, when allowed to mate with a single cow until he has allowed 30 minutes to elapse without mounting, will achieve approximately 10 ejaculations. The interval between successive ejaculations will have shown a steady increase. If the first cow is then removed and replaced by a second one, the bull will begin to copulate again, and typically, will achieve at least a few ejaculations at a rate approximating his initial performance with the first cow. If a new cow is introduced whenever the male reaches the 30-minute criterion of satiation, the male can continue to exhibit ejaculatory behavior with as many as seven cows. The introduction of new females can result in the bull exhibiting as many as 70 ejaculatory responses. In addition to emphasizing the importance of environmental control over copulatory performance, these results are of practical importance in that they produce a method for increased collection rates of semen (by the use of artificial vaginas) for eventual artificial insemination (Schein and Hale, 1965).

The influence of manipulating the identity of the female on the male's performance has been studied systematically in sheep (Bermant, Beamer, and Clegg, 1969; Beamer, Bermant, and Clegg, 1969). In one experiment rams mated with single ewes for an hour or until 20 minutes had elapsed without a mount, whichever occurred sooner. At that time the first female

[3]Domestic roosters are another good example. A droll anecdote involving the comments of a U.S. president at a chicken farm serves as the basis for the label given to the phenomenon: "the Coolidge effect."

was removed and replaced by a second, fresh female, until the criterion for satiation was reached again. Thirty rams were tested a total of 152 times. Under these conditions the rams ejaculated about four times with the first female and two or three times with the second. Twenty-eight of the 30 rams consistently ejaculated more often with the first female than with the second. These results may be interpreted as follows: the introduction of the second female did facilitate the male's performance, for males would not have achieved these additional ejaculations had the females not been switched; and the facilitation could not be considered a complete restoration of the male's copulatory competence, for the rate of ejaculation was decidedly slower with the second female than with the first.

In the case of the ram, it is not necessary to wait until several ejaculations have occurred in order to achieve a facilitating effect by switching females. If a new ewe is introduced after every ejaculation the male approaches her and ejaculates almost immediately. Some rams can continue this ejaculatory performance through the introduction of at least 12 different ewes, without any appreciable increase in the time taken between the ewe's introduction and the ejaculation. Experiments of this sort are typically terminated by the experimenter's fatigue, not the ram's. Figure 3.8 illustrates the magnitude of the effect through the introduction of the fifth female. This effect is not attributable to the general disruption caused by the experimenter entering the pen and moving animals around, for if the original ewe is removed and replaced after each ejaculation, the male's performance is no different than if no interruption had taken place. Thus the identity of the ewe exerts an influence on the ram's ejaculatory rate from the very outset. The decreasing ejaculatory rate observed with continued exposure to a single ewe is not an indication of physiological capacity, but is a measure of the ram's habituation to the sexually stimulating characteristics of the environment.

One would like to know exactly what it is about the female that the male responds to, or fails to respond to, during the course of his exposure to her. One experimental approach to this problem is to attempt to confuse the male by masking the females' identities in various ways, for example, by making them look alike or smell alike. Beamer et al. (1969) tried to hide the identity of different ewes by dressing them in identical body coats and face masks before introducing them to the ram after each ejaculation. The rams were not fooled by the visual similarities thus produced. Moreover, when a single dressed female was removed after each ejaculation and re-dressed in a new bag and mask, the rams showed the same habituation to her as they would had she not been dressed. The visual and presumably partial olfactory masking utilized in the experiment was totally ineffective in preventing the ram's discriminations of individual identity. Similar experiments using bulls and cows have also failed. The lack of success in these experiments implies that rams and bulls can identify ewes and cows by reference to several sensory modali-

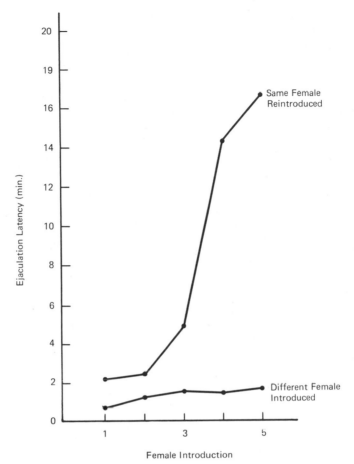

FIGURE 3.8 Average ejaculation latencies for rams (*Ovis aries*) sequentially reintroduced to the same female or introduced to different females. From Beamer, W., Bermant, G., and Clegg, M. (1969) Copulatory behavior of the ram, *Ovis aries*. II: Factors affecting copulatory satiation. *Animal Behavior* 17: 706–711.

ties and complicated combinations of information from these modalities. The only way to prevent the males from obtaining this information is to denervate them, that is, render them blind, anosmic, and so on. But this procedure disrupts the copulatory functioning of the males in ways more severe than simply abolishing their response to new females (Schein and Hale, 1965). Thus it is difficult to eliminate this portion of the behavioral repertoire without damaging the rest.

Male rats show relatively weak copulatory response to novel females. Most of the published studies indicate that males that have ejaculated several times with one female and hence have reached a postejaculatory period lasting about 30 minutes, will achieve only one or two more ejaculations with a new female (Bermant, Lott, and Anderson, 1968; Cherney and Bermant, 1970). Moreover, when new females are introduced after each ejaculation, the male rat does not maintain the rapid ejaculatory rate that the ram does (Wilson, Kuehn, and Beach, 1963). As we have already seen, the copulatory behavior of rats and other rodents is relatively complicated, consisting of a series of brief, discrete intromissions that culminate in the ejaculatory response. The quantitative characteristics of this preejaculatory behavior are changed by the introduction of new females, even though the overall ejaculatory rate is affected to a relatively slight degree. Some research has been directed at the question of what the male rat responds to; the question of particular interest has been whether he distinguishes a female that has not mated recently from one that has mated recently with another male. The results of the experiments are not in complete agreement, and the differential effects in all cases are relatively small (Fowler and Whalen, 1961; Wilson et al., 1963). Attempts to exclude relevant information by denervation have run into the aforementioned problem for other species (Bermant and Taylor, 1969).

Males of several other mammalian species have also been observed to respond positively, to one degree or another, to the introduction of a new female at the end of a period of copulation (Wilson et al., 1963; Schein and Hale, 1965). In order to assess the general biological significance of the phenomenon we would need to have good information on the relation between its occurrence in a given species and the normally occurring social-sexual system operating in that species. As far as we can tell, the effect is most pronounced in those species (bovine, ovine) in which one male naturally tends, guards, and typically mates with several females. On this basis we might predict that other similarly organized species, for example, elephant seals, would demonstrate a strong response in experimental conditions (LeBoeuf and Peterson, 1969). Similarly, we would predict that the effect would be smaller, or nonexistent, in species organized differently, for example monogamously. If one is tempted to interpret this prediction in terms of the human case, it is important, at the very least, to keep the distinction between sexual desire and sexual performance or competence firmly in mind.

ENVIRONMENTAL AND EXPERIENTIAL DETERMINANTS
OF ABNORMAL SEXUAL BEHAVIOR

In this final section of the chapter we will consider the roles of environmental circumstances in shaping and maintaining abnormal sexual behavior. We must of course be clear about what we take "abnormal" to mean. When we talk about human sexual behavior, there is a strong tendency for the word to carry two distinct messages: first, that the behavior so labeled is rare, and second, that it is bad or wrong. On reflection we realize that these two messages ought to be kept distinct, because one of them refers to matters of fact ascertainable simply by enumerating what people do, whereas the other refers to matters of value, and hence has as much to do with the characteristics of the person making the comment as with the behavior of the person being evaluated.

Most of us can learn to refrain from applying our value judgments about human sexual conduct to nonhuman animals, so this confusion about the meaning of abnormal may be less likely when applied to them. However, there is another problem that needs to be dealt with carefully. Even if we attempt to restrict our notions of normality and abnormality to a descriptive base, so that normal means frequent and abnormal means rare, we have to be very sure that we understand the circumstances under which our data were collected. In part this means only that enough individual cases have been observed to guard against random sampling errors; this problem is common to all research. But, more important in regard to animal sexual behavior, we must also be sure that we comprehend the extent to which the general conditions of observation and testing relate to the conditions in which the behavior evolved, both in the individual and the species. What animals do normally (frequently) in a setting such as a zoo or laboratory, they may do only rarely (abnormally) in their original environment, and vice versa. Hence comments about what is abnormal for the species cannot be made without reference to environmental circumstances.

This leads us to think in terms of normal and abnormal environments as well as normal and abnormal behaviors, and thus takes us one step away from purely numerical concepts of normality and abnormality, for it introduces a norm of *naturalness*. What animals do naturally, we believe, is what they do in the environment in which the species evolved. Behavior observed under these conditions of *ecological validity* is normal behavior in the sense of being natural. If qualitatively and quantitatively similar behavior is measured in members of the species when they are placed in a different environment, they may be said to behave normally under these new conditions. But if behavior appears in the new setting that never appears in the natural one, then that behavior is abnormal, in the sense of not naturally occurring, even if it is observed in every animal in the new environment and if there are more animals in the new environment than there are in the natural environment. For example, it may well come to pass in the next 50 years or so that there will be more orangutans in zoos than in jungles. But the normal behavior of orangutans will still be defined by the behavior of the animals in the jungle.

The sense of "natural" that we have been using is clearly related to the notion of "removed from human influence." To the extent that these two ideas are overlapping, it makes little sense to be concerned with what is natural in the behavior of domesticated species such as farm animals, standard lines of dogs and cats, and indeed, man himself. The concept of normality cannot rely on the concept of naturalness for support in these cases. Confusion about this can, and has, resulted in the misapplication of the ethological point of view to human affairs (Ardrey, 1966).

We will turn now to some of the better-known data on the environmental determinants of abnormal sexual behavior in animals. To begin with, the most typical problem is the failure of courtship and copulation when apparently healthy animals are placed in new environments. This is a problem that has long plagued zoo-keepers; 50 years ago expert opinion maintained that the artificial zoo environment produced physiological sterility. But it is now known that the failure to breed may be remedied by simulating the critical physical and social features of the natural setting. The details of the setting will vary from species to species, but Hediger (1965) has pointed out some general rules. For example, it is often important to allow the members of a potential breeding pair to smell each other before they can see each other. Hediger followed such a plan of gradual sensory exposure when he introduced two Indian rhinoceroses into a breeding situation, and thereby provided the first example of this species mating successfully in captivity. Careful attention to these kinds of details, guided by knowledge of behavior in natural settings, has substantially increased the number of species that have been bred under the "unnatural" conditions of the zoo.

A good deal of research attention has been paid to the effects of prepubertal experiences on subsequent sexual behavior in adulthood. There are several historical foundations for this research, including Lorenz's 1935 formulation of the function of imprinting (Lorenz, 1957), and the general psychological interest in infantile and childhood determinants of sexuality stemming from Freudian theory. As we shall see in Chapters 7 and 8, this interest has now expanded to include consideration of prenatal and neonatal hormonal conditions as well.

IMPRINTING

Almost a hundred years ago D. A. Spalding pointed out that newly hatched domestic chicks would follow virtually any available moving object just as soon as they could walk. If they were prevented from seeing anything for about the first three days of life (by tying hoods over their heads) then, rather than approaching a moving object, they would attempt to flee from it. In the ensuing decades several investigators added corroborating data for other bird species; this early history has been reviewed by Sluckin (1965). This early period of research was capped in 1935 when Lorenz reviewed available information and concluded that the "following" response was a reflection of the process by which the young animal comes to recognize members of its own species. The German word that Lorenz used as a label for this process, *Prägung*, has been translated as *imprinting*. Normally the first moving object a chick, duckling, or gosling sees is its mother; and it is this initial contact, so the theory went, that stamped in or imprinted the stimulus characteristics that would determine the object of the bird's social and sexual responses in adulthood.

Lorenz's theory of imprinting has generated considerable research in the last 25 years, some of which has dealt with the validity of the prediction that young birds imprinted on inappropriate objects (members of a different species, inanimate objects) would attempt to court and mate with these and not with their own conspecifics. In his review of the literature, Sluckin (1965) reports observations of "courtship fixations" on inappropriate objects in approximately 20 species of birds; unfortunately many of the data are anecdotal. He also points out that in some cases the tendency to court the inappropriate object does not preclude normal courtship and mating by the same animal, that the best evidence for cross-species mating tendencies occurs in birds that do not show the typical imprinting responses (e.g., "following") as hatchlings, and that hatchlings that show obvious imprinting are not particularly likely to exhibit courtship fixations. All of these points substantially weaken the validity of Lorenz's theory on the function of imprinting. However, this is due in part to the posited special nature of the imprinting process. We are still left with a reasonable body of evidence that for a number of avian species some forms of early social experience can predispose the adult to abnormal sexual behavior. Lorenz's fundamental point, that exclusive sexual responsiveness to conspecifics is not innate and unalterable, remains valid.

SOCIAL ISOLATION IN MAMMALS

The data on imprinting stressed the effects of early experiences on the object toward which adequate sexual responses would be directed in adulthood; the research that has been done with mammals, by contrast, has emphasized the importance of early experiences on the development of adequate copulatory responses per se. In particular, the concern has been with the effects of infantile and juvenile social isolation on adult behavior. The effects of social deprivation on nonsexual behavior has been reviewed recently by Mitchell (1973); here we will give a brief overview of the data on sexual behavior.

Initial experiments with rats and guinea pigs suggested that there might be a difference between the species in their sensitivity to social isolation. Beach (1942b, 1958) concluded that contact with adults or age peers after weaning was not necessary for adequate adult copulatory behavior in male rats, whereas Valenstein, Riss, and Young (1955) showed that male guinea pigs did not achieve intromission and ejaculation if they had been raised in postweaning social isolation. Kagan and Beach (1953) reported that male rats raised in all-male groups were less likely to show a completed ejaculatory pattern in adulthood than males raised in isolation, and they argued that this occurred because the socially reared animals developed nonsexual play responses that they carried over into adulthood and that competed with copulatory responses in the presence of an estrous female.

Subsequent experiments have complicated the picture substantially, for at least three independent studies have demonstrated that male rat copulation *can* be adversely affected by prepubertal social isolation (Folman and Drori, 1965; Gerall, Ward, and Gerall, 1967; Gruendel and Arnold, 1969). The results of these experiments have suggested that the effects of social deprivation in rats and guinea pigs are much alike if not identical. Gerall et al. (1967) also raise objections to the competing-response hypothesis of Kagan and Beach. A close reading of the available published reports does not allow one to determine the important variables behind the discrepant results. In other words, we cannot say precisely when "social isolation" will disrupt the adult male rat's copulatory behavior and when it will not. But it does seem clear that rats and guinea pigs need not be considered fundamentally different in their responsiveness to prepubertal social deprivation.

Female rodents are essentially unaffected by early social deprivation, at least with respect to the eventual occurrence of the lordosis response. It might be, however, that social deprivation would alter the voluntary characteristics of the female's copulatory behavior; the relevant experiments have not been done.

Male dogs and cats that are isolated from other animals during approximately the first year of life typically are disoriented in their copulatory attempts in adulthood (Beach, 1968a; Rosenblatt, 1965). At least for dogs, in order for the deprivation to be effective it must be complete; Beach (1968) reported that as little as 15 minutes of daily exposure to other animals during the first year was sufficient to prevent aberrant behavior in adulthood.

Whatever ambiguities remain in our understanding of the effects of isolation on rodents, there is no doubt at all that similar deprivations imposed upon rhesus monkeys, whether male or female, produce profound disruptions of adult copulation. Harlow (1965) has described the consequences of raising rhesus males and females without maternal or peer contact for the first 6 to 12 months of life. When the females reached puberty they were placed with previously isolated males. These pairs never copulated and often fought viciously. Pairing of isolates with normally reared individuals was ineffective in overcoming the isolate's deficiencies, as was allowing a group of isolates to live together for a summer in a large outdoor space. Numerous other techniques have been tried and found wanting. An independently conducted study of Missakian (1969) confirmed Harlow's earlier findings. In this experiment none of the adult rhesus monkeys that had been socially deprived as infants and juveniles were able to mount successfully as adults; and the deficit in copulatory behavior could not be remedied by extensive social contact during adulthood.

Chapter 4
Neural Determinants of Sexual Behavior

A very general hypothesis in physiological psychology is that every act of an animal can be explained by reference to a set of events occurring in the nervous system: the task of the physiological psychologist is to discover and demonstrate the details of this relationship for as much behavior as he can. For a number of reasons, this is a difficult task.

First, there is a problem of exact specification of the behavior to be explained by reference to the neural events. Behavioral descriptions such as "courting," "aggressing," "copulating," or "learning" are insufficiently precise; they need to be analyzed into their component parts. Later in this chapter we will have examples of how careful analysis of sexual behavior aids our understanding of its neural organization.

Second, it may be difficult to separate the neurophysiology that is *specifically* involved in the regulation of a particular act from that *generally* involved in maintaining the physiological integrity of the animal so that it may perform the act. This problem arises particularly when portions of the nervous system are destroyed experimentally in order to test their relevance in regulating the incidence or frequency of an act. Basic biological functions such as respiration and temperature regulation may be affected adversely by a relatively small lesion in certain brain areas. The sick animal produced by this lesion shows a variety of behavioral deficiencies, but we would not be justified in concluding that the damaged brain area was specifically involved in the normal regulation of all the affected behaviors. Careful experimental control procedures must be used to guard against incorrect generalizations.

A third, related difficulty stems from the obvious fact that the nervous system is indeed a *system*, a set of interconnected components and functions. Our natural tendency in studying the nervous system is to search out sites of particular neural activity and relate them to single acts or behavioral tendencies; this can lead to talk of "sex centers" or "hunger centers" or "pleasure centers" in the brain. These conceptualizations are meaningful to the extent that they point out the substantial degree of control over behavior that may be achieved by direct intervention at particular nodes of neural activity; but they can also mislead us into imagining the nervous system as a set of discrete components hooked together by relatively simple circuitry, each "in control" of a particular behavioral process. Particularly in the case of sexual behavior, which is typically the expression of an integrated response of both nervous and endocrine systems, a search for the controlling site of behavior is likely to be a fool's errand. In general the best model or theory for understanding the operation of the brain, particularly the human cortex, is very much an open problem (Eccles, 1966, 1970; Sperry, 1969).

We have been speaking as if there were only one kind of nervous system in the animal kingdom. But as most readers realize, there is a great diversity of forms of neural organization among animal phyla; a brief, informative overview of different forms may be found in Dethier and Stellar (1964). And within our own class, the mammals, and our own order, Primates, the different evolutionary histories experienced by different species have produced demonstrably different patterns of neural organization (Diamond and Hall, 1969). One consequence of this, for which examples may be found in the study of sexual behavior, is that damage done to the same location in the brains of two mammalian species does not necessarily produce the same effects in both species (e.g., Sawyer, 1960, pp. 1232–1233; see also Figures 6.2 and 6.3). We will provide examples of this point in more detail later, but we have introduced it here to indicate that nervous systems, like all other physiological and behavioral characteristics of a species, have been shaped and differentiated by the pressures of their separate evolutionary histories.

These introductory comments emphasize some of the difficulties inherent in understanding the neural correlates or determinants of sexual behavior. But none of the difficulties is insuperable, and there are now some facts and principles in which we may have confidence. We will limit our discussion almost entirely to a consideration of evidence from studies of laboratory mammals, for these are the animals we know best. Because our knowledge is closely connected to our techniques for acquiring it, we will turn now to a brief survey of the basic methodologies employed in this area.

METHODOLOGIES IN NEURAL PSYCHOLOGY

TISSUE DESTRUCTION

One way to try to understand how a machine works is to remove one part at a time and observe what happens. Physiologists have been employing this logic on the nervous systems of living animals for at least 300 years (Brazier, 1959), but only during the last 30 years or so have results relating tissue destruction to sexual behavior been collected systematically. Because many of the neural networks directly involved in the regulation of sexual behavior lie deep in the brain, particularly in the areas just above the pituitary gland, investigation depended on the development of techniques for destroying them selectively, without ablating or damaging all the tissue between them and the point of surgical entry on the skull. The first instrument to accomplish this form of surgery was invented by Victor Horsley and R. H. Clark in 1908; it is called the *stereotaxic instrument*. A modern instrument is illustrated in Figure 4.1. The stereotaxic instrument allows slender electrodes or other devices

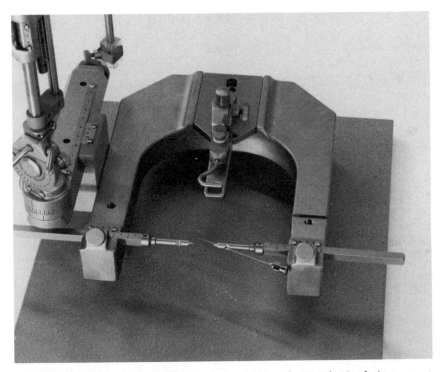

FIGURE 4.1 Stereotaxic instrument. Courtesy of David Kopf Instruments, Tujunga, California.

to be inserted accurately deep into the brain without undue damage to overlying tissue. Once the electrode is properly inserted, electrical current is passed between it and another electrode attached to the body surface. The type and amount of resulting tissue destruction depends on the size and composition of the electrode and the values of several electrical parameters. The routine use of stereotaxic surgery requires an atlas of the brain of the species being investigated; the atlas relates the location of brain areas to the three-dimensional coordinate system of the stereotaxic instrument (e.g., DeGroot, 1959; Pellegrino and Cushman, 1967). A single frontal section of the rat brain as it appears in an atlas is shown in Figure 4.2.

The brain damage produced by electrical current is permanent. Although the behavioral deficit initially produced by the lesion may show recovery, the damaged tissue does not. However, there are several methods by which "reversible lesions" may be made in the nervous system. For example, one may place a small blob of topical anesthetic on the end of the stereotaxic needle (Hoebel and Teitelbaum, 1962) or temporarily freeze a portion of the brain by passing a refrigerant through small stereotactically implanted tubes (Dondey, Albe-Fessard, and LeBeau, 1962). Or, if one is interested in temporarily suppressing the electrical activity of the brain surface, one may open a flap in the skull and apply a solution of potassium chloride to the tissue, coating the cortex; this results in a spreading depression of cortical electrical activity.

ELECTRICAL AND CHEMICAL STIMULATION
Evidence that neural transmission had an electrical character became available and partially accepted during the eighteenth century, but it was not until approximately the middle of the nineteenth that there were convincing demonstrations of electrical excitability in the nervous system (Brazier, 1959). The combination of modern electrical and electronic technologies with stereotaxic procedures now permits us to pass very small, nondestructive amounts of current through well-defined subcortical areas. The logic of this procedure, in relation to behavior, is that stimulation is the obverse of destruction: stimulation increases the amount of electrical activity in or exiting from a given area, whereas destruction reduces the activity to zero. With regard to some aspects of food-intake regulation in the hypothalamus, this logic has been used successfully to account for a good deal of experimental data (e.g., Teitelbaum, 1967). As we shall see later, the same approach has also been partially successful in relation to understanding the regulation of sexual behavior.

A major technique in the study of sexual behavior is the chemical stimulation of small brain areas by the direct implantation of hormones. This procedure allows the localization of neural sites directly involved in monitoring blood-hormone levels. The neural activity generated locally by the hormone begins sets of related neural events that eventually are reflected in sexual behavior.

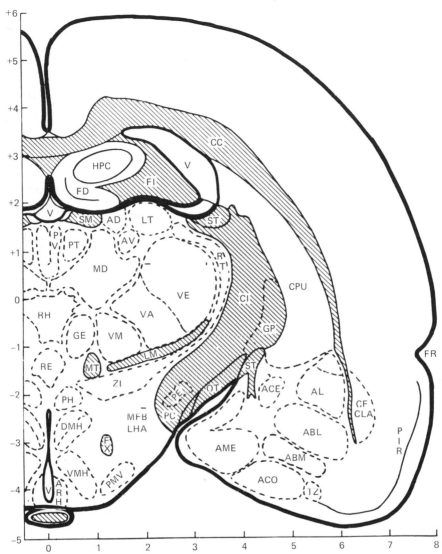

FIGURE 4.2 Stereotaxic coordinates of a frontal section of the rat brain. Abbreviations refer to various brain areas. From DeGroot, J. (1959) *The rat forebrain in stereotaxic coordinates.* Amsterdam: North Holland.

RECORDING BRAIN ACTIVITY

In 1929 the psychiatrist Hans Berger published the results of the first successful experiment to record the ongoing electrical activity of the human brain (Brazier, 1959). It was Berger who invented the name *electroencephalogram* (EEG), which now refers to recordings made from the surface of the brain or the scalp. The EEG gives an electrical picture of the combined activities of an immense number of neurons, and in that sense it is a rather gross measure of the brain's electrical activity. Nevertheless it has been widely and successfully employed in both experimen-

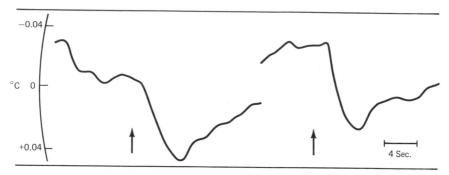

FIGURE 4.3 Cerebral temperature changes during intromission. Arrows indicate beginning of penetration. Downward slope of curve represents an increase in temperature. From Hull, C. D., Garcia, J., and Cracchiolo, E. (1965) Cerebral temperature changes accompanying sexual activity in the male rat. *Science* 149:89–90. Copyright 1965 by the American Association for the Advancement of Science.

tal and clinical contexts (e.g., Glasser, 1963). In regard to sexual behavior, we have good evidence that certain characteristics of the EEG are correlated with sexual receptivity in female cats and rabbits (Sawyer, 1960).

Subcortical electrical activity can be recorded by the stereotactic placement of recording electrodes. This procedure has been used to map different areas of the hypothalamus in terms of their sensitivities to hormone administration and stimulation of the genitalia (e.g., Lincoln and Cross, 1967). When very small recording electrodes are used it is possible to pierce the cell body of a single neuron and measure its individual rate of firing. This procedure has produced strikingly interesting results when used in studying the role of the hypothalamus in the regulation of food intake (Anand, Chhina, and Singh, 1962) and has also been useful in the analysis of hormone action on the brain (Ramirez, Komisaruk, Whitmoyer, and Sawyer, 1967).

Finally, we note that temperature changes in the brain may be recorded and, in some cases, related to behavioral changes. Figure 4.3 shows the cortical temperature changes recorded in male rats during a copulatory sequence (Hull, Garcia, and Cracchiolo, 1965). To the best of our knowledge, this interesting result has not been followed up and related to other known physiological changes that occur during copulation.

THE LOGIC OF NEURAL ORGANIZATION

In this section we want to introduce four ideas about the neural control of behavior that we will use throughout the rest of the chapter. We will not attempt to review basic neuroanatomy; readers who want to refresh themselves in the fundamentals will find good sources readily available (e.g., Thompson, 1967).

First, the simplest unit of behavioral organization is the *reflex*, and the simplest reflex is one that is organized at a single level of the spinal cord.

The simplest conception of the *reflex arc* includes three components: a receptor, a conductor, and an effector (Sherrington, 1947). The actual neural anatomy and physiology of even the simplest reflexes turn out to be surprisingly complex (e.g., Thompson, 1967, Chapter 12). For our purposes it is sufficient to point out that a movement or set of movements can be identified as a spinal reflex if it occurs, under conditions of proper stimulation, in an animal whose spinal cord has been severed from its brain. We shall see that a number of movements normally observed in copulatory behavior are spinal reflexes.

The second idea is that of *inhibition*. This word has many different uses in psychology. Here we will use it to mean an action of one part of the nervous system that reduces the likelihood that another part will function under conditions of stimulation. For example, we can say that a spinal sexual reflex is inhibited by neural influences descending from higher in the nervous system.

A third, related idea is that of *hierarchical organization* in the nervous system. The basic idea of a hierarchy is that parts of a system are organized so that "subordinate" parts are partially autonomous in their operation but are still under the final control of "superordinate" parts. A number of subordinate parts may be under the final control of a single superordinate one. To use a military example, there are several platoons under the control of a company commander, several companies within a batallion, several batallions within a regiment, and so on. Within the nervous system, spinal reflexes serve as units with some autonomy that nevertheless may be controlled by higher brain centers. Inhibitory and disinhibitory influences operate within the hierarchical organization of the nervous system. One consequence of this is the "release" or disinhibition of a spinal reflex as the result of higher nervous activity. If a neural component that normally inhibits the occurrence of a spinal reflex is itself inhibited, then the spinal reflex becomes disinhibited.

Fourth and finally, we introduce the idea of *neuroendocrine integration*. The normal operation of the nervous system in regard to sexual behavior depends on an adequately functioning endocrine system, which, in turn, normally depends on regulatory influences from the nervous system. Hence one of the key problems in understanding the physiological determinants of sexual behavior is the specification of how and where these two systems affect each other in the normal course of sexual behavior. For purposes of organization and clarity, we will reserve our substantive discussion of neuroendocrine integration for Chapter 6.

INHIBITION AND DISINHIBITION OF SEXUAL REFLEXES

A very interesting example of inhibition and disinhibition involves an insect, the male praying mantis, *Heirodula tenuidentata* (Roeder, 1963). Although the insect nervous system is in many ways quite different from the vertebrate system (see Figure 4.4), the case of the mantis highlights by analogy the relevant principle of behavioral organization.

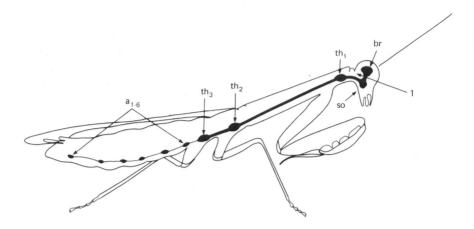

FIGURE 4.4 The central nervous system of the praying mantis: a_{1-6}, abdominal ganglia; br, brain; so, subesophageal ganglion; th_{1-3}, thoracic ganglia. From Roeder, K. (1963) *Nerve cells and insect behavior.* Cambridge, Mass.: Harvard University Press.

Mantids are carnivorous and as adults will cannibalize members of their own species. The tendency to cannibalism places the mantid male, who is considerably smaller than the female, in a position of risk during courtship and copulation. Male mantids have evolved a very cautious strategy to cope with this danger: their first response upon sighting a female is to freeze and remain motionless for minutes or even hours. The male's subsequent approach to the female begins quite slowly and may be interrupted, if the female's behavior warrants it, by additional periods of motionlessness. When the male finally gets within about a body length of the female, the tempo of his behavior picks up dramatically: he mounts the female's back, clasps her with his forelegs, and begins to beat her about the head with his antennae. His abdomen begins a series of copulatory movements at about the same time, but it may take as long as 30 minutes for actual coupling to occur. The male then remains mounted for as long as several hours, during which time sperm is transferred to the female. Following sperm transfer the male dismounts, and the encounter is ended. He has copulated and survived.

However, males do not always survive. At almost any time during the encounter, excluding only the time of sperm transfer, the female may attack him and cannibalize his head and upper thorax (see Figure 4.5). One might imagine that this inhospitable act would terminate his copulatory behavior; but, as it turns out, quite the contrary is true. The headless male mantis displays a series of movements that the intact male never

FIGURE 4.5 Cannibalism during copulation in the praying mantis, *Heirodula.* (A) Pair in normal upside down copulatory position. (B) Female attacking male and cannibalizing head. (C) Male continuing copulation after decapitation. From Roeder, K. (1963) *Nerve cells and insect behavior.* Cambridge, Mass.: Harvard University Press.

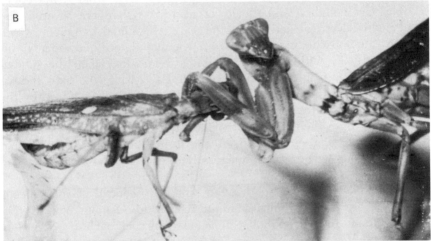

shows. First, the copulatory bending movements of his abdomen become especially intense. And second, these motions are accompanied by walking movements of the legs that draw the male's body into a position parallel with the female and covering her back. The behavioral changes following head loss have two important consequences: the male's upper body is driven away from the female's grasp so that she cannot easily continue to eat it, and the position of the male's body makes successful sperm transfer likely.

Put briefly, then, the female's cannibalism releases behavior in the male that leads to a high probability of sperm transfer. Roeder and his colleagues (Roeder, 1963) have done a number of experiments to work out the neural logic underlying this remarkable behavioral adaptation. They have found, for example, that the inhibitory "control center" for the vigorous locomotor and copulatory movements of the headless male is in the subesophageal ganglion (see Figure 4.4); when connections between this ganglion and the lower nervous system are severed, the spontaneous electrical activity of the sixth abdominal ganglion (which controls the musculature involved in the copulatory movements) increases markedly within about 5 minutes. Normally, when the subesophageal ganglion is intact, the intense neural activity in the sixth abdominal ganglion is either completely inhibited or selectively released by the various stimuli related to copulation: the sight of the female, perhaps by feedback from the genitalia after mounting, and so on. But when the subesophageal ganglion is removed, the disinhibition of the abdominal ganglion is complete, and the unnaturally intense copulatory motions are released.

Roeder (1963) emphasized that the inhibitory control of spontaneous neural activity is by no means unique to mantids. Similar neural processes are found in the copulatory movements of cockroaches, where they do not serve the obvious adaptive function that they do in mantids. Similar processes of inhibition are found to serve in the control of nonreproductive behaviors as well.

Roeder points out that the evolutionary basis for this form of control system in insects is in their historical relationship with the segmented worms, in which each generalized segment contains an almost completely independently operating set of neural mechanisms. Be this as it may, the copulation of the headless mantis and Roeder's elegant exposition of its neural control serve as a clear introduction to the concepts of neural inhibition and disinhibition of copulatory behavior. We can turn now to the more complex examples of sexual inhibition and disinhibition in mammals.

FEMALE SEXUAL REFLEXES

There have been studies of sexual reflexes in spinal[1] females of several mammalian species (Bard, 1940; Beach, 1967). Bard (1940) reported that spinal cats would respond with typically sexual postures and movements

[1]"Spinal" refers to the surgical separation of the spinal cord from the brain.

when their genitalia were stimulated. Hart (1969) obtained similar results with female rats, but emphasized that these responses were not reliably produced by artificial stimulation.

For at least some species of male and female mammals, then, the evidence suggests that some of the postures and motions directly involved in copulatory behavior are organized at the spinal level. It might be that these reflexes are inhibited by descending neural influences from the brain. If these brain areas could be located and selectively destroyed, their inhibitory influence on the lordosis reflex should be removed, and females should exhibit lordosis with greater intensity and/or estrous frequency than they otherwise would.

Law and Meagher (1958) have reported results of experiments with female rats that support this hypothesis. Females with small lesions in the posterior hypothalamus just anterior to the mammillary bodies, or in the preoptic area,[2] exhibited lordosis during the diestrus as well as the estrus phase of the normal estrus cycle. Goy and Phoenix (1963) performed a similar study with female guinea pigs. Unlike the results with rats, lesions in the posterior hypothalamus did not disinhibit lordosis in guinea pigs; however, some females with lesions in the midventral hypothalamus exhibited the reflex even after their ovaries were removed. This is evidence that the subcortical structures that inhibit the expression of sexual receptivity may vary from species to species. In Chapter 6 we will return to this issue and also place these and related data into the context of neuroendocrine integration.

Beach (1944) reported that some female rats demonstrated intensified strength and duration of lordosis following surgical removal of the cerebral cortex. This result has been confirmed and extended by the use of the spreading cortical depression procedure. Clemens, Wallen, and Gorski (1967) discovered that by the application of potassium chloride directly on the surface of the brain, the frequency of lordosis could be enhanced in spayed female rats injected with estrogen. As described earlier, this procedure temporarily suppresses electrical activity in the cortex. Under these conditions females increased their frequencies of lordosis by an average of 150 percent. Clemens et al. related this effect of cortical depression to the normal role of progesterone in facilitating the action of estrogen on lordosis during the normal estrus cycle (see Chapter 6). It may be that progesterone acts partially to inhibit the cortical influences on the lordosis reflex. Unlike the effects of subcortical lesions just mentioned, cortical damage or suppression does not facilitate lordosis in the absence of estrogen. And finally, as we shall see, cortical damage has a very different effect on the sexual behavior of male animals.

MALE SEXUAL REFLEXES

A clear picture of the role of spinal reflexes in the copulatory behavior of male dogs has been presented by Hart (1967a). Hart began his anal-

[2]Locations of brain areas mentioned in the discussion will be displayed later in Figure 4.10.

ysis with a detailed description of the copulatory behavior of intact animals; an appreciation of some of these details is necessary for an understanding of the results using spinal animals.

Figure 4.6 shows four stages of copulation in dogs. The male mounts the receptive bitch and clasps her. He then begins a series of shallow thrusting motions that eventuate in penile insertion. When insertion is complete the male's thrusting movements become much more intense, his forelegs pull back closer to the bitch's groin, and his hindlegs show characteristic stepping movements. The expulsion of semen begins during this reaction, which Hart has termed the *intense ejaculatory reaction* (IER). The male's erection is not complete until this point in the behavioral sequence. The IER continues for approximately 30 to 45 seconds. Then, unless the bitch throws the male off of her back, he dismounts with his penis still inserted and steps over it with one hind leg. This maneuver puts the two animals tail to tail, facing in opposite directions.

The penis remains inserted during the male's dismount because a pronounced swelling of its proximate part, the bulbus glandis, has locked it into the vagina (see Figure 4.7). The penis does not rotate within the vagina as the male steps over it; rather it twists through 180 degrees. Some seminal expulsion continues while the animals are in this position, which is called the *copulatory lock*, and is common to both domestic and wild canines. In Hart's sample of purebred beagles the average duration of the lock was approximately 14 minutes.

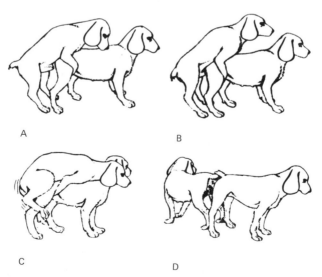

A

B

C

D

FIGURE 4.6 Copulatory behavior of the male dog. (A) Male mounting and clasping female. (B) Shallow pelvic thrusting leading to insertion. (C) Intense ejaculatory reaction. (D) Copulatory lock. From Hart, B. (1967a) Sexual reflexes and mating behavior in the male dog. *Journal of Comparative Physiological Psychology* 64:388–399. Copyright 1967 by the American Psychological Association, and reproduced by permission.

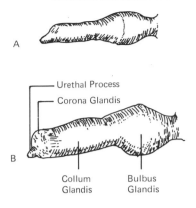

FIGURE 4.7   Diagram of the *glans penis* of the dog. (A) Nonerect state. (B) Erect state. From Hart, B. (1967a) Sexual reflexes and mating behavior in the male dog. *Journal of Comparative Physiological Psychology* 64:388–399. Copyright 1967 by the American Psychological Association, and reproduced by permission.

After the animals separate the male will not engage in additional copulation attempts for a period of time that ranges, across different males, from approximately 30 minutes to 48 hours. These numbers for the copulatory refractory period will figure importantly in our subsequent discussion of spinal reflexes.

Having established this sufficiently precise behavioral baseline, Hart (1967a) next prepared a series of spinal male beagles. Surgery consisted of sucking out (aspirating) a portion of the spinal cord in the midthoracic region. Thus, the spinal cord below this region was removed from any direct brain influence. Because the nerves that innervate the genitalia and hindquarters of the animal enter and exit the spinal cord in the lumbar and sacral regions, one can be confident that all the behavior exhibited by these portions of the operated animals reflect neural activity in only those sections of the spinal cord. Of course spinal animals are not capable of maintaining normal posture, locomotion, or coordination. Therefore all tests of sexual reflexes were made by stimulating the penis manually while the dog was comfortably cradled in a specially designed support apparatus.

Using these procedures, Hart (1967a) found that he could elicit four distinct spinal sexual reflexes. Each reflex was elicited by stimulation of a different portion of the penis. First, when the foreskin was moved back and forth over the penile collum (see Figure 4.7), the penis became partially erect and the hindquarters began shallow thrusting movements. Second, stimulation of the penile corona alone resulted in a strong downward arching of the back accompanied by leg extension and the loss of penile erection.

Third, if the penis was grasped just behind the bulb and friction applied there, the hindquarters responded with intense thrusting movements, the hindlegs began stepping or kicking, and within a few moments seminal fluid was expelled. This is just the pattern of behavior that characterizes the intense ejaculatory reaction of the intact animal. In the spinal animal it had a duration of 30 to 45 seconds and a refractory period of 5 to 30 minutes.

Finally, if the penis was grasped behind the bulb and the urethral process simultaneously squeezed gently, or if this stimulation was applied with the penis moved back between the hind legs and rotated through 180 degrees, the penile musculature responded so as to maintain the phasic expulsion of seminal fluid, but the hindquarters did not exhibit thrusting movements and the hindlegs did not step or kick.

The roles that these four reflexes play in the copulatory behavior of the intact animal are clearly definable. As the male first mounts the female and begins to move against her, the shaft of the penis will experience friction against the foreskin. This stimulates the first reflex of partial erection and shallow thrusting movements. The second reflex may also be elicited during the prepenetration phase of the sequence, when the male achieves an initial erection too large to permit penetration; the stimulation of the corona resulting from shallow thrusting would result in detumescence and withdrawal and thus allow the male to begin again. Normal erection plus thrusting leads to penetration, which, when deep enough, produces stimulation behind the bulb. The bulb begins to swell, the intense thrusting and stepping movements begin, and the initial ejaculations occur; these are the characteristics of the third spinal reflex, and the IER.

The behavior of the penile musculature during the copulatory lock closely resembles that observed when the penis is squeezed behind the bulb and the urethral process is stimulated (i.e., the fourth reflex); it is important to note that this reflex is elicited when the penis of the spinal dog is reflected and twisted into the normal copulatory lock position.

Thus we may analyze the copulatory behavior of dogs into a sequence of discrete reflexes organized in the spinal cord and elicited by stimulation of different portions of the penis.

Two additional points should be made. The first has to do with the durations of refractory periods following intense ejaculatory reactions in intact and spinal dogs. The minimum behavioral refractory period for intact animals was greater than 30 minutes, yet this was the maximum refractory period following IER in the spinal males. This difference shows that behavioral refractory periods are not measures of simple physiological fatigue of spinal sexual mechanisms; the spinal cord "recovers" from sexual activity more rapidly than the brain does. Evidence from "Coolidge effect" experiments in several species (Chapter 3) has shown that under certain circumstances behavioral refractory periods may be reduced by the appropriate social stimuli; the results of Hart's experiments give us direct evidence concerning the different rates of recovery from sexual activity operating at different levels of neural organization.

Finally, Hart (1967a) discovered that he could not elicit the intense ejaculatory reaction in an intact dog by manual stimulation of the penis in the absence of a receptive female. From this we may conclude that the IER reflex is normally inhibited by higher brain centers and remains so unless "unlocked" by appropriate visual and olfactory stimulation (the sight and smell of a receptive bitch) or removal of all neural input to it (spinal transsection). The effect is identical in the two cases: the disinhibited reflex can now be elicited by appropriate tactile stimulation of the penis.

In another experiment, Hart (1968a) searched for organized sexual reflexes in the spinal cord of the male rat. As in the case of the dog, the spinal rat's penis and hindquarters respond in a reasonably predictable fashion when the penis is stimulated. However, with one exception, it was not possible for Hart to draw clear parallels between the behavior of the spinal preparation and the copulatory behavior of the intact animal. The exception involved the possible role of inhibitory influences from the brain on a strong leg-kicking reflex that occurs in the spinal animal when the penile tip is stimulated. Hart suggested that this response is the foundation for the typical dismount of the male rat after an intromission. After a series of intromissions, however, the leg kicking is inhibited by descending influence from the brain, and the male continues his pelvic thrusting without dismounting. The additional genital stimulation thus achieved produces the ejaculatory response.

A Model of the Ejaculatory System

We turn now to the process by which stimulation of the penis produces changes in the nervous system that lead eventually to ejaculation. The discussion will be based on data collected with rats, but there is no reason to believe that the general features of the process will turn out to be different in other mammalian species. The rat turns out to be an admirable animal to use in this analysis because the pattern of the male's normal copulatory behavior permits us to do certain experiments that would be difficult or impossible with other species.

The schematic diagram in Figure 4.8 serves as a summary and review of the basic features of copulation in rats. Three behaviors are recorded: a mount without penile intromission, called simply a *mount*; a mount with intromission, called an *intromission*; and a mount with intromission and ejaculation, called an *ejaculation*. The number of intromissions preceding an ejaculation is called the *intromission frequency*, and the number of mounts preceding an ejaculation is called the *mount frequency*. Certain temporal features of these variables are also measured: the time from the beginning of the test to the first intromission (*intromission la-*

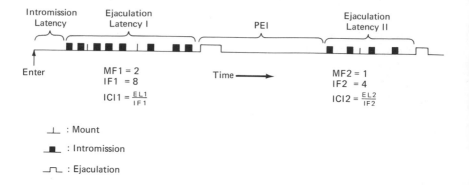

FIGURE 4.8   Schematic representation of rat copulatory behavior: MF, mount frequency; IF, intromission frequency; ICI, intercopulatory interval; EL, ejaculation latency; PEI, postejaculatory interval. See text for complete explanation.

tency), the time from the first intromission to ejaculation (*ejaculation latency*), the average interval between intromissions (*intercopulatory interval*), the duration of each intromission (*intromission duration*), and the time from ejaculation to the next intromission (*postejaculatory interval*).

The rat shares with a few other mammalian species a characteristically spaced-out pattern of intromission (see also Chapter 3). Each intromission lasts less than a half second but there is, typically, an interval of 30 to 60 seconds between intromissions. The first ejaculation is preceded by 8 to 15 intromissions, whereas the second and subsequent ejaculations in a single session are preceded by 4 to 7 intromissions. The first postejaculatory interval lasts approximately 5 minutes; subsequent delays after ejaculation are longer (Beach and Jordan, 1956; Bermant, Parkinson, and Anderson, 1969; Larsson, 1956).

This brief description is sufficient to highlight our major question of interest: How does the nervous system store the stimulation received during each intromission so that ejaculation is eventually achieved? That it must be stored somehow is particularly obvious for the rat, in which the interval between intromissions is 60 to 120 times longer than the duration of each intromission. Some trace or residue of stimulation from each intromission must remain over that period of time in order for ejaculation to occur.

Beach (1956) was the first to formulate explicit models of how the several intromissions lead to ejaculation. He presented two alternative accounts. First, it might be that the initial intromission triggers a timing mechanism, like a clock, which must run for a given time before ejaculation can occur; the first intromission after that period triggers ejaculation. Intromissions between the first and the last serve to keep the clock running. Alternatively, it might be that stimulation from each intromission builds upon stored stimulation from earlier intromissions (which we will call *excitation*) until sufficient excitation has been accumulated to cross the *ejaculatory threshold*; when the threshold has been crossed, the next intromission will release the ejaculatory response. At first glance these two accounts seem to be quite different. In fact it is difficult to collect experimental results allowing us to accept one and reject the other. In what follows we will emphasize the model of accumulating excitation.

Larsson (1956, 1959a) provided a good deal of experimental information that could be accounted for by a simple extension of the model of accumulating excitation. He found that when after each intromission a male was prevented from approaching and mounting a female, thus artificially extending the usual intercopulatory interval, the number of intromissions required for the first ejaculation was decreased. The extent of reduction depended on the duration of the enforced intercopulatory intervals, with a maximum reduction at a delay of 3 minutes. For example, in one experiment the mean intromission frequency for the first ejaculation under normal conditions was 15.6, but when a 3-minute interval was enforced after each intromission the mean intromission frequency was 4.9. Longer delays caused intromission frequency to begin to grow again (Larsson, 1956).

This so-called enforced interval effect on intromission frequency is quite reliable and has been replicated in several laboratories. It may be related to the model of accumulating excitation by postulating that excitation first grows, then decays after each intromission; maximum accumulation occurs at approximately 3 minutes after intromission. Normally the male doesn't wait that long, hence each intromission in normal copulation builds on less accumulated excitation than does an intromission during an enforced interval test. If intervals are enforced much beyond 3 minutes then each intromission is building on less than maximal excitation, and intromission frequency begins to grow again.

This form of the model of accumulated excitation is presented in Figure 4.9. The time axis is arbitrary and not meant to reflect actual data. However, an important feature of the model is clear from the figure: intromission frequency is assumed to be dependent on nothing but time, in particular the time between intromissions. No other factors are included. There is now very strong evidence that the single factor of time is insufficient to account for changes in frequency observed under certain experimental conditions (Bermant, 1964). In order to account for these changes additional factors must be introduced. Examination of the model as drawn suggests two obvious variables: the level of the ejaculatory threshold and the amount of excitation generated by each intromission. Changes in either or both of these would produce changes in intromission frequency at a constant value of intercopulatory interval.

It is much easier to speculate on the significance of these presumably important variables than it is to provide direct physiological evidence of their existence and operation. However, as we shall see, some of the following results of lesion and stimulation experiments can be considered

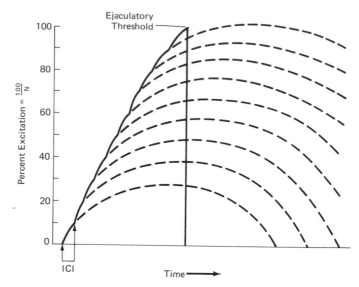

FIGURE 4.9 Model of accumulating excitation postulated for rat copulatory behavior. Each intromission builds on more excitation until the ejaculatory threshold is reached. ICI = intercopulatory interval.

within this framework. Finally, to relate the model to our earlier discussion of spinal reflexes, we repeat Hart's (1968a) conjecture that the part of the mechanism involved in the ejaculatory threshold is the inhibition of the intromission reflex, which normally involves a sharp extension of the hindlegs. When this leg reflex is inhibited the male continues to thrust until he ejaculates. According to this conjecture, inhibition of hindleg extension reflex constitutes a necessary but not sufficient condition for crossing the ejaculatory threshold.

Disinhibition of Ejaculation and Seminal Emission by Electrical Stimulation and Subcortical Lesions

Adult male rats spontaneously emit seminal fluid once or twice each day, and normally they eat the ejaculate. To gain an accurate count of spontaneous semen output, a collar is placed around the rat's neck; this prevents access to the genitalia (Orbach, 1961). The reflex that leads to seminal emission under these circumstances is only part of the entire set of reflexes that produce the complete ejaculatory response during copulation. Although male rats regularly exhibit a good deal of genital-cleaning behavior, it is not the resulting stimulation that produces the spontaneous emissions. It is that the tonic inhibition operating on the emission reflex oscillates during the course of the day, even without direct genital stimulation. When this inhibition falls below threshold level, the reflex operates to pump semen out through the penis.

The normal rate of seminal emission without erection or other behavioral indices of ejaculation can be increased by electrical stimulation of the brain. For example, Herberg (1963) reported that alternating current stimulation of one portion of the median forebrain bundle in the hypothalamus (see Figure 4.10) of male rats regularly produced seminal oozing without concomitant erection. Stimulation in this brain area was positively reinforcing: rats would learn and maintain a lever-pressing response that produced the brain stimulation. It would be interesting to know the extent to which the reinforcing properties of this situation depended on feedback from the genitalia during emission. The proper

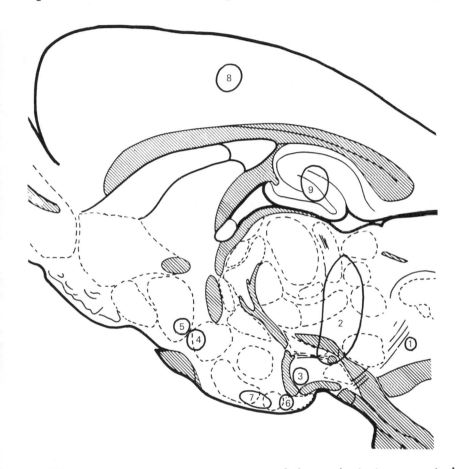

FIGURE 4.10 Diagrammatic representation of the rat brain in parasagittal section (from DeGroot, 1959). The numbers label areas in which lesion or stimulation studies have been performed, as indicated in the text: (1) reticular formation (Goodman et al., 1970); (2) medial forebrain bundle (Caggiula, 1970; Caggiula and Hoebel, 1966; Herberg, 1963; Hitt et al., 1970); (3) diencephalic-mesencephalic border (Heimer and Larsson, 1964a; Lisk, 1966); (4) anterior hypothalamus (Vaughn and Fisher, 1962); (5) preoptic area (Law and Meagher, 1958); (6) mammillary bodies (Law and Meagher, 1958); (7) midventral hypothalamus (Goy and Phoenix, 1963); (8) cerebral cortex (Beach, 1944; Clemens et al., 1967); (9) dorsal hippocampus (Bermant et al., 1968).

experiment to answer this question, which would involve deafferentation of the genital area, appears not to have been performed.

The simplest procedure for producing repeated seminal emissions is electroejaculation: rather intense electrical current is passed through the animal from an electrode inserted in the rectum. This stimulation triggers the release of semen directly, without involvement of the higher nervous system. Experiments with rats (Arvidsson and Larsson, 1967) and domestic sheep (Pepelko and Clegg, 1965) have shown that repeated electroejaculations do not affect the ability of males to copulate and ejaculate normally immediately afterward. In another experiment with domestic sheep, Beamer, Bermant, and Clegg (1969) found that when the males wore aprons that prevented penetration but not emission, they did not exhibit the normally observed decrement in sexual response with a single ewe (see Chapter 3). Taken together, the results indicate that in itself the occurrence of emission neither fatigues the normal ejaculatory process nor reduces the male's sexual arousal.

There have been two reports of greatly increased ejaculation frequency in copulating male rats when they were stimulated in or near the medial preoptic region of the hypothalamus (Van Dis and Larsson, 1971; Vaughn and Fisher, 1962). These authors reported that only a small proportion of their animals showed this effect; hence we are limited in what we may conclude. However, two points can be made on the basis of these data. First, the increased ejaculatory frequencies of affected animals reflect more than a simple increase in the rate of normal copulation; the number of intromissions required to achieve the ejaculatory response and the duration of postejaculatory refractory periods were both reduced to minimal values. This suggests that the effect of the stimulation is to lower the threshold of disinhibition of the ejaculatory reflex directly (see Figure 4.9). As we shall see, stimulation in other hypothalamic areas can increase the rate of ejaculation by increasing the rate at which the male achieves intromissions; in this latter case we need not assume that the ejaculatory threshold itself has been lowered. Second, the effective site of stimulation appears to be in an area of the brain that, when destroyed by lesion, abolishes copulation completely. This suggests that the site of effective stimulation is within the general complex of neurons that is involved in regulating various aspects of copulatory behavior, including initial sexual arousal.

A striking disinhibition of ejaculatory behavior can follow the placement of large lesions at the diencephalic-mesencephalic border (see Figure 4.10) in the brains of male rats (Heimer and Larsson, 1964a; Lisk, 1966). The numbers of ejaculations achieved in hour-long tests may more than double as a result of the operation. However, as in the aforementioned case of anterior hypothalamic stimulation, this effect does not occur in every operated animal, and to date no convincing evidence has been reported that relates the occurrence or magnitude of ejaculatory disinhibition with the exact sites of tissue damage. At present we can say only that the locus of greatest disinhibitory effect lies posterior to the mammillary bodies of the posterior hypothalamus (Heimer and Larsson, 1964b) and anterior to the midbrain reticular formation (Goodman, Jansen, and Dewsbury, 1971). To put the problem of precise localization in perspective, we should point out that in the rat brain this locus is a space in the anterior-posterior direction of approximately 1 to 2 mm (see Figure 4.10) that contains numerous nuclei and fiber tracts. A good deal of combined neuroanatomical and behavioral work will be required to elucidate the mechanisms underlying the large effect of lesions in this area on copulatory behavior.

Less dramatic increases in the rate of copulation in male rats have been observed following destruction of the dorsal hippocampus (Bermant, Glickman, and Davidson, 1968; Dewsbury, Goodman, Salis, and Bunnell, 1968; Kim, 1960). The effect is observed whether the lesions are made stereotaxically or by surgical ablation; in the latter case all the cortex overlying the hippocampus is also removed (see Figure 4.10). When both the dorsal and ventral aspects of the C-shaped hippocampus are destroyed, the rate of copulation is not increased; animals with these large lesions copulate at a normal rate (Bermant et al., 1968).

CONTROL OF COPULATORY PACE BY CENTRAL OR PERIPHERAL STIMULATION

Caggiula and Hoebel (1966) were the first to demonstrate that electrical stimulation of the male rat hypothalamus could produce an increased rate of normal copulatory behavior. As shown in Figure 4.10, the site of effective stimulation was in the posterior hypothalamus, in approximately the same region as that found by Herberg (1963) to be related to seminal emission. Unlike the effects of anterior hypothalamic stimulation as reported by Vaughn and Fisher (1962), posterior hypothalamic stimulation did not reduce the number of intromissions required to achieve ejaculation. Rather, the effect of stimulation was to reduce markedly the time intervals between intromissions and between an ejaculation and the next intromission. The enhanced rate of copulation was closely coupled to the stimulation; when current was discontinued the males' sexual behavior immediately fell to normal rates or below. Finally, the males will press a lever to receive electrical stimulation in this area of the brain. This is one of a number of examples that lead to the general principle that brain areas that are involved in regulating the incidence or rate of species-specific behaviors are also directly involved in neural systems of pleasure or reward (Glickman and Schiff, 1967).

An intermittent, mildly painful electric shock applied to the neck or back of male rats can produce increases in the copulatory rate that are very similar to those produced by direct hypothalamic stimulation (Barfield and Sachs, 1968; Beach, Conovitz, Goldstein, and Steinberg, 1956). The similarity of behavioral results obtained with central and peripheral stimulation might lead one to question whether the same neural processes or mechanisms are involved in both cases. For example, could it be that the posterior hypothalamic stimulation represents the same kind of generalized stimulation that skin shock does? Given a male rat in the presence of a receptive female, perhaps *any* intermittent, mildly painful or aversive stimulus will lead to an increase in the rate of copulatory activity. It is known, for example, that simply handling male rats during copulation, or systematically removing and then replacing them with the female, can increase the rate of copulation (Bermant, 1964; Larsson, 1963). However, there are three important reasons to doubt that direct brain stimulation is identical to peripheral stimulation in terms of the neural processes involved. First, the brain stimulation must be precisely localized if it is to produce increased copulatory rates; this would not be expected if it functioned as a "nonspecific" stimulus to copulation. Second, direct brain stimulation in this area is positively reinforcing, whereas peripheral stimulation is aversive. And third, Caggiula (1970) identified some points of stimulation that affected only sexual behavior, whereas peripheral stimulation can modify other behaviors as well, in particular the tendency to fight (Ulrich and Azrin, 1962). It may be, as Caggiula (1970) has suggested, that peripheral stimulation results eventually in electrical activity in the medial forebrain bundle equivalent to that produced by direct brain stimulation. At present we do not know.

HYPOTHALAMIC AND CORTICAL LESIONS THAT RETARD OR INHIBIT COPULATION

There are several separate areas in the hypothalamus that appear to be essential for the occurrence of copulation in one mammalian species or another. Surgical destruction of these areas leads to a marked deterioration or total cessation of sexual behavior. However, the reasons why the lesions are effective in inhibiting copulation are not the same in all cases. Lesions in the basal hypothalamus around the stalk of the pituitary gland produce cessation of copulation in both males and females of several species (e.g., Sawyer, 1960). Ordinarily the behavioral deficit is accompanied by degeneration of the gonads and/or sexual accessory structures, showing that the lesion has interrupted the neuroendocrine pathways between the pituitary gland and the gonads. In Chapter 6 we discuss the neuroendocrine mechanisms involved in these cases.

There is a second class of hypothalamic lesions that appear to be involved in maintaining sexual behavior independent of hormone action. Damage to these areas does not lead to deterioration of sex accessory structures or gonads, and the behavior deficits produced by the lesions are not remedied by treatment with exogenous hormone.

Preoptic and Anterior Hypothalamic Areas

Large lesions in the preoptic or anterior hypothalamic areas (see Figure 4.10) result in the total cessation of copulation in male rats (Heimer and Larsson, 1966/1967). No lesioned animal in this experiment displayed mounting behavior postoperatively, although most males approached the female and sniffed her genitalia. Injections of testosterone were totally ineffective in restoring copulatory behavior. Smaller lesions in the same areas produced partial and temporary deficits. Heimer and Larsson concluded that there were no specific locations within the preoptic-anterior hypothalamic continuum that were essential for maintaining copulatory behavior. As previously discussed, electrical stimulation of the preoptic area at least occasionally increases the frequency of ejaculation. Thus the preoptic area may provide one example of a brain site in which lesioning and stimulation produce opposite effects on copulatory behavior. In female rats anterior hypothalamic lesions, but not preoptic area lesions, have been shown to prevent the occurrence of lordosis. Preoptic lesions in the female left lordosis intact but markedly reduced the amount of malelike mounting that receptive females normally exhibit (Singer, 1968). Taken together, the data suggest that the preoptic area in the rat is involved in regulating mounting responses in both male and female rats, but is not involved in the regulation of lordosis. The anatomical distinction between anterior hypothalamic area and preoptic area is rather fine, and we do not yet know precisely what nuclear groups are involved. In species other than the rat, anterior hypothalamic lesions have also been shown to inhibit sexual receptivity independent of any action on the gonadotropin-steroid control mechanism (see Sawyer, 1960; Goy and Phoenix, 1963). On the other hand, in the rabbit the effective area for lesions inducing this effect was more posterior (see Chapter 6).

Medial Forebrain Bundle

We have already seen that electrical stimulation of some portions of the medial forebrain bundle in the posterior hypothalamus can produce either seminal emission (Herberg, 1963) or rapid copulatory pacing (Caggiula, 1970). Conversely, large lesions of the bundle abolish all mounting behavior in both male and female rats, but leave the female's lordosis response unimpaired. Smaller lesions in the bundle produce a partial decrement in responding (Hitt et al., 1970).

Cortical Damage and Male Copulation

Early experiments on the effects of cortical damage on copulation in male rats (Beach, 1940) suggested that there was a direct relationship between the amount of tissue removed and subsequent copulatory failure, regardless of the exact location of the removed tissue. This purely quantitative relationship between brain damage and behavioral deficit stood in contrast to results collected later in the male cat that showed that lesions in the frontal motor cortex were substantially more effective in producing behavioral deficits than were lesions in other cortical areas (Beach, Zitrin, and Jaynes, 1955; Zitrin, Jaynes, and Beach, 1956).

However, in a more recent series of experiments using male rats, Larsson (1962a; 1964) found that a certain degree of cortical localization does exist in the regulation of copulation. As in the cat, the single most effective location of damage was in the frontal cortex: 40 percent of the males with frontal lesions showed permanent postoperative suppression of copulation. Figure 4.11 is a diagram of the cortical surface showing the areas under consideration.

There is no doubt that lesions of the cortex can have quite different effects in males and females of the same species. Lesions that suppress copulation in male rats may enhance lordosis in females. Similarly, temporary suppression of cortical function by the direct application of potassium chloride suppresses male copulatory behavior but enhances lordosis in females (Clemens et al., 1967; Larsson, 1962b). This difference may be at least partially understood in terms of the greater dependence of the male on cortically mediated, fine-grained muscular coordination for the execution of copulatory responses. Also, it may be that an intact frontal cortex facilitates the integration of incoming sensory information regarding the estrous condition of the female. Sexual arousal and performance in male rats is partially dependent on visual and olfactory cues (e.g., Bermant and Taylor, 1969) and the suppressive effect of frontal cortex damage may reflect an inability to synthesize incoming information properly. The lordosis response, on the other hand, is elicited directly by tactile stimulation of the genital area or flanks; although severe cortical damage may cause some lack of general muscular coordination, the lordosis response itself remains unimpaired.

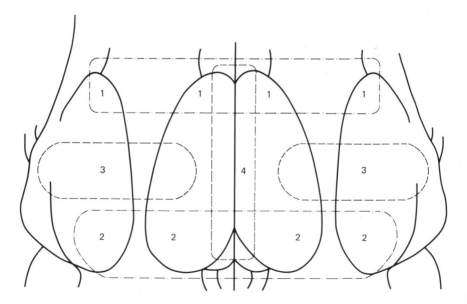

FIGURE 4.11   Top and side views of areas of the cerebral cortex of male rats: (1) frontal lobes, (2) posterior, (3) dorsolateral, (4) medial.

TEMPORAL LOBE DAMAGE AND MISDIRECTED COPULATION

Some of the most dramatic examples of altered sexual behavior following brain damage occur when lesions are placed in certain areas of the temporal lobes in male cats, rhesus monkeys, and several other mammalian species, but not including rats (Bermant, Glickman, and Davidson, 1968; Schreiner and Kling, 1956). Lesioned animals will mount other males, animals of other species, stuffed toys, and the arm of the experimenter if given the opportunity. One lesioned male cat, which had been given a lethal injection of anesthetic because of recurring epileptic seizures following surgery, continued to attempt copulation with a teddy bear until the anesthetic took effect. When the experimenter tried to remove the teddy bear the cat revived and continued his mounting and thrusting for a few seconds, at which point he died (Green, Clemente, and DeGroot, 1957).

These results in experimental animals have been compared to changes in sexual arousability and behavior in humans who have undergone temporal lobe surgery to correct severe epilepsy. For example, Terzian and Dalle Ore (1955) reported the case of a 19-year-old boy whose frequency of masturbation increased to two or three times daily following temporal lobe removal. Blumer (1970) has pointed out that only a relatively small percentage of surgically treated epileptic patients exhibit unusual sexual behavior postoperatively, but when such episodes occur they appear to share some characteristics with the behavior of lesioned monkeys. Importantly, he also notes that prior to surgery the patients show little or no sexual interest or activity, and that the surgery typically is followed by sexual behavior within the acceptable range.

SUMMARY

It will be clear to the reader that much of our information is fragmentary, spread between the sexes and across several species, and involves the use of different techniques. Nevertheless, the following general scheme seems to be plausible, particularly for male rats.

Sexual reflexes, organized at the spinal level, are modulated by the action of higher nervous centers. Areas of the anterior hypothalamus, and in particular the medial preoptic area, appear to be necessary for the initiation and maintenance of mounting and thrusting responses. Activity in the preoptic area is transmitted, via the medial forebrain bundle, to collections of cells in the anterior midbrain that are directly involved in inhibiting the release of spinal reflexes. The function of this activity may be to inhibit these midbrain inhibitory influences, thereby disinhibiting the spinal mechanisms. The hypothalamic area immediately surrounding the stalk of the pituitary gland is involved in monitoring the flow of gonadal hormones and regulating it by influencing the output of gonadotropic hormones from the pituitary. Lesions in this area lead to decreases in sexual behavior by producing a functional castration (see Chapter 6). Large lesions at the border of the posterior hypothalamus and the midbrain lead to increased ejaculation frequencies. In some but not all of these brain areas electrical stimulation has an opposite effect to that produced by lesioning.

The model of the ejaculatory mechanism presented here may be used to organize some of the data on relations between brain function and sexual behavior. For example, the greatly increased frequency of ejaculation observed during medial preoptic stimulation may reflect a drastic reduction of the ejaculatory threshold. Medial forebrain bundle stimulation, on the other hand, does not reduce the frequency of intromission preceding ejaculation; although we do not have enough data to be certain, it may be that intromission frequency is increased under stimulation. It is also clear that the stimulation-bound intromission occurs at a greater than normal rate. In this case, then, it may be that the amount of excitation available per intromission remains normal but that the males accumulate less after each because of their increased rate of response. Additional insight into these possibilities could be gained by coupling lesioning experiments with measurement of the durations of individual intromissions, for it has been shown that these durations increase as intromission frequency decreases (Bermant, Parkinson, and Anderson, 1969). To the extent that duration of insertion is correlated with penile stimulation, it is also correlated with the amount of available excitation. An experiment that found intromission duration to remain constant while intromission frequency varied would have demonstrated a shift in the ejaculatory threshold.

Large lesions of the rat cerebral cortex produce decrements in male sexual responding, but not necessarily in female responding. In some species temporal-lobe lesions lead to misdirected sexual responses. The complete significance of all these findings for the neural regulation of sexual behavior in humans remains to be understood.

Finally, we need to emphasize again that this chapter has treated neural determinants of sexual behavior out of their normal context of interaction with the endocrine system. In the next chapter basic sexual endocrinology is provided, and in Chapter 6 we describe the major properties of neuroendocrine integration.

Chapter 5
Hormonal Determinants of Sexual Behavior

It is not difficult to justify the decision to devote a considerable portion of this book to the relationship between hormones[1] and behavior. No single physiological variable or group of variables is as important in determining the occurrence or level of sexual responsiveness as the amount of gonadal hormones in the blood. This kind of relationship is unusual if not unique in behavioral physiology. In no other area of behavior do hormones appear to occupy such a commanding role, and no hormones have nearly as important an influence on sex behavior as the gonadal hormones. Brief amplification of this last statement will help to place in perspective the ensuing discussions.

EFFECTS OF HORMONES

HORMONES IN NONSEXUAL BEHAVIOR
The gonadal hormones affect behaviors other than courtship and mating. For example, progesterone stimulates maternal (nest-building and retriev-

[1]A hormone is defined as a chemical compound secreted by a specific gland and transported in the bloodstream to its site of physiologic action.

ing) behavior in rabbits and rodents and parental behavior in both male and female birds (Lehrman, 1961; Zarrow, 1968); and estrogen increases the general activity level of females. In males of many species, androgen seems to play a part in activating aggressive behavior. All of these behaviors may be viewed as parts of the entire reproductive function. Thus aggressive behavior plays a role in the defense of territory and the struggle for mates; like general activity and parental behavior, it is therefore related to courtship and mating. The gonadal hormones have not as yet been clearly shown to have any importance in learning, although it has been suggested that prenatal exposure of humans to steroid hormones may elevate IQ (Money, 1961a).

NONGONADAL HORMONES IN SEXUAL AND NONSEXUAL BEHAVIOR

Despite a variety of generally unconfirmed and even conflicting reports,[2] there is no reason to believe that under normal conditions the nongonadal hormones play important and specific roles in sexual behavior. However, as we shall see, the adrenal cortex produces hormones similar to those of the gonads, and the possible involvement of these hormones has to be considered; we will return to this topic. Of course, a variety of nongonadal hormones play nonspecific roles in maintaining sexual behavior in that they maintain the normal health and metabolism of animals.

To what extent are nongonadal hormones important in nonsexual behavior? Clinical medicine provides us with dramatic examples such as the psychotic manifestations of hyper- and hypoadrenalism; "myxedema madness" or, less dramatically, symptoms of slow mentation that may result from hypothyroidism; and the various mental aberrations that can accompany hyperthyroidism, from "emotional liability" to frank psychosis. But we have little systematic information on the role of the thyroid and adrenal in normal human behavior. Experimentally the thyroid hormones have been shown to have marked effects on various types of learning behavior in rats (Eayrs, 1968), and the adrenal corticoids and of adrenocorticotropin (ACTH) can clearly be shown to influence fear-motivated behavior (Levine, 1968). It should be emphasized that one reason why these effects seem less important than those of the gonadal hormones on sexual behavior may be simply that they have been less studied.

HORMONES AND BEHAVIOR IN PERSPECTIVE

Consideration of the significance of the extraordinary dependence of sexual behavior on gonadal hormones soon leads to very difficult problems relating to the physical basis of behavior. How does a simple change in the level of a hormone in the blood specifically and profoundly redirect

[2]For example, it has been reported that prolactin and ACTH decrease male sexual behavior in rabbits (Hartmann et al., 1966; Koranyi et al., 1966); but it has also been claimed that ACTH stimulates male sex behavior in this species (Bertolini et al., 1969).

the behavior of an animal? It is not likely that a complete answer will be found in the foreseeable future—there are philosophical problems as well as biological ones implicit in the question. But study of the interaction between hormones and sexual behavior is a particularly appropriate platform from which to plumb the depths of the biobehavioral mysteries, if only because in this area causal relationships are so clearly demonstrable between physiological variables and behavioral manifestations. Furthermore, the gonadal hormones in their relationship to sexual behavior have one characteristic that gives this relationship an advantage over other phenomena in the area of physiological psychology: the same hormones have vital, well-known effects on reproductive system anatomy and physiology, as well as on behaviors. These two classes of effects are analogous in some ways. Among those areas of analogy that we shall discuss are the following phenomena: (1) The cyclic rise and fall of gonadal steroid levels simultaneously initiates both behavioral and physiological events relating to reproduction. (2) The feedback mechanism that controls the levels of gonadal steroids depends in part on receptor cells in the hypothalamus. Similarly, the effects of these hormones on sexual behavior is mediated by hypothalamic receptors. (3) Synergistic and antagonistic relationships among the two major classes of ovarian hormones occur in relation to both physiological and behavioral phenomena. The hope and belief that our relatively advanced understanding of the physiological phenomena may help us build a model of the hormonal regulation of sexual behavior motivates the ensuing discussions. It also explains the attention we direct to concepts of hormonal physiology in the next section.

ENDOCRINE CONCEPTS

THE STEROID HORMONES

A thorough understanding of the biological basis of sexual behavior requires careful consideration not only of the direct involvement of gonadal hormones in activating and maintaining behavior, but also of the mechanisms by which secretion of the hormones is controlled. This in turn requires mastery of a few basic concepts of the chemistry and physiology of steroid hormones.

All known hormones produced by the male and female vertebrate gonads are steroids; that is, they share a common basic molecular skeleton (see Figure 5.1) that is common also to the hormones of the adrenal cortex. The hormones produced by the ovaries belong to two classes: the *estrogens* (most important example, *estradiol 17β*) and the *progestational compounds* (of which the only representative known to have major physiological importance is *progesterone*). The main hormones of the testes are known as *androgens*, and the most important in this class is *testoster-*

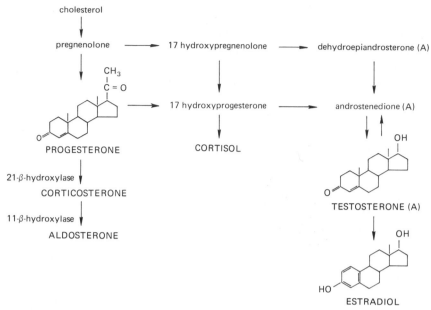

FIGURE 5.1   An abbreviated version of the basic pathways of steroid biosynthesis. The main physiologically important steroid hormones are shown in capitals, and complete formulae are given for the major gonadal hormones. Two important enzymes involved in production of adrenal corticosteroids are shown on the left. Absence of these enzymes results in accumulation of precursors, causing the synthetic "machinery" to shift toward increased production of the androgens dehydroepiandrosterone and androstenedione. Very little testosterone or estradiol is actually secreted from the adrenal cortex. A = androgen; double arrows indicate a freely reversible reaction.

one. The adrenal cortex secretes a variety of steroid hormones, of which the best known are *cortisone* and related *glucocorticoids*; it also produces androgens and small amounts of estrogens.

There is considerable overlap in the functional activity of different steroid hormones. For instance, although the corticosteroid hormones are divided into two classes according to whether their effects are mainly exerted on carbohydrate or electrolyte metabolism, each hormone has some of the activity of the other class. Similarly, one might expect a certain degree of overlap in the behavioral actions of the different gonadal steroids, if only because the differences in molecular structure between male and female sex hormones are so small (Figure 5.1). This overlap does indeed occur and will be discussed in Chapter 7.

It is also relevant that all the steroid hormones are linked through their biosynthetic pathways, which are believed to be qualitatively identical in ovaries, testes, and adrenals. Thus, in the chain of chemical conversions whereby estrogen is synthesized in the ovary, one important precursor is testosterone, the "male hormone"; and progesterone is a common precursor of testosterone, and therefore of the estrogens too (see Figure 5.1).

Of course for these precursors to be able to affect behavior they have to be secreted into the bloodstream in sufficient quantities, and we shall see that this may happen in certain cases.

Another important consequence of the common manufacturing process of the steroid hormones is that changes—pathological, experimental, and perhaps also physiological—in the chemical (enzymatic) apparatus of the cells can result in a redirection of the biosynthetic pathways, with the result that the steroid-producing gland begins to secrete inappropriate amounts of a given hormone or hormones. Thus, in certain pathological situations the adrenal cortex can shift its "emphasis" from manufacturing primarily adrenocortical hormones to begin a large-scale production of male hormones. This condition, known as the adrenogenital syndrome, can cause extreme genital and extragenital masculinization in human females. In male children it can lead to such unpleasant consequences as precocious sexual maturation (the "infant hercules" condition).

Control of Circulating Steroid Levels

Like all hormones, the circulating level of gonadal steroids (their concentration in the bloodstream) is determined by a balance between production of the hormones and their removal from the circulation. The removal is effected by metabolic transformation or destruction of the molecule (often by the liver), and by subsequent excretion of the hormone itself or of its metabolic products. These processes are primarily carried out by the liver and kidney. Under certain circumstances changes in liver function can have serious effects on the circulating level of sex hormones. Thus, men with liver disease (as in alcoholism) may have gynecomastia (overdevelopment of the breasts); this is due to failure of the liver to dispose of the small amount of estrogen normally produced in men. Under normal circumstances, however, by far the most important factor determining how much of the sex hormones are present in the blood at any one time is the activity of the cells that produce (secrete) the hormones. These are primarily the leydig cells (interstitial tissue) of the testis for testosterone. In the ovary the theca interna of the follicle and the corpus luteum secrete estrogens, and the corpus luteum is the primary source of progesterone.

ANTERIOR PITUITARY

The steroidogenic (steroid-producing) function of both male and female gonads is absolutely dependent on the secretion by the anterior pituitary of two protein hormones known as *gonadotropins*. These are follicle-stimulating hormone (FSH) and luteinizing hormone (LH). A third pituitary hormone, prolactin, although involved in control of the ovary in rodents, is of questionable importance in this regard in other mammalian species. The structure of these tropic hormones, whose molecular weights are about 30,000, is only now being elucidated after many years of effort by biochemists (Hafez and Evans, 1973). We have learned enough about their structure and actions, however, to be fairly certain that there is no major difference in the hormones produced by males and females.

In addition to controlling steroidogenesis by the gonads, FSH and LH are also responsible for maintaining both the structure of the testis and ovary and their gametogenic functions—production of spermatozoa and ova (see Figure 5.2). In the male, LH activates testosterone production by the leydig cells. The androgen probably diffuses to the neighboring cells of the seminiferous tubules, thereby helping to maintain spermatogenesis. FSH stimulates spermatogenesis by a direct action on the tubules. No physiological function has been assigned to prolactin in male animals.

In females, FSH is responsible for maturation of the ovum-containing follicles. Both FSH and LH are secreted throughout the cycle, and both are released in a burst at midcycle. The LH secreted at that time in great quantities is primarily responsible for causing ovulation within 24 hours, and subsequent luteinization of the follicle (corpus-luteum formation). It also plays a role in the production of progesterone by the newly formed corpus luteum.[3] The two gonadotropins appear to act synergistically in the control of estrogen secretion. Prolactin is important for growth of the corpus luteum and secretion of progesterone in rodents, although its gonadotropic function in other animals is not established. Its increased production and the resulting high levels of progesterone secretion function to maintain pregnancy following fertile matings. Following an infertile mating, similar hormonal events produce the (shorter) state of pseudopregnancy, in which ovulation and cycling are absent. Additionally, prolactin is the most important pituitary hormone in the control of mammary development and lactation in rodents and other species.

Although we can safely assume that the general principles of gonadotropin regulation are similar in all mammals, the great variety of different types of reproductive cycles found in different species suggests that there are considerable differences of detail in the mechanisms controlling the gonads. The length of the female cycle is primarily dependent on the duration of activity of the corpus luteum. Thus in rats, mice, and hamsters, the corpus luteum is functional only briefly, and the cycle is only 4 to 5 days long. The corpus luteum of guinea pigs, farm animals, and primates, on the other hand, secretes progesterone (and estrogen) for a prolonged period, and the cycle is several weeks long (see Chapter 3).

In several species, notably the rat, the changes in the ratio of circulating estrogen and progesterone throughout the cycle are reflected

---

[3]That LH stimulates progesterone secretion from other parts of the ovary also was suggested by the recent discovery that progesterone levels rise immediately after the midcycle LH release that causes ovulation but before ovulation (and therefore corpus-luteum formation) has actually occurred.

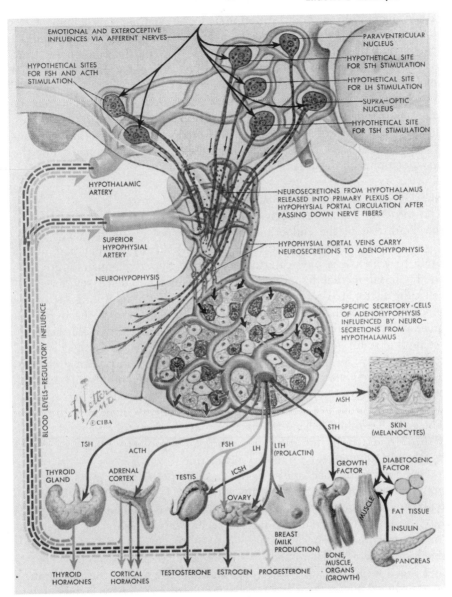

FIGURE 5.2   The hypothalamo-anterior pituitary system and the systems it controls. Note arrows from gonads, adrenal and thyroid, indicating feedback influences either directly on the pituitary (via its blood supply) or at the hypothalamic level, to regulate production of the releasing factors (neurosecretions). From Netter, F. H. (1953) *The CIBA Collection of Medical Illustrations*, vol. 4, *Endocrine System*. Summit, N.J.: CIBA Pharmaceutical Products Co. Copyright 1965 CIBA Pharmaceutical Company, Division of CIBA-GEIGY Corporation. Reproduced, with permission, from *The CIBA Collection of Medical Illustrations* by Frank H. Netter, M.D. All rights reserved.

rather accurately in transformations of the vaginal epithelium. Cells sloughed off from the vaginal walls are easily removed from the vaginal lumen, and the precise stage of the cycle can be determined by brief microscopic inspection of the cell types present in the vagina (the vaginal smear).

Another hormone of the anterior pituitary, ACTH, is responsible for maintenance of the adrenal cortex and the secretion of all its hormones, including progesterone, androgens, and estrogen. Under normal circumstances ACTH does not have significant effects on the gonads, nor the gonadotropins on the adrenal cortex.

NEUROENDOCRINE MECHANISMS

The pituitary gland (hypophysis), despite its traditional designation as the conductor of the endocrine orchestra, is not an independent agent, but is itself controlled by the brain, and specifically by that part of the brain to which it is physically attached—the hypothalamus (Figure 5.2). In the basal medial hypothalamus, or median eminence region, FSH-releasing factor, LH-releasing factor, and a prolactin-inhibiting factor are produced by specialized neurosecretory cells. This area of the hypothalamus is linked to the anterior pituitary by a specialized vascular arrangement—the small *hypophyseal portal vessel system* that collects blood draining from the median eminence and perfuses the pituitary with it. This private communication system between the hypothalamus and the anterior pituitary enables the small quantities of releasing factors produced in the brain to reach the pituitary in relatively high concentration. The gonadotropin-releasing factors stimulate both production and release of FSH and LH by the pituitary, whereas a prolactin-inhibiting factor has the opposite effect on pituitary prolactin.

This controlling mechanism in the basal hypothalamus provides a link between the environment and the endocrine system that is vital to the reproductive process. Many environmental factors, such as light, temperature, trauma, and sexual stimuli, are received by sensory receptors; and the information is conveyed, often via complicated neural pathways, to the hypothalamus. There they are converted to neurosecretory signals that are transmitted to the pituitary gland. At this level the information is transduced to produce hormonal signals to the gonads, which effect changes either in hormone secretion or gamete production, or both. An example of the various "neuroendocrine reflexes" that are important in the reproductive life of animals will suffice to illustrate the general principle.

Ovulation in the rat occurs in response to gonadotropin release occurring once every 4 or 5 days. The environmental factor responsible for the timing of this event is the pattern of illumination. This control is so precise that if illumination is supplied from 5:00 A.M. to 7:00 P.M. each day, it can be predicted that a neural event precipitating LH release will occur between 2:00 and 4:00 P.M. on the day of *proestrus*, the portion of the cycle closest to the onset of behavioral receptivity. Ovulation will follow soon after midnight. By shifting the light cycle in one direction or another by several hours, the time of ovulation can likewise be shifted in the corresponding direction. Although all the links in the chain from the light stimulus to the release of ova have not yet been elucidated, the events from the production of the hypothalamic "neurohormone" onward are beginning to be fairly well understood. We still know all too little, however, about how the brain codes information fed in from the environment and transmits it to the basal hypothalamic region. This is true also of other neuroendocrine reflexes such as the milk-ejection reflex, elicited on stimulation of the breast by sucking, and the thyroid-induced increase of metabolic heat production elicited by decreased environmental temperature. An additional complexity will be added to this scheme in Chapter 6, when it will be shown that behavioral events can themselves constitute important stimuli for eliciting neuroendocrine reflexes.

There is one other group of processes that plays an important role in the regulation of circulating steroid levels, namely the "feedback" self-regulatory mechanisms. A generalization that is true of all the important steroid hormones is that sustained increases in blood levels of these hormones have an inhibitory effect on the secretion of the appropriate pituitary tropic hormone, with the result that there is a decrease in further production of the steroid. Thus, if androgen or estrogen is administered regularly over a period of time, an inevitable consequence is the eventual atrophy and cessation of function of the testes or ovaries. Likewise, a decrease in the level of circulating hormone provides a stimulus for increased production of the appropriate pituitary tropic hormone, with the consequent increase in activity of the steroid-producing gland. An example of this aspect of the function of these "negative-feedback" mechanisms is seen following unilateral removal of one steroid-producing gland. Thus in bulls, if one testis is surgically removed, the other approximately doubles its output of androgen. Obviously, this "closed-loop" automatic control system can operate to maintain the blood level of steroids in the face of stimuli that tend to change them.

Unfortunately, however, the feedback control of steroid hormone concentrations is more complicated than is suggested by this simple schema. It has been found that in certain circumstances injections of estrogen provoke increased LH release, which produces in turn a further increase in estrogen secretion by the ovary. This "positive-feedback" mechanism seems to play an important role in precipitating the great augmentation of LH secretion required for ovulation. Obviously the operation of such positive-feedback mechanisms must be limited in time and extent, otherwise the organism would rapidly become completely flooded with the particular steroid and gonadotropin. We do not yet know the precise conditions under which steroids operate in positive or negative fashion, but the dose and duration of exposure are important factors.

It is also clear that estrogen and progesterone interact in their effects on gonadotropin secretion. We cannot detail here the complexities of the interactions, but a few brief examples taken from the experimental control of ovulation in the rat will serve to illustrate the point. In rats showing a 4-day cycle, estrogen administered on the first day after ovulation (metestrus) will not shorten the next cycle. But if it is followed by progesterone administration on the next day, gonadotropin release and the consequent ovulation will be advanced by 24 hours. Progesterone administered on the third, but not the second, day of the 5-day cycle also advances ovulation by 24 hours. Probably the important factor determining whether progesterone will have a stimulatory effect in both these examples is the level of preexisting estrogen in the blood, which conditions the gonadotropin response to progesterone in a manner as yet undefined (Brown-Grant, 1969).

Prolonged exposure to the high levels of estrogen and progesterone produced by the corpus luteum does, however, have an invariably negative feedback effect on FSH and LH, and thus inhibits ovulation. This is presumably responsible for the long cycles in species with long-lived corpora lutea, which prevent development and ovulation of new follicles. Similarly these two events are absent during pregnancy, when greater amounts of progesterone (and estrogen) are produced.

This consideration of some concepts of reproductive endocrinology is intended to lay a basis for our discussion of behavioral endocrinology. The dual role of the testes and ovaries in gametogenesis (germ-cell production) and steroidogenesis is but one of many examples of the amazingly intricate coordination whereby the many processes involved in the physiology of reproduction are interwoven and made to function in concert. The gonadal steroids themselves have the dual role of maintaining many functions of the reproductive system[4] and of regulating sexual behavior. For a true understanding of the latter role one has to bear in mind how behavioral function meshes into the complex scheme of hormone-dependent reproductive mechanisms.

THE RELATIONSHIP BETWEEN STEROID HORMONES AND SEXUAL BEHAVIOR

A first step in elucidating the function of any endocrine gland is to remove (ablate) the gland and observe the effects on the animal's physiology and behavior. Second, crude extracts of the gland, and later its pure hormonal products, are administered to the ablated animals in the attempt to counteract the effects of the ablation (replacement therapy). Finally comes the period of careful investigation of the quantitative re-

[4]Estrogens and progesterone have extensive regulatory effects on most reproductive events, for example, gamete transport, ovulation, implantation, and pregnancy. Testosterone, apart from being necessary for the function of the sex accessory glands, is probably also essential for spermatogenesis. The gonadal steroids also control the various secondary sexual characteristics and affect various phases of metabolism.

lationship between replacement dosage and the animals' response, the factors affecting this dose-response relationship, and the mechanisms of hormonal action. Because the endocrinology of sexual behavior is still not far advanced into the third phase, we shall be discussing in large part information from the experiments on ablation and replacement therapy. At this point, a few words of caution are in order about the interpretation of this kind of data.

Endocrine-gland ablation is performed to observe the resulting deficit when the hormone is absent from the circulation. The results of this procedure may not, however, necessarily be relevant to the normal situation because a *complete* lack of circulating hormone may never be found under "physiological" conditions. In the case of seasonally breeding animals, a situation does periodically recur in which the gonads may be virtually or completely inactive, but most laboratory mammals are not seasonal breeders (see Chapter 3). As for replacement therapy, it is difficult to reproduce the normal patterns of change, periodic or otherwise, in the circulating levels of a given hormone. To do so would require prior knowledge of the normal circulating levels of the hormones and their patterns of change. Moreover, the experiment would require the animal to be continuously attached to the external source of hormone, for example via chronically implanted intravenous catheters. Such experiments have not yet been performed,[5] and precise information on the changing levels of gonadal steroids in blood has only begun to appear in recent years. Thus, it must be borne in mind that experiments do not reproduce the dynamic pattern of changing hormone secretion existing in physiological conditions.

These reservations concern mostly the detail of the picture we have to draw, not its general outline. As to the latter, we can commence by making one secure generalization: the sexual behavior of at least the subprimate mammals is always dependent on the present or prior exposure of the animal to gonadal hormones. The rest of this chapter and the following one are essentially devoted to an amplification of this statement. The term *present exposure* means the immediate "activational" function of the hormones. *Prior exposure* means both the "organizational" effects of these hormones at critical periods in development (see Chapter 7) and the residual effects of hormones administered days, weeks, or months previously in the animal's adult life.

OVARIAN HORMONES AND FEMALE SEXUAL BEHAVIOR

The ability of the female to reproduce is a cyclic phenomenon; the various forms which this cyclicity takes in different species were described in Chapter 3. It was pointed out that the necessary coordination between cyclic sexual behavior and cyclic reproductive ability could be achieved

[5]In studies on sexual behavior, hormones are usually administered subcutaneously in oil. This method results in a relatively slow but not necessarily stable release of the hormone into the bloodstream.

in one of two ways: either the female is sexually "receptive" only close to the time of ovulation (spontaneous ovulators), or the coital events themselves precipitate ovulation (reflex ovulators). Rape is apparently unknown among nonhuman mammals, so coitus is limited to periods of natural female receptivity. How absolutely essential is close temporal coordination of the behavioral and physiological events becomes obvious when we consider the following facts. The sperm and egg are viable only for a day or less. Behavioral estrus[6] is usually brief. It lasts only a few hours in the rat, commencing on the afternoon of proestrus and waning that night. In this and other species, if mating occurs early in the period of sexual receptivity, there will be considerable delay before ovulation occurs. Freshly ejaculated spermatozoa are however often infertile. Changes of a chemical and perhaps an anatomical nature take place in the sperm cell during its first few hours of residence in the oviduct. This process, called *capacitation*, renders the spermatozoon capable of fertilizing the egg. The point is, that about the time such capacitation is completed ovulation will have occurred, making ova available to the now fertile sperm.

Ovariectomy and Replacement Therapy

Because reproductive cyclicity is a direct result of cyclic changes in hormone secretion, it is reasonable to assume that behavioral cycles are also related to these changes. This suggestion of a causal relationship between hormones and sexual behavior is overwhelmingly confirmed by the results of endocrinologic experimentation. Over 2000 years ago Aristotle wrote "the ovaries of sows are excised with a view to quenching their sexual appetites." In fact, ovariectomy in most subprimate mammalian species that have been studied results in virtually complete disappearance of sexual receptivity. The exceptions appear to be of very minor significance; for example, in some rabbit strains spayed females occasionally show receptive behavior.

The effects of ovariectomy in subhuman primates have not been extensively studied until recently. The results seem to show a somewhat lesser dependence of sexual receptivity on ovarian hormones than in the case of "lower" mammals. In the remainder of this section we will discuss the

---

[6]Some confusion arises from the use of the term *estrus* to describe female cycles. This term was originally coined to describe the behavioral manifestation of heat. It has subsequently been used by physiologists to describe the state of vaginal cornification that occurs in many species around the time of ovulation. As a result of this dual use, peculiar semantic anomalies have arisen. Thus the term *constant estrus* is used to describe animals in which, due to a variety of experimental or pathological causes, the vagina is constantly cornified, although the animal may not be sexually receptive. Another source of confusion arises from the fact that behavioral heat may not coexist with full vaginal cornification, but, as in the rat, may precede it and terminate before cornification does. The reason for this is that although these processes are both dependent on prior exposure to estrogen, the vaginal changes take longer to complete than the behavioral ones.

characteristic influences of the ovarian hormones on behavior without reference to the sites or mechanisms of their action. After presenting the relevant experimental facts, we shall attempt to draw conclusions about the hormonal control of female sexual behavior under natural conditions.

From experiments on the induction of estrous behavior in ovariectomized animals of many species it appears that both estrogen[7] and progesterone usually are involved in the control of female sexual behavior (Young, 1961). It is generally accepted, for example, that both hormones are essential in guinea pigs, rats, mice, hamsters, and cows. The usual method for experimental induction of estrous behavior in these species is one or two daily injections of estrogen, followed by one administration of progesterone a few hours before the mating tests. The commonly held concept is that in these species estrogen activates some process that requires at least a day or two to have its full effect. Progesterone has a subsequent facilitating effect on appearance of the behavior, which is manifested a few hours after subcutaneous administration of this hormone.

A number of reservations have to be made about this scheme. First, in a variety of species such as cats, dogs, ferrets, goats, rhesus monkeys, and rabbits, estrous behavior is commonly induced after ovariectomy with estrogen alone, and it is assumed that progesterone is not needed in these species. Although progesterone may have a facilitatory effect after estrogen treatment in the rabbit (Sawyer and Everett, 1959) and dog (Beach and Merari, 1968), a single low dose of estrogen produces excellent results in the rabbit (McDonald et al., 1970).

Even in classical estrogen-progesterone species such as rats and guinea pigs, it can be shown that repeated but low doses of estrogen (in the "physiological" range) can induce levels of lordosis responding that, at least in rats, are entirely similar to those found in natural heat (Davidson et al., 1968a, b). A linear relationship can be shown to exist between the logarithm of the dose of estradiol benzoate and the mean lordosis quotient (see Figure 5.3). The response is not dependent on progesterone or any other steroid from the adrenal, because similar responses to estradiol were found in ovariectomized rats as in ovariectomized-adrenalectomized ones (Davidson et al., 1968b).

In relation to our later discussion on the natural hormonal events responsible for estrous behavior, it will be advantageous to summarize three possible treatments with ovarian steroids, all of which are experimentally effective in ovariectomized animals: a single large dose of estrogen, one or two small doses of estrogen followed by one of progesterone, and daily injections of estrogen for about a week.

How long does it take for estrogen to activate female sexual behavior? The point is of some importance in relation to attempts to understand the mechanism of the hormone's action. No matter which of the three

---

[7]Although *estrogen* is a generic term, it is used here, for the sake of simplicity, as if it referred to a single hormone.

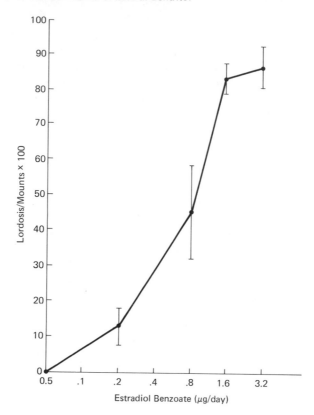

FIGURE 5.3 Dose-response relationships between estrogen (estradiol benzoate) administered daily by subcutaneous injection to ovariectomized rats and female sexual behavior (as measured by lordosis quotient) eight days after onset of treatment. From data of Davidson, J. M., Rodgers, C. H., Smith, E. R., and Bloch, G. J. (1968a) Relative thresholds of behavioral and somatic responses to estrogen. *Physiology and Behavior* 3:227–229. ˙

methods is used, the time from initial estrogen administration to onset of receptive behavior cannot be brought to much less than one to two days in the rat. The optimum latency after estrogen treatment in the guinea pig is 36 hours (Dempsey, Hertz, and Young, 1936), whereas the effect of progesterone is fully manifested in about 4 hours after subcutaneous injection in both species. Interestingly enough, estrogen has to be present in the circulation for only a fraction of the latent period. If guinea pigs are exposed to estrogen (in a removable capsule) for as little as 30 minutes, lordosis can still be induced with progesterone. Even when estrogen is injected subcutaneously in oil (as is usual) the hormone cannot be detected in the tissues (including brain) by the end of the latent period. Clearly, a chain of events intervenes between some trigger-like action of estrogen and actual precipitation of the behavioral response (Bullock, 1970). Those events are still shrouded in mystery (see Chapter 6).

An important complication in this story is the surprising fact that progesterone can also be shown to have clearly inhibitory effects on female sexual behavior. Thus, appropriate administration of progesterone will terminate estrous behavior or reduce estrogen-induced sexual activity in female rabbits, ferrets, sheep, goats, rhesus monkeys, pigs, and cows (Young, 1961). Whether progesterone will have an inhibitory or a stimulatory effect depends on judicious choice of experimental conditions, namely the doses and precise timing of the administration of both ovarian steroids (Zucker, 1966). Two explanations have been proposed to reconcile the apparent contradiction between experiments showing facilitatory or inhibitory effects of progesterone. First, low levels of progesterone facilitate receptivity, whereas high levels suppress it. This kind of dose-dependent reversal of effects is not unfamiliar to physiologists and pharmacologists. Second, there is a biphasic temporal pattern: initially progesterone facilitates female sexual behavior but later suppresses it. Experimental evidence can be found to support both proposals.

The biphasic effect of progesterone on sexual behavior has been demonstrated in the rabbit (Sawyer and Everett, 1959) and the guinea pig (Zucker, 1966). In the latter species, progesterone administered after estrogen has no inhibitory effect, although it does when administered simultaneously or before estrogen. The inhibitory effect is easier to demonstrate in guinea pigs or rabbits than in rats or mice, where it only appears when the prior estrogen level was low (Edwards, 1970; Edwards et al., 1968; Lisk, 1969; Powers, 1970; Zucker, 1967). A similar situation is to be found in the experimental control of ovulation. Progesterone may advance or retard ovulation, depending on the time of its administration during the rat's cycle. A stimulatory action is best obtained when estrogen levels in the blood have risen (Brown-Grant, 1969).

We may summarize our conclusions on the experimental conditions for activation of sexual behavior by ovarian steroid administration in ovariectomized subhuman mammals as follows: circulating estrogen is always a necessary condition. In some species it appears to be a sufficient condition; this is probably the case in all species only if the period of estrogen administration is sufficiently prolonged. Progesterone clearly facilitates the behavior-activating effect of low estrogen doses in most species. Progesterone can also inhibit estrogen-induced receptivity, an effect more prominent in some species than in others. A variety of factors seem to determine the direction of the progesterone effect. These include dose, time after hormone administration, and preexisting level of estrogen. These experimental facts will eventually have to be explained before we can make valid generalizations about the hormonal control of female sexual behavior.

Hormonal Control in the Behavioral Cycle

The ultimate goal of research on the effects of the ovarian steroids on female sexual behavior is to elucidate fully the behavioral phenomena

occurring during the normal cycle in different species. Because the physiological and behavioral events are best understood in the rat, we will
first consider this species before attempting to draw analogies to others.
Can we explain in endocrinologic terms the rapid waxing and waning
of the female rat's sexual receptivity, which reaches its peak on the evening of the day of proestrus (about the time of full vaginal cornification)?

Recent improved methods of steroid analysis have made possible the
description of estrogen and progesterone levels throughout the estrus
cycle. In the rat, estradiol secretion rises rapidly to reach a peak on the
morning of the day of proestrus. Progesterone levels do not begin to rise
until the afternoon of proestrus, most likely as a direct result of the large
release of LH at that time, although this point is still debated. Peak blood
levels of progesterone appear to coincide fairly closely with the appearance of a peak in sexual receptivity in rats and guinea pigs (Feder, Resko,
and Goy, 1966; Feder, Goy, and Resko, 1967). The estrogen peak occurs
around ten hours or less before the behavioral change.

We see therefore that the hormonal conditions preceding the onset of
sexual receptivity in the rat appear to approximate those found to be
effective in artificially induced estrus—initial high levels of estrogen followed by a sharp progesterone peak. One apparent discrepancy, how-

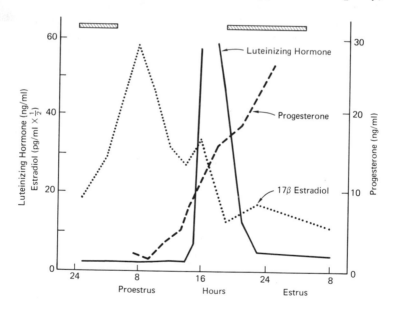

FIGURE 5.4 Circulating levels of luteinizing hormone (LH), estradiol, and
progesterone during proestrus and early estrus in the rat. Ovulation occurs
around 1:00 to 2:00 A.M. on estrus. The bars at the top of the figure represent
periods of darkness; ng = nanograms (grams $\times$ 10^{-9}), pg = picograms (grams
$\times$ 10^{-12}). From Brown-Grant (1971), The role of steroid hormones in gonato-
tropin secretion in adult female mammals. In *Steroid Hormones and Brain Function*. C. H. Sawyer and R. A. Gorski (Eds.). Berkeley and Los Angeles: University
of California Press, pp. 269–288.

ever, is that the time lags between appearance of peak levels of both estrogen and progesterone in the blood, and the appearance of lordosis, seem shorter than would be predicted on the basis of the experimental work. However, it must be remembered that these events occur on a background of prior stimulation with ovarian steroids, a situation different from that in the experimentally treated castrate. Furthermore, as pointed out earlier, experimental hormone administration may not duplicate the precise pattern of natural secretion. Lisk (1960) suggests that the latency from progesterone administration to lordosis may be much shorter when the hormone enters the bloodstream directly, following intravenous injection, than in the usual experimental situation, when it is administered subcutaneously.

These considerations strongly suggest, though they do not prove, that both the early exposure to estrogen and the proximal exposure to progesterone are causally related and in fact essential for induction of behavioral receptivity.

But are we to believe that the sole function of the proestrus secretion of ovarian hormones is to produce sexual behavior? This seems unlikely in light of the many important physiological events taking place at this time in the estrus cycle. Much could be said of the requirements for estrogen and progesterone in preparing the various parts of the female reproductive tract for gamete transport and ovum implantation, but a particularly relevant analogy in this context is the role of these steroids in the control of ovulation.

An impressive body of evidence establishes the preovulatory estrogen peak as being the essential trigger for the ovulation-inducing release of gonadotropins (the so-called positive-feedback effect of estrogen). Thus, a rise in estrogen secretion precedes ovulatory gonadotropin secretion in all species that have been studied. Furthermore, obliteration of the effect of estrogen by ovariectomy or administration of specific antiestrogen antibodies prevents that gonadotropin "surge," which can then be restored if estrogen is replaced. Finally, early estrogen administration can advance the occurrence of ovulation. The ovarian progesterone output, on the other hand, does not seem to be involved in precipitation of ovulatory gonadotropin secretion. Progesterone levels remain low until after the LH surge commences, and antibodies to progesterone do not affect ovulation (Ferin et al., 1969). Although hormone levels in the cycle vary among different species, the temporal ordering of estrogen, gonadotropin, and progesterone secretions are similar in all species studied to date including the human (see Figures 5.4 and 5.5).

How close is the relationship between these physiological events and the control of sexual behavior in female rats? There is little doubt about the analogous roles of estrogen in the two situations. Ovariectomy before the estrogen rise prevents mating (Schwartz and Talley, 1965); estrogen administration early in the cycle advances its appearance (Aron et al., 1966). We can conclude that the estrogen peak on early proestrus has the

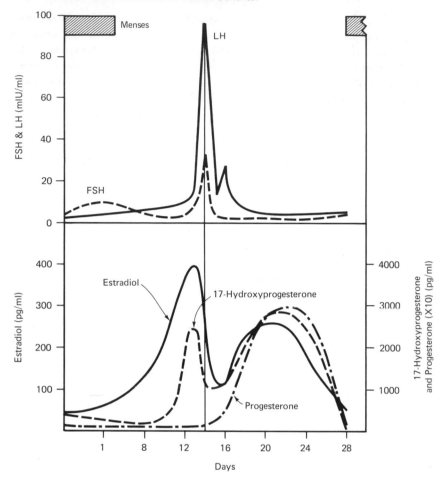

FIGURE 5.5 Schematic representation of the changes in circulating follicle stimulating hormone (FSH), luteinizing hormone (LH), and several steroid hormones during the normal menstrual cycle of women. Because progesterone levels are much higher than those of the other steroids, their true concentrations were divided by 10 in this diagram. mIU = milli international units; pg = picogram (grams $\times 10^{-12}$). From Odell, W. D., and Moyer, D. L. (1971) *Physiology of Reproduction*. St. Louis: Mosby.

simultaneous dual action of stimulating both the ovulatory release of gonadotropin and the onset of behavioral receptivity.

What of progesterone? Our previous discussions have indicated its probable importance in facilitating receptivity on the basis of both the experiments on castrates and the time of its secretion in the cycle. But its apparent unimportance in ovulation and the fact that experimentally it is not essential for the behavioral response differentiate its role from that of estrogen. We are thus led to seek more definitive evidence of its role in activation of cyclic behavior.

Although the effects of antibodies to progesterone on mating have not yet been studied, a recent investigation found that ovariectomy just before, but not just after, the expected peak of progesterone production markedly inhibited lordosis (Powers, 1970). In earlier experiments in a different laboratory (Schwartz and Talley, 1965), ovariectomy on the morning of proestrus seemed compatible with receptivity later that day. However, subsequent analysis (Schwartz, 1969) suggested that the lack of inhibition following ovariectomy was due to the stress of the operation resulting in ACTH-induced adrenal progesterone secretion. This seems reasonable in the light of the demonstration that lordosis may be facilitated in chronically ovariectomized estrogen-primed animals by stress (Goy and Resko, 1969) or ACTH injection, apparently due to increased adrenal progesterone release (Feder and Ruf, 1969). This subsidiary mechanism for behavioral activation is probably not entirely reliable; noneffectiveness of adrenal removal was reported by Powers (1970).

In summary, there is excellent evidence that what the experimentalists have been doing for many years does in fact roughly duplicate the sequence of events that in the normal cycle produces the state of sexual receptivity necessary for successful fertilization. A "priming" influence of estrogen is followed by an acute exposure to progesterone, facilitating the full onset of behavioral receptivity. Two "fail-safe" mechanisms seem available to ensure normal receptivity: more prolonged exposure to estrogen may elicit the behavior in the absence of progesterone, and adrenal progesterone can substitute for the ovarian steroid under certain conditions.

Recent measurements of ovarian steroid levels in the blood of the sheep have allowed for the resolution of a previously puzzling experimental situation. In that species it has been known for a long time that the usual order of estrogen-progesterone administration has to be reversed in order to induce early receptivity. We now know that levels of progesterone remain high in the ewe until shortly before estrus, when a rapid fall occurs and this immediately precedes an abrupt rise in estrogen production. Here again we find that the natural situation duplicates the conditions previously empirically found by investigators to be optimal for induction of sexual receptivity.

How can we fit into the picture of natural behavioral cyclicity the experimental information on the second, inhibitory phase of the progesterone effect that follows the initial facilitatory action? Biological significance for this effect might be sought in terms of the limitation of mating to the period when the female is fertile. As mentioned before, progesterone inhibition of estrogen-induced lordosis is easier to demonstrate in long-cycling animals such as the guinea pig and the monkey than in short-cycling species such as the Mongolian gerbil (Kuehn and Zucker, 1968) and the rat and mouse, where it seems to depend on special circumstances. Because progesterone is present only relatively briefly in the short-cycling animals, it is not surprising that a more prominent role for this hormone in limiting the mating period has not evolved in these species.

Of course, rats as well as the long-cycling animals are behaviorally unreceptive during pregnancy. The chronically high levels of progesterone seem to be responsible for this refractoriness despite the presence of adequate estrogen (Powers and Zucker, 1969). High progesterone levels probably also have an inhibitory ("negative-feedback") effect on gonadotropin secretion in pregnancy and pseudopregnancy—another parallel between the physiological and behavioral roles of ovarian steroids.

What of the situation in reflex ovulators like the cat and the rabbit? In these animals, LH release and progesterone secretion follow rather than precede mating. Accordingly, the experimental observations show the unimportance of progesterone in these species, in which acute (single) administration of small amounts of estrogen alone suffices to induce receptivity.

That progesterone nevertheless does facilitate the response in rabbits, when receptivity is submaximal, merely demonstrates what seems to be a general principle—the existence of a potential pool of responses to hormones that may not be called for under normal circumstances. Previously mentioned possible examples of this include the inhibitory effect of progesterone in rats and mice, and the facilitatory role of adrenal progesterone. Other examples will appear later when we discuss the endocrinologic consequences of mating (Chapter 6) and the effects on behavior of hormones of the opposite sex (Chapter 7).

We have seen that a large variety of experimental investigations have led us to a fair understanding of the main features of the hormonal control of female sexual behavior in mammals. These include the predominant, stimulatory role of estrogen in all the subprimate species; the secondary, facilitatory role of progesterone in a number of "spontaneously ovulating" species; the inhibitory role of progesterone, particularly in long-cycled species and in pregnancy (real and pseudo); and the existence of potential hormonal responses that are not normally operative but may become so under special circumstances. Less understood are the mechanisms whereby hormonal effects are exerted and how the behavioral patterns and specific responses to hormones of the female become established during development. These difficult problems and that of the effects of behavioral stimuli on endocrine function will be the subjects of later discussions.

TESTICULAR STEROIDS AND MALE SEXUAL BEHAVIOR

At first sight the hormonal regulation of sex behavior in males appears to be less complex than in females. For one thing, only a single principal hormone, testosterone, appears to be of major significance,[8] so that the difficult questions of interpreting synergistic and antagonistic interactions of different hormones can be avoided. Second, we do not have the prob-

[8]In addition to testosterone, the testes secrete smaller amounts of a precursor possessing androgenic activity—androstenedione. Although this steroid has action similar to testosterone, it is less potent; hence in this discussion it will not usually be mentioned specifically.

lem of estrus cycles. Reproductive capacity is a seasonal phenomenon for male animals in the wild, although not usually in domesticity. But during the breeding season itself the male seems to have a constant capacity for sexual behavior. He is not subject, so far as we know, to major cyclic fluctuations from day to day.

Despite this apparent lack of complexity, however, there are several rather puzzling aspects of hormonal regulation in the male. The close temporal correlation between changes in hormone levels and sex behavior in female animals is not nearly so obvious in males. When, for example, behavioral responses are stimulated by administration of testosterone to castrated males, the latency to onset of behavior is considerably longer than in females. The difference is even more striking when we consider the effects of castration. Very often castration in males does not lead to a rapid loss of all elements of mating behavior.

Male Sexual Behavior Without Androgen

Unless replenished by new secretion, testosterone is rapidly removed from the bloodstream by metabolism and excretion. In a matter of hours after castration, there is no testosterone in the circulation except the insignificant amount that comes from the adrenal cortex. It is of some interest to look briefly at the quality of the behavior that continues for a variable but often prolonged period of time in the absence of testicular hormone, and to compare it with that seen in the intact subject.

When we say that rats can ejaculate weeks or months after castration we are referring to the observation of the behavioral event seen at the culmination of the mating pattern, which consists of a long intromission and slow withdrawal and dismount. The physiological process of seminal emission that accompanies ejaculatory behavior in intact subjects is absent because semen is not formed in castrates.[9] The most general and striking behavioral effect of castration is a steady decline over time in the proportion of subjects that will initiate or complete mating. This decline is illustrated for rats in Figure 5.6. The ejaculatory pattern may disappear long before mounting behavior, which may continue for prolonged periods despite the total absence of intromissions. The extent to which this is due to failure of erection is unknown.

If we wish to quantitatively analyze changes in male behavior patterns we must first distinguish between two sets of parameters. The first reflects readiness to copulate: intromission latency (time from onset of test to first intromission) and postejaculatory interval, or the refractory period

[9] As discussed by Beach et al. (1966), seminal emission may be separated from "orgasm" or "ejaculation" in a variety of situations. Examples of seminal emission unrelated to coitus are nocturnal emission in man and cats; spontaneous emission, which is a daily occurrence in rats; stimulation of the central nervous system; treatment with various drugs, including epinephrine, ether, mecholyl, morphine, and methamphetamine. Other drugs, such as guanethidine, prevent seminal emission without preventing behavioral ejaculation in man and rats.

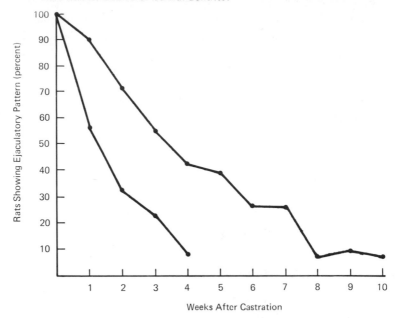

FIGURE 5.6   Percentages of two groups of rats (46 in upper curve and 63 in lower curve) that showed the ejaculatory pattern of behavior on successive tests after castration. Note the variability in postcastration behavior as indicated by the difference in the two curves, and the prolonged behavioral maintenance in series 1. From Davidson, J. M. (1966b) Characteristics of sex behavior in male rats following castration. *Animal Behavior* 14:266–272.

between the end of one mating sequence and the beginning of the next. The second set of parameters reflects the rate of consummation of the mating act: frequency of intromissions and latency from first intromission to ejaculation.

In the initial postcastration period the pattern of changes is different insofar as the two types of measure are concerned. Thus intromission latency and postejaculatory interval show steady increases from the time of castration to that of disappearance of sexual behavior. The effects of castration on intromission latency and postejaculatory interval can be almost entirely prevented by daily administration of 25 µg testosterone propionate (see Figure 5.7).

Surprisingly, however, the mating pattern is consummated with increased efficiency[10] in the first weeks following castration, in the sense that ejaculation is achieved in less time and with fewer intromissions (see Davidson, 1969b). These changes in "consummatory" elements of the mating pattern do not, however, appear to be as hormone-dependent as the other changes, because they cannot be prevented by testosterone except, to some degree, when very large doses are given. The reason is that

[10]Of course, it is not implied that this behavior pattern is more efficient in any biological sense. In fact, as we shall show, high intromission frequencies are conducive to fertility.

decreases in intromission frequency and ejaculation latency following castration are not due merely to androgen removal but occur in response to the operative trauma itself. They have been observed following such diverse stressful stimuli as surgical trauma and audiogenic stimulation. Apparently, however, they may be produced also by decreasing testosterone levels in the absence of trauma (Bloch and Davidson, 1968). We do not as yet understand the mechanism that mediates between these diverse environmental events and this change in copulatory behavior.

Responses to Testosterone Treatment

Replacement therapy with testosterone restores the mating patterns in castrates of all species that have been studied. We know of no evidence that any other hormone plays an important specific role in male sexual behavior. One would therefore expect that a dose-response curve could be drawn to express the relationship between the hormone administered and the behavior elicited, just as can be done for other hormone-dependent physiological effects. As we have seen, such a dose-response

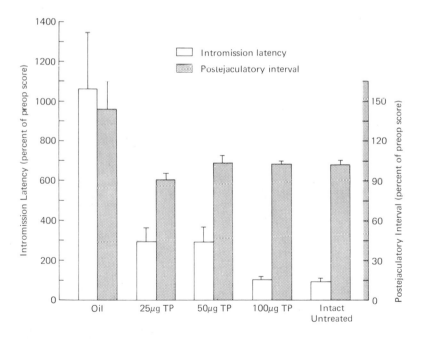

FIGURE 5.7 Effects of castration and replacement therapy with testosterone propionate (TP) in an oil vehicle on intromission latency and postejaculatory interval in rats. The postoperative scores were averaged over six tests during the three weeks following castration. Injections were commenced on the day of castration and continued daily thererafter. Columns indicate means for the groups, and vertical lines show standard errors of the mean. Oil alone was injected as a control condition. From Davidson, J. M. (1969b) Hormonal control of sexual behavior in adult rats. *Advance in Bioscience* 1:119–169.

curve is easily obtained for the effect of estradiol on lordosis in ovariectomized animals. For the male rat, too, similar curves can be constructed to describe the probability that groups of castrates will complete the mating pattern in terms of amount of testosterone administered. Both this parameter and the latency to first intromission can be shown to be more or less linearly related to the logarithm of hormone dosage (see Figure 5.8).

Obviously these curves do not tell the whole story relating levels of testosterone to masculine sexual behavior. Thus we may ask: what of other parameters of male sexual behavior? Are individual differences in behavioral performance correlated with differences in circulating androgen among individuals? Does administration of testosterone to normal males increase the level of performance?

Part of the difficulty in obtaining clear-cut answers to these questions arises from the complexity of the behavior itself. No single behavioral response, like that of lordosis in the female rodent, is sufficiently representative in the male, but rather we must depend on a complex of different responses that are mutually interdependent and may be differently affected by androgen. Intromission frequency (number of intromissions occurring before each ejaculation) is not directly related to androgen

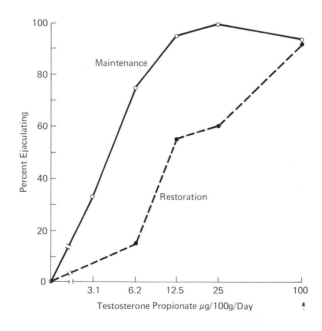

FIGURE 5.8   Effects of long-term daily administration of testosterone on sexual (ejaculatory) behavior in male rats. Less hormone was required when treatment was instituted from the time of castration (maintenance) than when behavior was restored by treatment commencing two months after castration (restoration). From unpublished data of Smith, E. R., Weick, R. F., Rodgers, C. H., and Davidson, J. M.

dose, although total number of intromissions in a prolonged test period may be so related, because androgen shortens the periods before and between each series of intromissions. A particularly interesting phenomenon encountered in androgen replacement experiments is the long latency from onset of testosterone treatment to complete behavioral reactivation in rats or guinea pigs (several weeks in Figure 5.8). This contrasts sharply with the situation in castrated female rats, in which estrogen-progesterone treatment will invariably produce lordosis in 24 to 48 hours.

Apparently it is more difficult to demonstrate such quantitative relationships in male guinea pigs than in rats. Grunt and Young (1953) found that behavioral responsiveness (ejaculation) was restored in almost step-like fashion when 25 µg/100g body weight of testosterone was administered, and clear-cut dose relationships were not particularly obvious. This situation has not been studied extensively in other animals. Further research on different species is required before we can say whether behavioral responses in different species are activated in all-or-none fashion, that is, whether they appear fully on attainment of a threshold level of circulating androgen.[11]

Whether variations in behavioral performance can be accounted for by differences in circulating androgen can best be answered by a direct attempt to correlate blood testosterone levels with inter- and intraindividual differences in performance. Only recently have practical chemical techniques for wide-scale measurements of this kind become available; hence the important studies have not yet been performed. However, the problem can be tackled by indirect approaches. Thus precastration behavior can be compared with that resulting from postcastration treatment with different levels of testosterone. Both in guinea pigs and in rats, several investigators have found that testosterone tended to restore the precastration level of performance regardless of the dose (see Figure 5.9). On the basis of this kind of information, Young (1961) has emphasized the importance of a somatic or constitutional factor that determines the individual level of behavioral responsiveness in males as well as females, thus discounting the importance of the precise level of circulating gonadal hormones. Of course this kind of explanation does not really settle anything; if the "somatic factor" is really so crucial it becomes the task of physiological investigation to define it and elucidate its nature.

Whether it is possible, by administering testosterone, to increase the level of behavioral performance above that normally seen in the intact animal has also been studied, albeit not extensively. Again differences appear between the results of experiments performed on guinea pigs and on other species. Increasing the dose of testosterone 20 times above that

[11]Even in female guinea pigs the situation may not be so simple. Goy and Young (1957) found linear relationships between estrogen dose and duration of heat, but not strength of response (duration of lordosis and frequency of mounting by the female). The latter seemed unaffected by above-threshold quantities of estrogen.

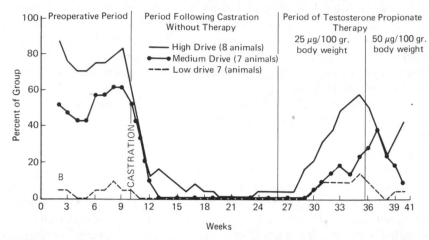

FIGURE 5.9 Effects of castration and daily treatment with 25 μg of testosterone on the mating behavior of male guinea pigs that showed high, intermediate, or low levels of sexual behavior before castration. From Grunt, J. A., and Young, W. C. (1953) Consistency of sexual behavior patterns in individual male guinea pigs following castration and androgen therapy. *Journal of Comparative Physiological Psychology* 46:138–144. Copyright 1953 by the American Psychological Association, and reproduced by permission.

required to maintain normal behavior in castrated guinea pigs did not affect performance in studies in Young's laboratory. On the other hand, administration of testosterone to intact rats (Kagan and Beach, 1953) and rabbits (Cheng et al., 1950) did result in increased performance on some measures of copulatory activity. In this, as in other types of experiments, dose-response relationships seem easiest to demonstrate with measures that can be regarded as manifestations of the state of sexual "arousal."

Another way to look at these questions is to relate the daily androgen dose necessary for behavior maintenance in castrates with that normally present. Table 5.1 compares doses of testosterone (administered over a two-month period) required for permanent maintenance of behavioral and several anatomical-physiological responses. Accessory sex gland growth is linearly related to amount of circulating testosterone from castration

TABLE 5.1. Minimum Approximate Doses of Testosterone Propionate (μg/day/100 g body weight, administered for 2 months) Required to Maintain Various Androgen-Dependent Parameters at the Level Found in Normal Intact Rats

	Sex Behavior[1]	Seminal Vesicle Weight[2]	Pituitary Weight[2]	Plasma LH Level[2]
Maintenance	12.5	25.0	25.0	25.0–100.0

[1]Dose required for 100% normal behavior.
[2]Dose required to reach level not significantly different from that in normal intact rats.
(From Davidson, 1972.)

levels, through the level normally found, to those that cause excessive hypertrophy. Thus we can take the amount of testosterone needed for normal seminal-vesicle and prostate weights as a rough approximation of the normal output of androgen over a period of time. It can be seen from Table 5.1 that considerably less testosterone (approximately 50 percent) is required to maintain sexual performance at the 100-percent level than to maintain normal growth of accessory sexual glands. This finding is consistent with the conclusion that variations in circulating androgen level within or above the normal range do not affect the performance of sexual behavior. In this respect, the amount of androgen secreted may be thought of as providing a "safety factor" such that performance level is unlikely to be affected adversely by fluctuations in secretion rate.

Although no final resolution of the problem of the relationship of testosterone levels to behavioral differences can be given until considerably more research is done on different species, we can summarize several conclusions to this point. Young's (1961) concept, which emphasizes individual responsiveness rather than hormone level,[12] is not adequate to explain the relationships between gonadal hormones and sexual behavior. Linear dose-response relationships are demonstrable in a number of circumstances. Yet it does appear that there are situations in which testosterone (and to a lesser extent estrogen) has a "threshold" type of effect on certain parameters of sexual behavior. In this respect the behavioral response is not essentially different from certain other endocrine responses. In what other ways can we show a similarity between sex behavior and other steroid hormone-dependent physiological responses?

It is known that when a hormone-dependent tissue is deprived of its trophic hormone, there is a developing insensitivity to that hormone such that a greater dose is required to elicit a response of given strength. This can be studied by observing the effects of treatment with androgen from the day of castration ("maintenance" condition) and comparing them with treatment following a period of postcastration regression or atrophy of the tissue ("restoration" condition). Figure 5.8 shows results of experiments in which animals were treated for two months starting either immediately after castration, or two months postoperatively. All responses had plateaued by the end of the experiment, when the values shown were obtained.

The difference between the two conditions is even more striking in the case of sexual behavior than for tissues of the reproductive system. This is not only true for the percentage of animals ejaculating; the intromission latency measure also shows a large discrepancy between maintenance and restoration conditions. One must presume that with increasing time of deprivation from androgen something akin to the atrophy and

---

[12]This concept resembles the idea of the "permissive" effect of adrenocortical hormones developed by D. Ingle, which states that small threshold amounts of the corticosteroid hormones have to be present in the circulation to "permit" many metabolic processes (such as glucose formation from protein) to proceed.

developing insensitivity to the trophic hormone that we find in the sex accessory (and other androgen-dependent) glands results in decreasing sensitivity of the behavior "substrate" to androgenic stimulation.

EVOLUTIONARY CONSIDERATIONS

It is not easy, on the basis of present knowledge, to generalize about evolutionary differences in the control of sexual behavior. Perhaps the only serious attempt to derive such a concept was the hypothesis of Beach (1947) suggesting an increasing degree of "emancipation" from dependence of sexual behavior on hormonal control with increasing development of the neocortex. The correlation was based on the observation that hormones seemed to assume lesser importance, and the cortex greater importance, with progression along the following evolutionary series: rodents, carnivores, primates. The decreasing role of hormones was assumed mostly on the basis of information on the duration of retention of sexual behavior patterns following castration in these different mammalian orders.

We have little precise information on the probability of persistence of sexual behavior in human castrates and the degree of its normalcy (see Chapter 8), and even less is known of the situation in subhuman primates. Most rats fail to copulate several weeks after castration. Cats, on the other hand, may show a retention of intromission behavior in some cases for years postoperatively (Rosenblatt and Aronson, 1958). Dogs have been observed to show intromission and copulatory lock over a period of up to five years after castration (Beach, 1970; Hart, 1968c). Of course in both cats and dogs there is a steady deterioration in sexual behavior after the operation, and some animals show very rapid loss of responses. The fact that some rats may persist in showing complete mating patterns for a period of up to 5 months (Davidson, 1966b) should be considered in relation to the lifespan of a rat, which is approximately two years (i.e., 3 percent of human life expectancy). Even in the "lowly" fish and amphibians, as well as in some birds, a slow decline in postcastration male sexual behavior has been reported (Young, 1961). In considering the degree of hormonal control in different species, one should of course also consider other factors. One of these is the extent to which the prepuberal animal demonstrates sexual behavior. As pointed out by Aronson (1959), the fact that prepuberal rats show more sexual behavior than prepuberal cats is not in line in Beach's hypothesis. Furthermore, even in the rat the neocortex seems to play an important role in the control of male mating behavior (Beach, 1940; Larsson, 1962a, 1964; see Chapter 7).

It seems, therefore, that Beach's pioneering attempt to construct an overall evolutionary concept in this area needs to be brought into line with more recently collected information and new theoretical viewpoints. The increasing corticalization of behavioral functions, which is a real evolutionary development, is more likely to be manifested in more subtle aspects of sexual behavior, such as the role of experience (see Chapter 6).

Chapter 6
Neuroendocrine Mechanisms and Integration

We have seen how the endocrine and nervous systems are both vitally involved in the regulation and control of reproduction in general and of sexual behavior in particular. The question before us in this chapter is how these two major communication systems of the organism interact. Consideration of this question will lead us to a further analysis of many of the phenomena discussed in earlier chapters, but our aim now will be to uncover the complexities of neural-hormonal interaction that underlie these phenomena. Because the types of neuroendocrine relationships involved vary as to direction as well as quality, it will be convenient to divide our discussion into two parts: the mechanisms of hormonal action on the nervous system, and phenomena involving effects of the nervous system on the endocrine system. In the first case we shall be examining mainly one problem, which is central to any future understanding of the mechanism of action of hormones on behavior: the site or "locus" of the hormonal action. In the second we shall discuss ways in which neural events are responsible for initiating hormonal changes within the behavioral context.

ACTIONS OF HORMONES ON THE NERVOUS SYSTEM

Given the assumption that cerebral events underlie all behavior, it is difficult to conceive of any way hormones may affect behavior other than by acting on the nervous system. Although this action could be indirect (via the production of other agents that in turn act on the nervous system), there is no evidence for this speculation. In searching for the location(s) of this action, several possibilities emerge. Hormones may act in a generalized way to change thresholds of functioning throughout widespread areas of the "centers" in the brain; they may affect spinal cord function; or finally, their action may be on the peripheral nervous system, presumably to affect afferent stimuli coming from sensory receptors in the genitalia or elsewhere. These possibilities are not necessarily mutually exclusive. We shall examine the data that bear on them as we describe the various experimental approaches used in their study.

In this discussion we shall not be constrained by the model of central neural inhibition-disinhibition presented in Chapter 4. We relied on this concept to organize a diverse and partially conflicting set of data dealing with effects of lesions and stimulation of the brain. The reader should not be misled into believing that the concept of cerebral inhibition of spinal reflexes is adequate to account for all the evidence relating to neural mechanisms of sexual behavior. Although it is one of the very few global concepts that are useful in simplifying available information in this area, the cerebral inhibition model leaves several vital questions totally unanswered. For example, although inhibition helps to account for the spinal sexual reflexes underlying the consummatory aspects of sexual behavior, it does not help us to understand the appetitive, arousal, or "motivational" components of this behavior. By this we mean those mechanisms that facilitate the initiation of the sexual encounter, as opposed to those that facilitate its completion once commenced. Second, the model does not help to explain why the destruction of some brain areas inhibits behavior and why the destruction of others stimulates it. At any rate, the present discussion of *where* the hormones act does not depend on the question of *how* they act—whether by suppressing an inhibition or, alternatively, by direct activation of neural mechanisms.

EXPERIMENTS ON SPINALLY SECTIONED ANIMALS

It was mentioned earlier that certain sexual reflexes in male and female animals are organized at the level of the spinal cord and can be elicited when all connections between the brain and the cord are severed. This phenomenon was reported over 30 years ago, but was not studied in such a way as to provide conclusive results (Bard, 1940; see Beach, 1967). More recent investigations by Hart (1967b, 1968b, 1969, 1970) have revealed interesting sex differences in relation to this question. In the spinally sectioned female rat, a response resembling lordosis was evoked, albeit with some difficulty. This response was not influenced by the presence or absence of estrogen. Similarly, in spinal bitches, although estrogen facilitated some responses to tactile stimulation of the genital region, the main sexual reflexes were not affected by hormone level. This corresponds to the failure of the early workers to find influences of gonadal hormones on sexual reflexes in spinal female cats, guinea pigs, and rats (see Hart, 1970).

In male dogs and rats, on the other hand, Hart found that testosterone was clearly facilitatory to the sexual reflexes evoked by tactile stimulation in spinal animals. This was most marked in the castrated spinal rat, where erections and a variety of penile motions accompanied by pelvic flexion waxed with testosterone administration and waned with its withdrawal (see Figure 6.1). In dogs he noted a quantitative increase in the frequency of "intense ejaculatory reactions" and of "simulated copulatory locks" resulting from testosterone injections. These results do not demonstrate direct effects of the hormone on the spinal cord, but rather that testosterone acts on some segment or segments of the spinal reflex arch, presumably on either receptors or spinal neurons. It must be emphasized at this point that these responses in spinally sectioned animals represent only fragmentary segments of the total pattern of sexual behavior. Thus their responsiveness to androgen supplies only a partial answer to the question of the site of hormonal action on behavior.

The Effects of Brain Lesions

There are two main experimental approaches used to look for possible cerebral sites of action of gonadal hormones. The first involves the combination of localized destruction of specific areas and administration of sex hormones. It was mentioned that the production of basal hypothalamic lesions (region of the pituitary stalk) results in cessation of copulation in both males and females of several species (see Sawyer, 1960). Ordinarily the behavioral deficit is accompanied in this case by degeneration of the gonads and sexual accessory structures, showing that the lesion had interrupted the neuroendocrine pathway comprising brain, gonads, and pituitary gland. When reproductive system atrophy accompanies the behavioral deficit, we hypothesize that the *hormone-dependent* portions of the neural process involved in sexual behavior are unable to function: it is as if the animal had been castrated. This hypothesis is tested in males by administering testosterone, and in females estrogen and/or progesterone. If copulation resumes after the appropriate treatment, then the hypothesis is confirmed. The lesion did not destroy neural tissue directly involved in regulating sexual behavior; rather it prevented the secretion of gonadotropin required to maintain production of hormones required for the behavior by the gonads. On the other hand, if copulation does not resume after hormone replacement therapy, and if the lesioned animals do not show signs of general debilitation, it is often concluded that the lesion removed tissue that directly responds to the hormone by activating behavioral responses. We will soon see that this argument is not entirely foolproof; for the moment, however, let us examine the evidence obtained using this approach.

The results of these experiments show in general that deficits in sexual behavior produced by *basal hypothalamic* lesions may be remedied by treatment with the appropriate hormone. (The basal hypothalamus contains the nerve terminals that produce releasing factors necessary for the secretion of pituitary gonadotropic hormones—see Chapter 5.) On the other hand, experiments on males and females of several mammalian species show that behavioral deficits resulting from damage in other parts of the hypothalamus are *not* reversible with exogenous[1] hormone treatment. The most commonly effective site of lesion production is the anterior hypothalamic-preoptic region (Heimer and Larsson, 1967; Singer, 1968). However, as we have pointed out before, damage to the same brain structures in different species may have different effects on sexual behavior. For example, lesions in the mammillary bodies (posterior basal hypothalamus) of female rabbits produced a hormone-dependent cessation of sexual behavior: the behavioral deficit produced by the lesion could be remedied by injections of estrogen (see Figures 6.2 and 6.3). Mammillary body lesions in female cats, on the other hand, produced a hormone-independent cessation of sexual behavior: the behavioral deficit could not be remedied by estrogen injections, nor did the lesions lead to any ovarian atrophy (Sawyer, 1960). Such species-specific neural organization makes a general account of the neural organization of mammalian sexual behavior very difficult.

Although the lesion-replacement therapy approach has been useful in pinpointing areas involved in the control of sexual behavior and in eliminating areas such as the median eminence region (medial basal hypothalamus) as targets or receptors for the behavior-inducing activity of hormones, its usefulness in defining hormone-sensitive structures is limited. Frequently it is not recognized that the aforementioned rationale

---

[1]*Exogenous*—administered by the experimenter, as opposed to *endogenous*—produced by the organism.

FIGURE 6.1 Influence of withdrawal and administration of testosterone on number of erections, quick flips, long flips, and response clusters per test. Tests were conducted at two-day intervals, and all animals were tested at two-day intervals throughout the experiment. There were six rats with transsected spinal cords in each group. When hormone withdrawal is indicated, there was no injection given on the day of the last test that was conducted while the animals were on testosterone (TP). When readministration of hormone is indicated, the first injection was given 48 hours before the first test that was conducted while the animals were on testosterone. When it was withdrawn from rats in group B, the number of long flips per test fell to approximately two per test (individual range of zero to nine) on tests 10 through 14. From Hart, B. L. (1967). Testosterone regulation of sexual reflexes in spinal male rats. *Science* 155:1283–1284. Copyright 1967 by the American Association for the Advancement of Science.

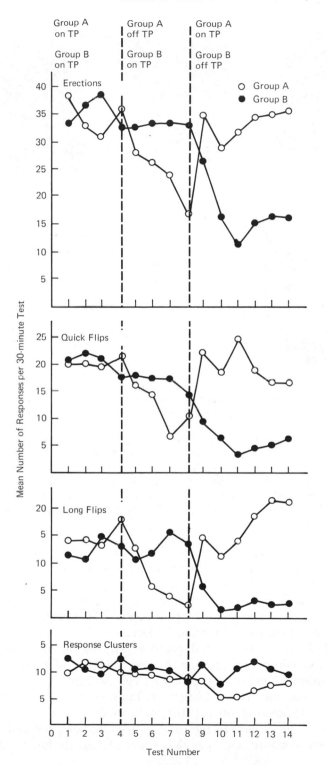

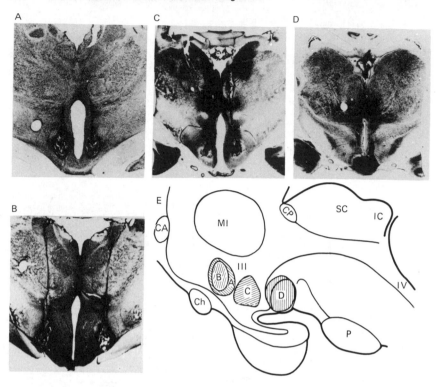

FIGURE 6.2 Anterior hypothalamic lesions (A and B), which induced permanent anestrus in the female cat in spite of treatment with exogenous estrogen. Stimulation at these sites had failed to cause ovulation and the lesions did not induce ovarian atrophy. (C) Ventromedial and (D) mammillary lesions, which induced ovarian atrophy but did not abolish mating behavior if exogenous estrogen was supplied. Stimulation at these sites had induced ovulation. (E) Midsagittal reconstruction showing anterior-posterior extent of lesions (A-D). From Sawyer, C. H. (1960) Reproductive behavior. In Field, J. (ed.), *Handbook of Physiology: Endocrinology*, II, p. 1232. Washington, D.C.: American Physiological Society.

concerning this type of experiment contains a logical fallacy. That replacement therapy fails to reverse the effects of a lesion only indicates that the lesion has removed tissue necessary for manifestation of the behavior. The area removed may be *either* a hormone-sensitive region *or* an essential structure without which a different hormone-sensitive region cannot produce its functional effects.

An additional, more general problem with lesion experiments stems from the difficulty of interpreting the effects of the lesion on brain tissue. Does it destroy neural "centers" (i.e., cell bodies), or does it destroy pathways from one area of integration to another? Are the results actually due to removal of neural structures, or to irritation of cells at the periphery of the lesion? In the cases of feeding behavior and ovulation, it has been demonstrated that deposition of ions during production of electrolytic lesions with steel electrodes may produce a chronic irritative locus that stimulates neurons for a long period thereafter. For these reasons we need additional information derived from other experimental methods for further elucidation of the neuroendocrine mechanisms of sexual behavior.

Hormonal Implantation

The second approach to locating the behavior-activating receptors of gonadal hormones (which we can refer to loosely as the "hormone receptors") is to implant small amounts of solid (crystalline) hormone directly into brain tissue at suspected sites of the receptor. The amount of hormone administered is less than the threshold amount necessary to induce behavioral changes if administered by any of the more conventional systemic routes. The rationale of this approach is as follows: because steroids are relatively insoluble in body fluids, a steroid implant provides a depot of hormone that diffuses to supply only a small region of brain tissue surrounding it. If this area contained the receptor to the implanted hormone, then the animal would behave as if subjected to high levels of hormone in its bloodstream.

Hormone Implantation in Females   The first experiments using this method were performed in the laboratory of G. W. Harris in London

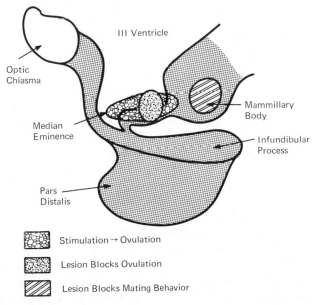

FIGURE 6.3   Sagittal section of rabbit hypothalamic hypophyseal region indicating sites of gonadotropic and behavioral "sex centers." From Sawyer, C. H. (1959) Nervous control of ovulation. In Lloyd, C. W. (ed.), *Recent Progress in the Endocrinology of Reproduction*, pp. 1–20. New York: Academic.

during the mid-1950s. Needles bearing solid estrogen fused to their tips were implanted chronically into the hypothalamus of spayed (ovariec- tomized) cats, and mating tests were conducted regularly thereafter. Complete sexual receptivity to the male was restored following hypo- thalamic exposure to estrogen. However, a variety of control procedures had to be performed before valid conclusions could be drawn as to the site of the behavioral receptor. Thus, it had to be shown that the results were not due to diffusion of the hormone from the site of implantation to other areas of the brain; that the results were not due to a mechanical lesion produced by introduction of the implant or to nonspecific irritation of the brain tissue; that the animals could not have shown the behavioral phenomena as a spontaneous reappearance of mating unrelated to the experimental procedures; and that the hormone was not released in sig- nificant amounts into the general circulation.

In this and similar studies, the first eventuality can be controlled by comparing the effects of the implant at one site to those at neighboring sites, and the second and third eventualities by implanting into the brain inert materials that should have no effect. The last possibility—absorption of the hormone and its release into the bloodstream—is somewhat more complex. The problem might seem to be resolved by using smaller quan- tities of hormone than the total dose that is effective on systemic admin- istration, so that even if the implanted hormone were completely absorbed it would be insufficient to produce the effect. However, there is a fallacy in this reasoning. The usual method of steroid administration— subcutaneous injection of the hormone dissolved in oil—is a less efficient method of delivering the hormone to the bloodstream than in the case of steady absorption from a depot of crystalline steroid. A better way to deal with this problem is to find some measure of the presence of hor- mone in the blood and use that to determine the extent of the release of hormone from the implant.

This approach was used in the studies by Harris and Michael (1964). They found that when an estrogen (diethyl stilbestrol) was injected into spayed cats, less hormone was required to produce vaginal cornification than to induce behavioral receptivity. When the hormone was implanted in the brain, however, it was possible to restore full mating behavior in the absence of estrogenic stimulation of the genital tract, showing that the effect of the implanted steroid was not due to its release into the blood. Furthermore, when inactive material was implanted in the hypo- thalamus or estrogen was implanted outside the hypothalamus, sexual behavior was not activated, which argues against the first three afore- mentioned criticisms. These carefully controlled experiments thus pro- vided solid evidence that the behavioral receptor for estrogenic hormone in female cats was located somewhere in the hypothalamic region of the brain, although more precise localization was not obtained.

Subsequent experiments by Lisk (1962) extended these observations to the female rat. An area just above the suprachiasmatic nucleus (an- terior hypothalamic-medial preoptic region) was found to be sensitive to

the behavior-activating effects of implanted estradiol, although the possibility that other areas were similarly sensitive was not eliminated by
these experiments. In rabbits, estrous behavior could be reliably evoked
by estrogen implants in the ventromedial-premammillary area (Palka
and Sawyer, 1966). Sawyer (1959) had previously found that mating behavior was most effectively inhibited by lesions just posterior to this area,
in the mammillary bodies; and it was concluded that the lesions eliminate behavior not by destroying steroid receptors, but by interrupting
"critical nervous pathways concerned with the expression of this behavior." In this species, therefore, the steroid receptor for behavior is close
to but apparently different from the area of feedback sensitivity for gonadotropic regulation.

Until very recently little or no solid information was available on the
location of the progesterone receptor in the brain. In 1971, however, Ross
et al. reported the results of experiments on the intracerebral implantation of solid progesterone. They used the double cannula technique,
whereby an outer intracranial tube is permanently fixed to the skull and,
following recovery from the operation, an inner tube containing hormone
at its tip can be introduced into the brain with no anesthesia or trauma
to the animal. In ovariectomized rats primed with estrogen, lordosis behavior was activated by intracerebral progesterone. The rapidity with
which behavioral effects followed progesterone application to the brain
was particularly impressive; significant results were obtained in 15 minutes. It will be recalled that following subcutaneous administration of
the hormone, the latency for lordosis activation is some two to four
hours, although quicker results can be obtained with intravenous injection (see Chapter 5).

A second interesting aspect of this study was that the effective area for
progesterone implantation was in the midbrain reticular formation, and
implants in the anterior or midhypothalamus or the preoptic area were
ineffective. If these findings are substantiated, the clear separation between estrogen-sensitive and progesterone-sensitive areas in the female
rat brain would suggest that these two hormones affect different components of the cerebral control of female sexual behavior. An important
function of the reticular formation is to produce "arousal" of the cerebral
cortex, and the same group of investigators (Clemens et al., 1967) previously found that inactivation of the cortex by application of potassium
chloride (spreading depression) substituted for progesterone in estrogen-
primed rats. The possibility that a progesterone-sensitive reticular activating system mechanism tied to this inhibition at the neocortical level
is involved in lordosis behavior merits consideration. It will be recalled
that lesions in an area just anterior to the one under discussion—at the
border of the diencephalon and midbrain—facilitate sexual behavior in
the male rat.

Hormonal Implantations in Males   As far as the male is concerned, on
the surface it appears less likely that sexual behavior could be activated

simply by the direct action of testosterone on a relatively circumscribed area of the brain. Mating in the male rat depends on a highly coordinated series of behavioral events involving considerable expenditure of energy, whereas in the female the only essential component of successful mating seems to be a relatively simple postural reflex. Moreover there is the general impression that male sexual behavior in animals is less directly dependent on hormones (see Chapter 5).

In 1966, studies were published on the reactivation of normal sexual behavior following chronic implantation of testosterone in male rats. Davidson (1966a) established that when testosterone propionate was injected into castrated male rats that had reliably ceased to show mating behavior, less hormone was required to restore the complete pattern of mating behavior, including the ejaculatory response, than to stimulate the sexual accessory glands (seminal vesicle and prostates). Implants of solid testosterone propionate were then placed in various cerebral locations in castrates and five mating tests were performed throughout the next three weeks. Animals with testosterone in the hypothalamus (but not the inactive steroid cholesterol) showed complete restoration of the mating pattern on approximately half of these tests in the absence of any detectable stimulation of the sexual accessory glands. Although the area from which behavior was more consistently obtained was the anterior hypothalamic-medial preoptic "continuum" (see Figure 6.4), responses were also obtained from other parts of the hypothalamus, and there were insufficient data to establish clearly whether or not the anterior region was the most effective site (Davidson, 1966a). Similar studies by Lisk (1967) and more recent (unpublished) results of Johnston and Davidson (see Figure 6.5) suggest that, indeed, implants in the anterior hypothalamic-preoptic region are most effective in restoring behavior in castrate male rats. A similar area of the brain was found to be androgen-sensitive in the male capon (Barfield, 1969) and in the Barbary dove (Hutchison, 1967).

CENTRAL VS. PERIPHERAL ACTIONS OF HORMONES

The results of these experiments on the implantation of gonadal hormones in the brain suggest that, given the appropriate stimulus object, hormonal action on a specific hypothalamic area of the brain is both necessary and sufficient for the manifestation of sexual behavior. However, closer inspection of the experimental results shows this to be an oversimplification. It is unlikely, for example, that any experiment on intracerebral implantation has resulted in completely normal sexual behavior. In the studies by Davidson (1966a), testosterone implants restored ejaculatory patterns in about 50 percent of the tests (although most of the implanted rats did respond on at least one test), and in subsequent experiments in the same laboratory the percentages were between 50 and 80 percent (unpublished). Qualitatively the behavioral responses were not entirely normal; for example, there was a tendency to prolonged latencies.

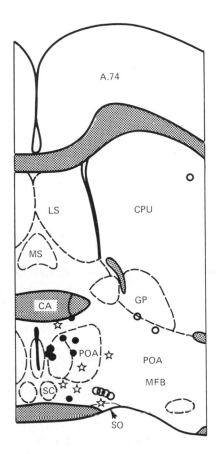

FIGURE 6.4 Intracerebral distribution of testosterone implants at frontal plane A7.4 of the DeGroot atlas (DeGroot, 1959). Implant locations include points projected from an antero-posterior distance of 0.4 mm. ○: no positive tests; ●: > one positive test; ☆: cholesterol implant (no positive tests). CPU: caudate-putamen; LS: lateral septum; MS: medial septum; GP: globus pallidus; CA: anterior commissure; POA: preoptic area; SC: suprachiasmatic nucleus; MFB: medial forebrain bundle. From Davidson, J. M. (1966a) Activation of male rats' sexual behavior by intracerebral implantation of androgen. *Endocrinology* 79:783.

What is missing in these animals? One possibility is that the steroid implant fails to deliver the hormone to the relevant cells in the appropriate manner, temporal pattern, and so on. An equally likely interpretation, however, is that extracerebral tissues may also require testosterone for full implementation of optimal behavior, a possibility to which we shall now direct our attention. Following castration in the adult rat and cat, there is a decline in the overall size of the penis as well as in certain of its structures, such as the disappearance of the cornified papillae ("spines") on the glans (Beach and Levinson, 1950). These changes can be reversed by testosterone treatment. Although it is not clear what the

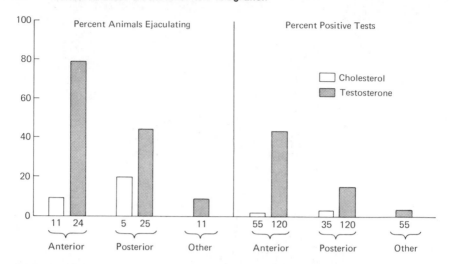

FIGURE 6.5   Restoration of sexual behavior in long-term castrated male rats by implantation of androgen in different parts of the hypothalamus and other areas of the brain (mostly thalamus). Each animal received five mating tests following implantation. Numbers beneath bars on left side of graph refer to number of animals tested and on the right to the total number of tests.

functional consequences of these morphological effects of androgen are in terms of behavior, the importance of information conveyed via sensory nerves from receptors in the genitalia has been emphasized earlier (see Chapter 4). We do not yet know to what extent these mechanisms are androgen-dependent. However, the work of Hart (see Figure 6.1) and clinical observations on humans show that erection may be stimulated by androgen. The most direct evidence for a spinal site of androgenic action is the finding of Hart and Haugen (1968) that implantation of testosterone in the lower spinal cord of rats stimulated certain of the spinal sexual reflexes, although no effect on other reflexes, including erections, was found in that study (see Figure 6.6).

What can we conclude about the site of the behavior-activating effects of gonadal hormones? It seems clear from work on both sexes in different species that the primary behavioral receptor is located in the hypothalamus, and particularly its anterior region. This does not imply for the male, however, that the supporting role of peripheral androgen is not important in ensuring maximally effective behavioral performance. The answer to the question posed at the beginning of this section seems to be that neuroendocrine integration operates via multiple actions of hormones at different levels of the nervous system, the most important receptor area being the hypothalamus. The net effect of these actions is to make the animal maximally responsive to incoming sensory stimuli.

We have seen therefore, that the anterior hypothalamic-preoptic area, whose integrity is essential for the manifestation of sexual behavior in both males and females, seems to depend for its action on receiving adequate levels on gonadal hormones. In a larger sense, the preoptic region plays a central role in neuroendocrine integration of female reproductive processes. Small lesions in this area abolish reproductive cycles in female rodents by eliminating the cyclic mechanism that controls release of the ovulatory surge of gonadotropin; and electrical stimulation here provokes ovulation. Analogous to the sensitivity of this region to estrogen in relation to female sexual behavior, estrogen implants in the preoptic region of prepuberal female rats can initiate puberty and normal ovulation (Smith and Davidson, 1968). Other aspects of the "feedback" actions of gonadal hormones may not, however, be mediated by cells in this region. Thus the positive-feedback effect of estrogen in adult rats, whereby rising estrogen levels trigger cyclic ovulatory gonadotropin release, seems to be due to a direct action on the pituitary gland (Weick et al., 1970). The negative-feedback effects, whereby rising levels of gonadal hormones inhibit the secretion of gonadotropins in both sexes, are mediated by the basal hypothalamus and/or pituitary. Thus we may say that particularly

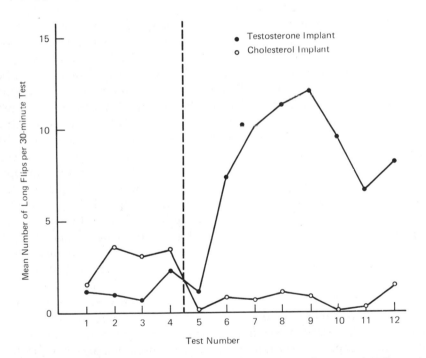

FIGURE 6.6 Number of long flips per 30-minute test for sexual reflexes in each of four tests before (1–4) and eight tests after (5–12) implantation (vertical dashed line) of testosterone or cholesterol. Points represent means of six subjects in the testosterone implant group and means of five subjects in the cholesterol implant group. From Hart, B. L., and Haugen, C. M. (1968) Activation of sexual reflexes in male rats by spinal implantation of testosterone. *Physiology and Behavior* 3:735–738.

in the female, implementation of the reproductive process requires a complex series of steroidal actions on different segments of the neuroendocrine reflex arc. The situation in the male seems to differ in having no cyclic gonadotropin release mechanism and therefore no positive-feedback effects, and in that the removal of the preoptic region does not suppress reproductive function. This area does, however, retain its behavioral significance as a steroid-sensitive region controlling the expression of sexual behavior. In both cases the end result is identical: a lower threshold of those neural mechanisms that respond to sexual stimuli by facilitating the appearance of copulatory behavior.

As mentioned at the beginning of the chapter, this discussion on the location of the behaviorally relevant receptors has not been formulated in terms of the model of cerebral inhibition of spinal reflexes described in Chapter 4. We avoided this because it was not essential for the argument: behavioral activation by a hormone acting at a given site may be conceived of either as a suppressive effect on inhibitory neurons (for detailed discussion see Beach, 1967) or as a direct activation of the neural mechanisms mediating the behavior.

In male rats the areas whose destruction speeds the course of sexual behavior (anterior midbrain or dorsal hippocampus) are remote from those where the androgen receptor is located. Similarly, in most cases where lesions have been found to have a stimulatory effect on female sexual behavior, the effective areas (midhypothalamus or cortex in rats; midhypothalamus in guinea pigs) are clearly differentiated from the anterior-preoptic estrogen receptor area. It has been argued that the steroids, acting on the preoptic region, suppress an inhibitory pathway to those areas that chronically inhibit spinal reflexes. This rather complex hypothesis is consistent with the finding that destruction of the steroid-sensitive preoptic region does indeed suppress behavior. However, it is equally possible that the steroid receptors activate mechanisms that are parallel to rather than in series with the disinhibitory areas of the brain. In fact, they may affect different components of the neural process underlying sexual behavior. It is also likely that there are neural structures in the brain that are essential for the activation of sexual behavior, but are not themselves steroid-sensitive, even though they may physically intermingle with steroid-sensitive neurons, so that a lesion will inevitably destroy both.

We need remind the reader of only one more complication—that in the female rat the estrogen receptor seems to be fairly remote from the progesterone receptor (in the midbrain reticular formation according to Ross et al., 1971)—to convince him that the time is not really yet ripe for a satisfactory unifying concept explaining the neuroendocrine mechanisms underlying the expression of sexual behavior. Nevertheless, the delineation of the hormone-sensitive as well as the hormone-insensitive areas of the brain opens the way for electrophysiological and neurochemical investigations that should eventually clarify this difficult problem.

EFFECTS OF SOCIAL·STIMULI ON HORMONE PRODUCTION

To this point we have discussed the effects of hormones on sexual behavior; that is, we have considered hormone level as the independent variable and the behavior as the dependent one. This type of relationship does not, however, suffice to describe the spectrum of endocrine-behavior interrelationships in the area of sexual behavior. Increasingly we learn that changes in endocrine secretion may be the consequence rather than the cause of behavioral events. The hormonal change can result directly and immediately from the behavior-derived stimuli, as in the case of the endocrinologic consequences of coitus. Here the sexual partner actually applies mechanical stimuli that activate the endocrine system.

Alternatively, however, the sexual partner or other individuals may activate the recipient's endocrine system without any tactile contact. This is accomplished for example by visual stimuli or by the release of chemical "messengers"—the pheromones. Pheromone-dependent phenomena in mice—the Bruce, Whitten and Lee-Boot effects—were described in Chapter 3, but their important relationship to the endocrine system requires further attention in the present context. As we shall see, mice are by no means the only mammals to show this type of phenomenon. A complete description of the reproductive process in any mammal would probably consist of chains of behavioral events linked by information transmitted by chemical means, hormonal or pheromonal. Figure 6.7 depicts in diagrammatic form some of the complexity of psychoneuroendocrine phenomena in terms of the interaction of two "closed loops"— one neuroendocrine and the other neurobehavioral.

Phenomena of the type we are discussing are often called *psychosomatic* in medicine. For instance, psychogenic amenorrhea—interruption of the menstrual cycle resulting from psychologic stress—is a case where behavior-derived stimuli seem to affect hormonal function. These phenomena are a mystery to medicine today; research on presumably analogous phenomena in animals might provide valuable insights.

We will begin the discussion with an example of external stimuli (produced by the social partner) that have an effect on neurocndocrine-mediated events: the case of reflex ovulation. This phenomenon, because of its relative simplicity, is better understood than most, and can serve as a model.

REFLEX OVULATION

The most parsimonious "solution" to the problem of coordinating copulation and ovulation is found in the so-called reflex ovulators. In cats, rabbits, ferrets, minks, ground squirrels, and many avian species ovulation does not occur unless the female is exposed to sexual stimuli—usually copulation itself. Stimuli arising from coitus activate a neuroendocrine reflex relayed through the central nervous system to cause the production

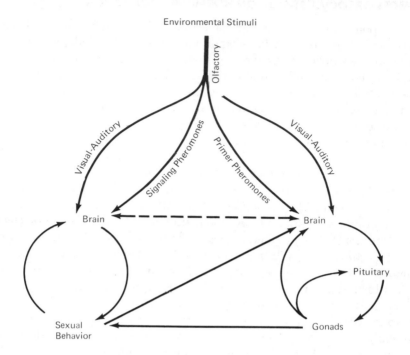

FIGURE 6.7 Environmental stimuli, including "social" stimuli from sexual and nonsexual partners, influence brain mechanisms controlling sexual behavior and the reproductive system, each of which is part of a "feedback loop." The two loops are interconnected via the effects of gonadal hormones on sexual behavior and effects of coitus-derived stimuli on neuroendocrine activity. The relations between the two brain systems themselves (depicted here by a broken line) are not yet clear. This diagram portrays many of the relationships discussed in this chapter as well as in Chapter 3. Based on a similar diagram in *Hormones and Behavior*, ed. R. E. Whelan. © 1967 by Litton Educational Publishing, Inc. Permission of D. Van Nostrand Company.

of LH releasing factor. As a result, LH is released from the pituitary within minutes of mating, and ovulation follows—in cats, rabbits, and other species—some 10 or 11 hours later (Asdell, 1964).

What is the nature of the adequate stimulus to provoke this reflex and by what sensory means is the information received by the female and transmitted through her central nervous system? In an effort to locate the sensory modalities involved in triggering reflex ovulation in the rabbit, early workers performed surgical deafferentation (destroying the sensory nerves) of the genital region. This procedure did not block copulation-induced ovulation, suggesting that direct stimulation of the genital organs was not necessary (Brooks, 1935).

The possibility that a variety of forms of stimulation may elicit ovulation in rabbits is supported by the more recent work of Staples (1967). This investigator found that ovulation could be produced in submissive does following mounting by other does, but the stimulus was only successful in the case of repeated mounting. Females that assumed the dominant role (i.e., the "mounters") did not ovulate. Males were more effective in provoking ovulation than females, because a single mount by the male (with intromission) constituted an adequate stimulus. A single mount without intromission was inadequate, even if lordosis occurred. Thus the direct stimulation of the genital tract appears to add something to the "nonspecific" stimulation of mere mounting. The (somewhat anthropomorphic) conclusion of Staples was that the "degree of excitement attained by the submissive female" was the operative factor in elicitation of ovulation. Of parenthetic interest is the observation that ovulation may also be produced in rabbits (Sawyer and Markee, 1959) and cats (Greulich, 1934) by vaginal stimulation with a glass rod, but only if exogenous estrogen is supplied, an interesting example of the lowering by hormones of a neuroendocrine threshold to sensory stimulation. A recent striking example is the demonstration that estrogen increases the sensory field around the rat's vagina from which responses to tactile stimulation can be recorded in the afferent nerve (Komisaruk, B. R., Adler, N. T., and Hutchison, J. 1972 Genital sensory fields: enlargement by estrogen treatment in female rats. *Science* 178:1295–1298).

Mechanical stimulation seems even less essential in birds. In pigeons, for example, this neuroendocrine reflex can be set in motion entirely by visual stimuli (perhaps due to conditioning processes?). Pigeons may ovulate on simple exposure of the female to a male pigeon or even to another female. Most intriguing is the demonstration that merely placing a mirror in the cage of the female pigeon can be effective in inducing ovulation (Matthews, 1939). This is a striking example of the activation of a neuroendocrine reflex by stimuli—heterosexual, homosexual, or narcissistic.

Reflex Ovulation in "Spontaneous Ovulators"

Recent work is increasingly calling into question the concept of a rigid separation between spontaneously and reflex-ovulating species. We now know that ovulation may be induced by mating in the "spontaneously ovulating" rat under a variety of special conditions in which natural ovulation would not otherwise occur. These include induction of receptivity early in the cycle by appropriate estrogen treatment (Aron et al., 1966); prior blockage of ovulation by treatment with the tranquilizer chlorpromazine (Harrington et al., 1966), or barbiturate (Ying and Meyer, 1969) or androgen treatment shortly after birth (Ericsson et al., 1966). Normal implantation and pregnancy can follow coitus-induced ovulation in barbiturate-blocked rats (Zarrow et al., 1968).

Probably the most effective way to make the spontaneously ovulating rat into a reflex ovulator is by exposing it to constant light, which results in cessation of cycling and of spontaneous ovulation. These animals become continuously sexually receptive, and (as shown in Figure 6.8) mating is followed by a very rapid release of LH into the circulation. Although the amount of LH released is less than in spontaneous ovula-

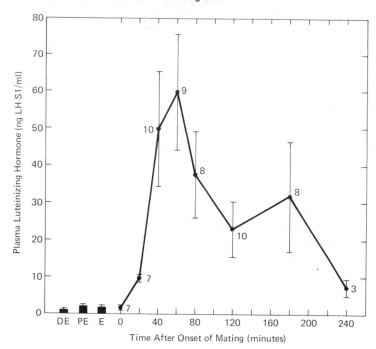

FIGURE 6.8   Rapid release of LH triggered by mating (20 minutes with an active male) in rats exposed to constant illumination. Bars on the left signify values on the morning of diestrus (DE), proestrus (PE), and estrus (E) in normal rats kept in cycling light. Numbers on the curve refer to number of tested animals. Vertical bars are standard deviations. From Brown-Grant, K., Davidson, J. M., and Greig, F., (1973). Induced ovulation in albino rats exposed to constant light. *Journal of Endocrinology 57:7–22.*

tion, it is more than adequate to produce ovulation, which almost invariably follows coitus in these constant light-exposed rats.

The claim that human beings may also show this phenomenon of reflex ovulation has been made by various authors (see Hartman, 1962; Clark and Zarrow, 1971). It has been reported that coitus under conditions of extreme stress, such as rape, might induce ovulation at unexpected times in the cycle. Most of the work on which this claim was based originated in Germany of the 1940s, and its validity is questionable. In view of the information in rats, however, this question should be reevaluated in humans, because it might provide an explanation for the occasional reports of conception occurring during or around the time of menstruation. The fact that oral contraceptive agents prevent ovulation does not belie this possibility. These antifertility steroids apparently inhibit brain mechanisms controlling gonadotropin release, and thus may block the neuroendocrine pathway whereby coitus activates the ovulatory mechanism.

The existence of an intrinsic mechanism for copulation-induced ovulation in spontaneous ovulators such as the rat and possibly the human could perhaps be explained as an evolutionary vestige (Zarrow and Clark, 1968). On the other hand, it might be thought of as a redundancy factor or "fail-safe" mechanism to help ensure ovulation under conditions where the normal mechanism is impaired. At any rate, this phenomenon in the rat provides the opportunity for examining the mechanism of a behaviorally activated neuroendocrine reflex in a species whose reproductive endocrinology is better known than that of the rabbit or the cat.

Studies on the adequate stimulus for reflex LH release in rats provide results not too dissimilar to those resulting from the work on the rabbit, a "true" reflex ovulator. Artificial mechanical stimulation of the vagina and cervix was found more effective in inducing ovulation than electrical stimulation of the cervix alone (Zarrow and Clark, 1968). Aron et al. (1968), on the other hand, claimed that intromission is not an essential part of the stimuli inducing ovulation because either genital stimulation or mounting by the male was partially effective in inducing ovulation. However, in constant light-exposed rats, in which ovulation can be elicited far more reliably by mating, Brown-Grant, Davidson, and Greig (1973) showed that intromissions were more effective stimuli for induction of ovulation than mounting alone (see Table 6.1). Ejaculation was shown to be unnecessary for this effect. Here, as in rabbits, the precise modalities, receptors, and CNS pathways involved are unknown.

As biologically useful as mating-induced LH release obviously is in the female, such an effect in males would seem, on the face of it, to be quite worthless. However, Taleisnik et al. (1966) found that LH is released in significant amounts following coitus in male rats. Although this report has not been confirmed, other data in both rabbits and rats suggest that mating in males increases either gonadotropin secretion or other factors that stimulate the testes. Thus, it has been found that in male rabbits mating is followed by increased testosterone production by the testes (Endroczi and Lissak, 1962; Saginor and Horton, 1968). Observations on rats by Drori and Folman (1964) suggest a biological function for this seemingly strange phenomenon. There is a tendency for degeneration of the testes, accessory sex glands, and other tissues affected by androgen

TABLE 6.1. Analysis of the Behavioral Stimuli Responsible for Evoking Ovulation in Rats Exposed to Constant Light

Treatment	Percent Rats Ovulating	No. Eggs Shed[1]
Controls: Cycling Light	100	14 ±0.5 (S.E.)
Constant light		
No treatment	0	—
Mating with ejaculations	87	8 ± 1.0
15 intromissions, no ejaculation	100	9 ± 1.3
20–40 mounts	47	9 ± 1.4

[1]Per ovulation per rat. (S.E. = Standard error of the mean.)

with advancing old age. This tendency toward atrophy of the reproductive system is lessened when male rats are allowed to cohabit with females for prolonged periods. The opportunity to engage in sexual activity results in higher weights of accessory sex glands, penis and penis muscles, and kidney, as well as a longer life span.

The changes observed by Drori and Folman are all androgen-dependent, and testosterone production is controlled by gonadotropin, particularly LH. The intriguing possibility arises that each time the male copulates, a "dose" of gonadotropin and thus of testosterone is "injected" into the bloodstream. The summation of these doses over time contributes to maintenance of the reproductive system, and perhaps enhances general health. Whether these findings are applicable to other species is as yet unknown, but their extension to the human species would provide food for thought.

The adequate stimuli for this reflex in male rats was studied by Thomas and Neiman (1968). They found that providing the male with mounts alone did not suffice to prevent the degenerative changes in the reproductive system. Intromissions alone were partially effective, but the effects were greatest when the full pattern, including ejaculations, was permitted. This work, if confirmed, provides one indication (others will be discussed) of the adaptive value of the complex mating pattern in rodents.

BEHAVIORAL STIMULI AND PREGNANCY

The endocrinologic events triggered by coitus are not limited to the phenomena described in the last section, whose biological significance is still speculative. In short-cycling species, a related set of events seems to be of crucial importance for the implementation of reproduction. The maintenance of conditions within the uterus that allow the growing fetus to develop to term is the function of the placenta, and perhaps of the fetus itself. However the *initial* event that makes ovum implantation and early embryonic development possible seems to be a direct consequence of the hormonal processes initiated by the mating stimulus, at least in these species. Following mating there is a release of prolactin (and probably LH), which contributes to formation of the corpus luteum and its continued secretion of progesterone. This keeps the uterus in a quiescent and otherwise appropriate condition for establishment of pregnancy.

This progonadotropic effect of mating should be distinguished from the ovulatory release of gonadotropin, which, as previously pointed out, may *also* follow mating in rats. Everett (1961, p. 532) has in fact shown that these two can be experimentally dissociated.

Which component of the male's mating behavior is involved in this effect? If pregnancy is to follow mating, ejaculation is required, so it was easier in the initial experiments on this question to study the phenomenon of pseudopregnancy for which insemination is unnecessary. When in the rodent fertilization does not follow mating, a period of corpus-luteum development, progesterone secretion,[2] and anovulation ensues, which lasts about half as long as pregnancy itself. Mating with a vasectomized male thus produces a condition resembling that of early pregnancy, which can be checked experimentally by observing the immediate cessation of vaginal cycling (smears), and the condition of the ovary and uterus.

Although luteal activation resulting in pregnancy maintenance or in pseudopregnancy can be separated from ovulatory gonadotropin release, the stimuli are similar in the two cases. Mechanical or electrical stimulation of the cervix can replace sterile mating as a means of inducing pseudopregnancy in rats. And yet the same kind of apparent sensory redundancy seems operative as was noted earlier in the case of ovulation induction. Thus, in an early investigation (Ball, 1934) it was found that although cervix removal decreased the response to single matings, pseudopregnancy still invariably resulted when multiple vaginal plugs[3] were deposited. Both the cervix and the vaginal wall therefore seemed capable of contributing to activation of this neuroendocrine reflex. Actually, this investigator believed that the vaginal plug supplied an essential stimulus for pseudopregnancy not replaceable even by many intromissions. Because these experiments were done before the characteristic differences between mount and intromission behavior were generally recognized, however, it has been suggested that Ball's rats might have been showing high-mount and low-intromission frequencies (Chester and Zucker, 1970). At any rate, Ball's finding suggests that the threshold for luteal activation is higher than for ovulatory LH release, which may follow mounting alone.

More recent work by Wilson et al. (1965) provides an interesting substantiation of the way behavioral stimuli contribute to physiological mechanisms essential for birth of the young. By appropriate switching of stud males, these investigators arranged for some females to be inseminated with vaginal plugs after three or fewer intromissions, and others after approximately nine intromissions. (It should be noted that between intromissions the rat dismounts and an interval of genital grooming ensues.) The rate of "successful" pregnancy in females that received three or fewer intromissions before ejaculation was only about 20 percent; those receiving normal numbers of intromissions had a pregnancy rate of about 90 percent.

How does the stimulus provided by intromissions influence the outcome of pregnancy? Adler (1969) has shown that multiple intromissions in the rat provide two necessary conditions: the induction of luteal

[2]Although it will be assumed here that the periods of prolonged progesterone secretion that characterize pseudopregnancy and pregnancy are the result of prolactin secretion from the pituitary, it should be recognized that even in rodents prolactin may not be the only stimulus activating progesterone secretion, but that other gonodotropins, especially LH, are most likely involved.

[3]In various rodent species the ejaculate coagulates inside the female tract, forming a solid "plug."

activation (prolactin secretion causing progesterone secretion) and the facilitation of sperm transport from the cervix into the uterus. It is not known how the latter effect is mediated, although possibilities include a neuroendocrine reflex involving oxytocin release (for which there is evidence in other species), relaxin[4] secretion, or some local reflex inducing uterine relaxation. Chester and Zucker (1970) have shown that these two phenomena of luteal activation and sperm transport facilitation can be separated. Under the conditions of their study, three intromissions with ejaculation sufficed for maximal sperm transport into the uterus, whereas luteal activation required more than five intromissions. As previously demonstrated by Blandau (1945), sperm transport requires a vaginal plug because the sperm do not reach the uterus following intravaginal artificial insemination or mating with males deprived of the ability to form plugs by seminal-vesicle removal. Luteal activation with subsequent pseudopregnancy is not, however, dependent on plug deposition. An indication of the separateness of these effects is that multiple intromissions could be separated by as much as five hours from the later ejaculation without adversely affecting the outcome of pregnancy (Chester and Zucker, 1970).

The stimulus requirements for activation of the corpus luteum, resulting in pseudopregnancy, have been studied in the mouse by McGill and co-workers (McGill, 1970). Although the mouse has the same general type of copulatory behavior as the rat—multiple discrete intromissions followed by a long type of ejaculatory mount—there are differences of detail. Thus the mouse may show more intromissions—with hundreds of pelvic thrusts —and a more prolonged ejaculatory reflex (see Figure 2.9). These differences apparently are reflected in differences in the stimulus properties required for luteal activation. As in the rat, deposition of a vaginal plug is not *necessary* for luteal activation. Thus pseudopregnancy followed mating with castrated males or males in which plug formation was prevented by drug treatment or seminal-vesicle removal. However, multiple intromissions apparently are not necessary in the mouse, unlike the rat, perhaps because of the large number of thrusts per intromission. McGill (1970) hypothesizes that the specific swelling of the penis during the mouse's ejaculation is the necessary stimulus. Using an artificial penis designed to stretch the vagina as does the penis during ejaculation, McGill was able to produce pseudopregnancy in over 80 percent of receptive females. This hypothesis leaves unexplained the role of the particularly high numbers of intromissions and thrusts that frequently occur during the mouse's copulatory behavior.

But regardless of the possible role of stimulation during ejaculation, pseudopregnancy in the mouse can be produced by stimulation that resembles that provided by penile intromission. Using carefully calibrated and controlled artificial vaginal stimulation, Diamond (1970) found that

---

[4]Relaxin is a little-known ovarian hormone that causes relaxation of the pelvic ligaments and cervical softening before delivery.

the successful induction of pseudopregnancy in the hamster and laboratory mouse, which, like the rat are short-cycle rodents, depends on both the number of intromissions received and the intervals between them. Different patterns of stimulation are required for the mouse and the hamster, and both appear to be different from that required for the rat. Hence Diamond suggests the concept of a "species vaginal code" for the proper stimulation of the neuroendocrine reflex required for the induction of events leading to successful pregnancy. Further experiments will be required to validate and extend this hypothesis.

Little is known of the biological role in reproductive physiology of the diverse mating patterns that may occur in mammalian species other than those discussed here. In man, for instance, the possibility that behavioral or at least "psychosomatic" factors affect fertility is suggested by the mysterious (albeit anecdotal) observation that infertile couples frequently become fertile after adoption of a child. Of course, in long-cycled species such as the human, luteal activation does not depend on a mating stimulus but is spontaneous. However, the efficiency of sperm transport through the cervix can be an important factor in human fertility (Masters and Johnson, 1966), and it would be of considerable interest to know what endocrine or psychoendocrine factors might influence this process, for instance by effecting relaxation of the cervix.

PHEROMONES

Another particularly intriguing example of an area where behavioral stimuli play upon the endocrine system is the case of the pheromones. Although these potent biological agents are certainly not hormones, like hormones they are chemical messengers. Furthermore, as we shall see, they often seem to affect hormone secretion, as well as being dependent for their production on hormones. Most of our knowledge of pheromones (including the chemical structure of a number of these agents) comes from work on insects, but research related to mammalian pheromones has grown considerably in recent years.

Primer Pheromones

Three well-established effects of social stimuli in mice, apparently all induced by pheromones, were described in Chapter 3. The precipitation of pseudopregnancy when female mice are housed in groups (the Lee-Boot effect) apparently is due to a stimulation of prolactin secretion by a factor in female urine. Male urine, on the other hand, contains a pheromone that induces estrus and synchronizes estrus cycles, as is seen when a male is brought into a group of females caged together (the Whitten effect). The same pheromone mediating the Whitten effect may also be the active factor in the Bruce effect, whereby exposure to the presence or odor of an alien male prevents implantation of the fertilized egg in a recently fertilized mouse. Here the evidence is particularly good that the mechanism is via inhibition of prolactin secretion. In short-cycling species, implantation is delayed for some five days after mating, during which time progesterone secretion (stimulated by prolactin) prepares the uterus for implantation. As mentioned in Chapter 3, the Bruce effect cannot be elicited after the fifth day because the blastocyst will have been successfully implanted by then. Thus the effect can be prevented with prolactin, progesterone, or reserpine (which activates prolactin release). It does not occur in lactating females or those bearing pituitary transplants, in both of which cases high prolactin levels are found (see Bronson, 1968).

Pheromones produced by male mice are dependent on androgen for their production; castration prevents and androgen restores both the male's effect on the estrous cycle of the female and his ability to block pregnancy (Bruce, 1969). The chain of hormone-behavior interactions seems to be fascinatingly complex: male hormone induces pheromone production, which in turn influences female hormone production, which subsequently affects the behavior of that female. And, most likely, this is not the whole story.

Phenomena related to mammalian pheromones have been researched primarily in mice, but an increasing amount of information is becoming available that suggests that these processes may be much more widespread within the class of Mammalia. This evidence is still rather indirect for the type of pheromones we have been discussing so far—the "primer" pheromones, defined as those that activate relatively long-term physiological (particularly endocrine) events. Introduction of rams into flocks of ewes has been reported to result in synchronization of births (Schinckel, 1954; Hulet, 1965). Stimulation or suppression of the olfactory system results in a variety of modifications of the female reproductive cycle in rats, cats, rabbits, and guinea pigs. There are also a number of reports indicating that olfactory bulb removal (in both sexes) can have quite severe inhibitory effects on sexual behavior as well as on gonadal maintenance. These effects tend to be variable, depending on species, strain, age, and other factors (Bronson, 1968).

There is little direct evidence for primer pheromone effects in males. However, information on effects of olfactory bulbectomy is most suggestive that chemoreception of pheromones is necessary to maintain male reproductive function in some cases. The stimulatory effect of mating on the reproductive system of aging male rats does not, however, seem to be a pheromone effect, because it cannot be duplicated by exposing the males to urine from estrous females (Thomas and Neiman, 1968; Drori and Folman, 1964; Folman and Drori, 1966).

The question of the biological significance of the Bruce, Whitten, and Lee-Boot-type effects is a difficult one. It has been suggested that the Whitten effect may be important in promoting fertility and the Lee-Boot effect in synchronizing births. The capacity to terminate a pregnancy started by another male, as in the Bruce effect, and then to inseminate the female, does confer a selective advantage (Bronson, 1968).

Signaling Pheromones

Quite different from the primer pheromones are the "releaser" or "signaling" (Bronson, 1968) pheromones, in which olfactory stimuli activate not a prolonged physiological response, but an immediate behavioral one. To this class belong the sex attractants of insects, and an increasing volume of information suggests that agents of this type aid in the mating process of a considerable variety of mammalian species. LeMagnen (1952) first demonstrated the ability of male rats to discriminate the odors of estrous from nonestrous female rats. Not surprising perhaps, males can also be shown to prefer the odor of estrous females in a choice situation where, for instance, they can approach samples of either estrous or nonestrous urine. The preference (but not the *ability* to discriminate) is androgen-dependent because it disappears after castration and can be restored by testosterone treatment (Carr and Caul, 1962).

Olfactory discrimination or preference for the urine of estrous females has also been reported in male dogs, stallions, bulls, and rams (see Michael and Keverne, 1968). In the case of the pig, however, no preference for the odor of estrous sows could be demonstrated (Signoret, 1970). Just as male primer pheromones seem to be generally dependent on testicular hormone, so also the signaling pheromones emanating from the female seem to be dependent on ovarian hormones. Thus, Beach and Merari (1970) showed that males were attracted to cotton balls soaked in vaginal secretions from estrogen-treated spayed bitches, and that the attraction increased to the point of licking and chewing when the secretion came from bitches treated additionally with progesterone. Merari (personal communication) once noted intense interest on the part of a male dog presented with a cotton ball on which progesterone solution was poured. This chance observation might not have any significance, but is worth mentioning in the light of some curious observations by Kloek (1961), who exposed police dogs to the scent of progesterone. Later the dogs were reportedly able to identify rods held by women who were in the luteal phase of the cycle or were pregnant. It is not inconceivable that some mammalian pheromones are steroid hormones. More concrete evidence for one such steroid deserves mention.

In some species an important olfactory attraction by the female toward the male has been observed. Sows react in the presence of boar urine by adopting the mating stance ("standing reaction"). The active principle is probably the 3-hydroxy-5-androst-16-one component of the boar submaxillary gland, which causes the "boar" odor or taint in pork (Patterson, 1968). This may be the first chemical identification of a mammalian pheromone.[5] Another similar case of an olfactory stimulus emanating from males is mentioned by Michael and Keverne (1968). Various elements of the mating pattern of female cats—crouching, treading, rolling, and rub-

---

[5] Recently Muller-Schwarze (1969) reported the chemical structure of the active ingredient of the scent from the tarsal gland of male Blacktailed deer which attracts females: cis-4-hydroxydodeca-6-enoic acid lactone.

bing—were reportedly released when estrous females were placed in cages "recently occupied and marked by an active male," but not if the cage was first washed with disinfectant.

In subhuman primates, recent research has uncovered a possibly important role for olfactory stimuli emitted by the female in activation of the male's sexual behavior. Estrogen treatment of female rhesus monkeys markedly stimulates the sexual excitability of males caged with them, who respond with prompt increases in mounting. Although estrogen also causes reddening of the "sex skin" in the genital region, this does not supply the stimulus to the male; estradiol applied directly to this area in spayed females reddens it without affecting the male's activity (Herbert, 1966). Attempts to block the sense of smell in male rhesus monkeys suggested, on the other hand, that olfactory stimuli from the female markedly stimulates the male's sexual behavior. Apparently, production of this pheromone(s) by the female macaque is facilitated by estrogen and suppressed by progesterone, since the latter hormone inhibits both female sexual activity and the attractiveness of the female to the male (Michael, Herbert, and Welegalla, 1967). Recently Michael and his coworkers (1971) claimed the first isolation and chemical identification of primate pheromones. Vaginal secretions collected from estrogen-treated monkeys, which increased sexual attractiveness when applied to the sexual skin of untreated spayed females, were subjected to chemical extraction and fractionations. Surprisingly enough, the active principles in the vaginal secretions were a mixture of simple short-chain fatty acids, such as acetic acid (vinegar). A synthetic mixture of the same acids that have been isolated from the vaginal secretions have the same effect on behavior as was found in the natural material.

Pheromones in Humans

Man is described as a microsmatic species (i.e., one in which olfaction is relatively unimportant), and Bronson (1968) believes that pheromonal function is only vestigial in humans. This view is not shared by Wiener (1966), who believes that people emit sex attractants. Relevant or not, the prodigies of olfactory discrimination described for some individuals are most impressive. The literature contains reports of people who could reliably recognize all individuals by their scents. In fact, this ability was apparently possessed both by Helen Keller and by the physiologist Bethe, who coined the term *ectohormone* (the predecessor of *pheromone*—see Wiener, 1966).

Of course, it is one thing to demonstrate the capacity of olfactory acuity and another to show that olfaction plays an important biological role in human beings or is related to the endocrine system. Although clear evidence on this score is lacking for man, there are a multitude of anecdotal accounts (Wiener, 1966, 1967) and a few experimental observations indicating that sex odors may be of considerable importance in humans. As far as signaling pheromones are concerned, much was written about the olfactosexual behavior of human beings long ago by Havelock Ellis (1905), and we need not dwell on this here. Comfort (1971) suspects the presence of pheromones in the sweat glands of the axillary region or in the smegma from the prepuce. He also points to the presence in human urine of musklike[6] odors, the susceptibility of humans to these odors, and their use in perfumes as indications of signaling pheromone in the human species. The possible signaling function of pheromones in humans is undoubtedly suppressed by psychosocial factors during ontogeny.

Recent findings by McClintock (1971) provide intriguing evidence for primer pheromone effects in women. The menstrual periods of girls living in a college dormitory showed significant synchrony among the individuals who had most contact (friends or roommates). Cycle length was found to be shorter among those "who spent more time" with males. The former observation is reminiscent of the Lee-Boot effect, the latter of the Whitten effect; both of these phenomena have been demonstrated in mammalian species other than mice, where they were initially observed. LeMagnen's classic studies (1952) on the "exaltolide phenomenon" showed the hormone dependency of certain olfactory perceptions in humans. Exaltolide is a synthetic compound with a musklike odor that can be perceived by adult women, but hardly ever by prepuberal girls or by males. The ability to perceive this synthetic analog of mammalian sex attractants varies tremendously throughout the menstrual cycle, reaching a maximum at the time of ovulation and a minimum during the luteal phase, when progesterone levels are highest. Similar hormone-dependent phenomena in human males are unknown. However, Michael (1969) is so impressed with the possible relevance of the aforementioned data on monkeys as to suggest that the (progestational) contraceptive steroids may adversely affect the sexual behavior of men via the effects of these agents on their female partners (see Chapter 8).

[6]"Musk" is derived from the preputial gland of the musk deer, a native of Tibet.

Chapter 7
Sexual Differentiation

THE EXTENT OF SEXUAL DIFFERENCES

The time-honored phrase *vive la difference* reflects the sentiments of most people about variations between the sexes. The exhortation is misleading, however, in that the difference is by no means limited to that small though crucial matter of anatomical detail, but includes many aspects of physiology and behavior. The combination of events that lead to establishing these sexual dimorphisms is referred to as the process of *sexual differentiation*. The elucidation of this process is a fascinating problem, both for its intrinsic interest and because its study can yield important insights into the physical basis of behavior.

Although we are concerned here with the anatomical, physiological, and psychological differences that are related to reproduction, it should be noted that the actual range of sex differences goes well beyond the areas of reproductive biology and behavior. In many organ systems there are sex differences in anatomy (e.g., skin, subcutaneous tissues, and bones). To take one small example, the adrenal glands are bigger in female than in male rats, although the situation is reversed in hamsters and in humans. This difference in size carries over also into the physiological area. Both the blood level of adrenocortical hormones and the secretion of these hormones in response to stress are higher in female than in male rats, and the converse is true in hamsters. Among the many other physiological differences between males and females of many species is the higher basal metabolism of males, which in turn results in other metabolic divergencies.

Of course there are also many behavioral differences between the sexes other than those related to copulation. An obvious example is that of aggressive behavior, which is influenced by sex hormones (Davis, 1964). Less obvious perhaps are various aspects of feeding behavior (e.g., taste preferences), spontaneous activity, and play and emotional behavior, which are clearly sexually dimorphic. In what follows we shall attempt first to define the sex differences in mating behavior and then discuss the processes of their differentiation and also those of other sexually dimorphic behavior patterns.

Although the sexes are differentiated in the areas of anatomy, physiology, and behavior, it is clearly impossible to consider these sets of differences as separate entities. Physiological function is dependent on structure; behavior is dependent both on structure and physiological function. We shall see that the actual processes of anatomical, physiological, and behavioral differentiation have much in common apart from the basic fact of the dependency of behavior on form and function. In fact, a theme that will pervade our discussion will be the effort to separate processes affecting anatomical differentiation from those determining behavioral differentiation. Expressed differently, to what extent are the basic differentiating processes "central" (i.e., affecting brain behavioral mechanisms) or "peripheral" (i.e., affecting structure, particularly of the sexual organs)? But first we must look purely descriptively at the differences between males and females in what pertains to sexual behavior.

LIMITATIONS TO THE SEX-SPECIFICITY OF SEXUAL BEHAVIOR

It is probably no exaggeration that virtually all the elements of copulatory behavior typical of either sex have at one time or another been observed in the other sex as well. This being so, any discussion of sexual differentiation should be preceded by a consideration of the extent and circumstances of these "anomalous" responses. In this discussion we shall refer to responses typical of one sex as *homotypical* when they are manifested by subjects of that sex and as *heterotypical* when they are manifested by subjects of the other sex. It should be emphasized that heterotypical behavor can in no sense be thought of as abnormal. This is particularly obvious in the case of mounting behavior in females, as will be discussed.

There are, of course, some rather obvious anatomical constraints on the extent to which the individuals of one sex can show the behavioral responses of the other. But the neuromuscular mechanisms required for mounting behavior and lordosis are present in both sexes, and we shall concentrate on these two important elements of mating behavior.

Mounting has been described in normal females of many species, including rats, hamsters, guinea pigs, rabbits, porcupines, sharp-tailed shrews, marten, dogs, cats, African lions, swine, cattle, and rhesus monkeys (Beach, 1947, 1968b). The ubiquity of mounting behavior raises the question of whether it is meaningful to regard it as a truly sex-differentiated or even a true sexual behavior. Mounting is often seen in prepuberal animals, and in this and other cases it might be more justifiably regarded as an element of play behavior, especially when not accompanied by pelvic thrusting. However, in all the species just mentioned, females do show mounts with thrusts, and often other accompanying manifestations of "male" behavior. For example, cows in estrus will sometimes paw the ground, bellow, and pursue other animals in the manner of the bull.

A point that emerges clearly from Beach's comprehensive analysis of mounting behavior (1968) is the importance of the nature of the animal that is mounted. In several species (as will be shown in the next section) the probability of mounting by females is greatest when estrous females are available as the recipients of this act. The stimulus properties of the recipient are thus a factor of prime importance, and more than for any other aspect of sexual behavior, mounting is relatively independent of the sex of the "performer." The degree to which it is dependent on the *hormonal* condition of the performer will be discussed in the next section.

The absence of male external genitalia precludes the performance by normal females of other elements of male sex behavior such as intromission and ejaculation. However, the behavioral elements that often accompany these events could conceivably be shown by females. For instance, on very rare occasions the typical backward lunge accompanying intromission by the male rat has been observed to occur spontaneously in females of this species (Beach, 1942a). We shall see later how certain hormonal manipulations can precipitate the appearance of intromission and ejaculatory behavior in females.

The prime manifestation of female receptivity, lordosis, is hardly ever seen in normal or castrated males, and certainly this form of heterotypical behavior is much less common than is mounting in females. There are, however, a few well-documented cases that have been reported in male rodents and primates (Beach, 1938; Bingham, 1928; Stone, 1924). Interestingly, lordosis is most likely to occur in normal male rats if they are vigorous copulators. When the female is removed in the course of mating, such males have been observed to show lordosis if mounted by another male (Beach, 1947, p. 268).

HORMONES IN RELATION TO HOMOTYPICAL AND HETEROTYPICAL BEHAVIOR

To what extent are these heterotypical behavior patterns dependent on hormones? Can they be elicited by the heterotypical gonadal steroids?

These are questions of some importance. If female behavior patterns could be elicited by ovarian hormones with equal ease in males and females, and male behavior patterns by testosterone, one would have to conclude that there is no differentiation of the central nervous "substrate" of sexual behavior. In such a case the only differentiation would be that affecting hormonal secretion by the two sexes. We shall define the differentiation of sexual behavior as differentiation of the capacity of central nervous mechanisms to respond to hormones. The question before us is to what extent differentiation as so defined exists.

If we consider the two classes of gonadal hormones, male and female, and the two corresponding classes of behavior, it is apparent that eight combinations exist in regard to the possibilities for the manifestation of both types of behavior in both sexes. Young (1961) has pointed out that each one of these eight possible results has been reported; that is, estrogen and testosterone have at one time or another been reported to elicit both male and female responses in both male and female animals. Nevertheless, in his view, only the two obvious relationships—testosterone induction of homotypical behavior in males and estrogen induction of receptivity in females, as well as the stimulation of mounting by estrogen —are "commonly encountered." The crucial question for this discussion, however, is the relative extent to which a given hormone produces "male" vs. "female" types of responses.

Ovarian Hormones in Females

As mentioned before, mounting behavior is very common in female animals. Does the manifestation of this behavior depend on hormones? In his authoritative 1961 review, Young was unable to arrive at a clear answer to this question, due to the apparent discrepancies in reports of experiments on different species. Young's own pioneering work on guinea pigs pointed to a rather clear dependence of this behavior on ovarian hormones because it was rarely found in anestrous or spayed animals and could easily be produced by estrogen and progesterone in the same dose range that elicits sexual receptivity. There are also many reports on the high incidence of mounting in cows and other animals in estrus. These reports often tend to be anecdotal, or at least do not clearly establish the relationship with hormones. In contrast there is the work of Beach (1968b) on rats, which seems to negate completely the importance of ovarian hormones because mounting was not affected by the state of estrus, the presence of the ovaries, or the administration of estrogen.

These apparent discrepancies appear to be at least partially resolved by Beach's recent work, which sheds considerable light on the whole question. It is now well established for dogs, rats, and cows that the probability that a female will mount another female depends primarily on the hormonal condition of the mounted animal. Estrous females, at least in these species, become highly attractive to the attentions of other females, probably in part because the estrous female is more cooperative under these conditions. Much less important is the hormonal condition of the mounting female, although estrogen may increase the intensity and perhaps the frequency of mounting. In Young's work on guinea pigs this point apparently was missed and, as the nature of the "mountee" was not defined, the situation in this species is not entirely clear (Beach, 1968b).

Apparently a different situation exists with regard to the mounting of males by females, a relatively common occurrence in dogs and rats, and one that is particularly noticeable in dogs following coitus interruptus. The attentions of the male are probably an important element in the elicitation of female "heterosexual" mounting. Male dogs are relatively "uninterested" in anestrous females, which accordingly show little heterosexual mounting. Again the stimulus properties of the mounted animal are of crucial importance. "Mounting cannot properly be viewed as a unilateral activity, but is more correctly conceived as a form of social or interindividual behavior which is only to be understood in terms of the qualities and activities of both participants" (Beach, 1968b).

Testosterone in Males

In view of the rarity of spontaneous lordosis in males, it is not surprising that testosterone treatment stimulates female behavior in males only rarely and under rather special circumstances (rodents: Stone, 1924; Fisher, 1956; Engel, 1942; Beach, 1941, 1945a; lizards: Noble and Greenberg, 1941). Generally in these studies the doses of testosterone are massive and the quality of the resulting behavior not very impressive. In a few rats found by Stone (1924) and Beach (1945a) to show a high incidence of spontaneous lordosis, castration inhibited and testosterone restored this behavior. It should be noted, however, that in those males testosterone also restored the capacity to respond to stimulus females with homotypical behavior. Males that spontaneously manifest lordosis also show vigorous homotypical behavior.

What can we conclude from the studies on the effects of homotypical hormones on heterotypical behavior? First, two rather obvious statements warrant assertion. Heterotypical behavior in males or females is never as intense or as frequent as homotypical behavior, and it is easier to produce homotypical behavior with homotypical hormones than to produce heterotypical behavior. Second, the stimulus conditions are probably of greater significance in producing heterotypical behavior than is the hormonal treatment. Thus if copulation is interrupted, heterotypical behavior may appear. This is seen when male rats show lordosis following removal of the female and introduction of another male, and when female dogs begin to mount the male following coitus interruptus.

The concept that emerges is that the potentiality to display various types of behavior patterns is "built in" in both sexes, but that with appropriate stimulus conditions the threshold for homotypical behavior is lower than that for heterotypical. This concept raises a number of interesting issues, such as the question of the differential response of the two built-in behavioral mechanisms to heterotypical hormones. Is the particular response to a given hormone more a function of the type of hormone or of the soma—the substrate on which the hormone acts? The next section describes the effects of heterotypical hormones in terms of homotypical and heterotypical behavior patterns.

Testosterone in Females

When androgen is administered to female animals, both male and female mating patterns may appear. Which type of behavior predominates? A great deal of data have been collected from many species, to, the effect that this treatment results in intensification of mounting and other masculine responses (Young, 1961). In dogs and rats, where mounting behavior in the female is not dependent on ovarian hormones, testosterone increases the frequency of mounting over that found in estrous females (Beach, 1967). In rats the type of mount is shifted to the more typically male kind with pelvic thrusts (Beach and Rasquin, 1942). On the other hand, the guinea pig, which is the only known species in which female mounting is entirely estrogen-dependent, does not respond as well to testosterone as to estrogen. Female hamsters rarely mount, with or without testosterone administration (Tiefer, 1970).

Studies on the rabbit present a somewhat different picture. Klein (1952) reported that administration of large doses of testosterone to female rabbits results in lordosis and receptivity to mating attempts by males. Much smaller doses were found by McDonald et al. (1970) to be effective, and they suggested that testosterone (small amounts of which are produced by the normal female's ovaries and adrenal glands) might synergize with estrogen in the normal activation of female receptivity. In these experiments, androgen was at least as effective in this respect as in producing mounting behavior. What is not clear is whether the effect on feminine responsiveness is really due to testosterone per se, because significant amounts of estrogen might be formed by metabolic conversion of the androgen (see Chapter 5 and West et al., 1956). In fact, when different androgens were tested in rabbits, lordosis was elicited only by those molecules intrinsically capable of this type of conversion (Beyer et al., 1970). Although sexual receptivity may be induced in female rabbits by implantation of small amounts of testosterone in the hypothalamus (Palka and Sawyer, 1966), the cells of the hypothalamus may themselves be able to convert testosterone to estrogen. In Chapter 8 we will discuss the evidence of clinicians that testosterone stimulates female sexual behavior in women and the difficulties of interpreting these clinical observations on human behavior.

Estrogen in Males

Less equivocal evidence comes from studies on administration of estrogen to males. A number of investigators have reported only weak masculine responses to this treatment. Even when massive doses of estradiol benzoate were given to castrated rats (see Table 7.1), only a few ejaculatory patterns resulted, although in some cases intromission without ejaculation was observed. On the other hand, much smaller doses of estradiol stimulated the manifestation of lordosis in response to mounting by vigorous males (see Figure 7.1). Considerable variability was encountered among animals in this experiment, but some animals displayed lordosis quotients quite close to those seen in the normal proestrus female, with doses which, although higher than those required in females, were not massive. Estrogen-induced lordoses in the male are not generally as complete as those seen in the highly estrous female. Another difference is that progesterone does not facilitate estrogen-induced lordosis in males (see Table 7.2). Hamsters are exceptional in that doses of estrogen similar to those needed to elicit lordosis in the female are also effective in the male, although the intensity (duration) of the males' response is less (Tiefer, 1970; Crossley and Swanson, 1968).

RESUME AND CONCLUSIONS

In view of the mass of sometimes contradictory data and the difficulties of interpretation it would be folly to attempt to draw dogmatic conclusions. It is clear that the forms of sexual behavior typical of both sexes are capable of manifestation in both males and females. The preponderance of homotypical over heterotypical behavior is caused by the following factors: homotypical hormones are present in much greater quantity than heterotypical ones; and homotypical behavioral mechanisms have lower thresholds of response to homotypical hormones than do heterotypical mechanisms.

The other problem we have considered is the relative responsiveness of the two mechanisms to heterotypical hormones. As mentioned previously, this question is important theoretically in defining the nature of the differentiation of sexual behavior, but it also has practical application in cases of endocrinologic aberrations. For convenience it will be dis-

TABLE 7.1. Effect of Estradiol Benzoate (200 μg/day) on Homotypical Behavior in Six Long-Term Castrated Rats[a]

	Days After Onset of Injections				
	—[b]	3	6	10	13
Intromissions	0/6	0/6	2/6	4/6	0/6
Ejaculatory patterns	0/6	0/6	2/6	2/6	0/6

[a]Figures are numbers of rats that showed the behavior on a given test.
[b]Results from 3 tests before estradiol benzoate administration.
(From Davidson, 1969a.)

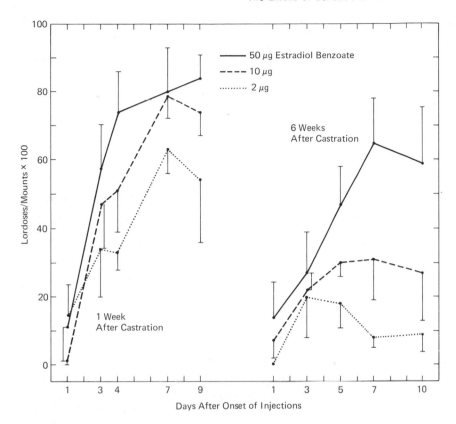

FIGURE 7.1   Lordosis quotients in castrated male rats treated with daily injections of estradiol benzoate. From Davidson, J. M. (1969) Effects of estrogen on the sexual behavior of male rats. *Endocrinology* 84:1365.

cussed in terms of two possible hypotheses. The hypothesis of *host-specificity* postulates that the predominant behavioral response to any hormone is the homotypical one. The hypothesis of *hormone-specificity*, on the other hand, implies that the predominant response to a hormone is the response typical to that hormone, regardless of the sex of the subject.

Although the results of some investigations on androgen administration to females seem to support the hypothesis of host-specificity, this may be due to conversion of androgen to estrogen within the animal. On the other hand, the experiments on estrogen administration to males much more clearly support the hypothesis of hormone-specificity. It appears that estrogen stimulates predominantly lordosis-type responses, regardless of the sex of the recipient animal. On the basis of their experiments on cats, Green et al. (1957) suggested that responses are hormone-specific in prepuberal animals and host-specific after puberty. Unfortunately they reported no details of their behavioral results, and it is not clear if careful quantification of the behavioral measures was applied. One implication of their work that merits further investigation is that (post-

TABLE 7.2. Effect of 0.5 mg Progesterone on Estrogen-Induced Lordosis in Male Rats[1]

Treatment	Number of Rats	Days After Onset of EB Treatment		
		2	4	8
Estradiol benzoate (EB)	6	2 ± 2	14 ± 7	34 ± 15
EB + progesterone	5	7 ± 5	30 ± 8	20 ± 12

(From Davidson and Levine, 1969.)
[1]Figures are lordosis quotients, means ±S.E.

puberal) sexual experience affects the potential quality of response to hormones. Nevertheless the findings on heterotypical responses of adults contraindicates any general validity for this hypothesis.

It should be noted at this point that heterotypical hormones can have heterotypical physiological effects analogous to their effects on behavior. Testosterone is quite active in stimulating uterine growth; likewise, estrogen stimulates the growth of the male sexual accessory glands. The action of the hormones of the opposite sex on the reproductive system is not, however, identical with that of the homotypical hormone. Thus, in seminal vesicles and prostate, estrogen tends to stimulate more of the fibromuscular elements of these glands, and less of the secretory epithelial tissue, than does androgen. Both androgens and estrogens are capable of inhibiting the secretion of gonadotropic hormones ("negative-feedback" effect) in both males and females (see Chapter 5). Indeed testosterone inhibits gonadotropin secretion more effectively in female rats than it does in males (Hertz and Meyer, 1937).

In conclusion, it is worth restating that sexual differentiation is not an all-or-none phenomenon; it should be thought of in terms of the relative probability of the occurrence of different patterns of sexual behavior and of the relative capacity to respond to gonadal hormones in specific ways.

THE CAUSES OF SEXUAL DIFFERENCES

Until quite recently virtually nothing was known about the control of the differentiation of sexual behavior. The mechanisms determining the anatomical differentiation of the reproductive system and genitalia have, however, been studied for many years. More recently, much information has become available on the sexual differentiation of the physiological processes involved in reproduction. Current concepts indicate that the mechanisms of differentiation of sexual behavior are similar in many respects to these two other areas of differentiation. Furthermore, because behavior is so dependent on sex hormones, one particular aspect of physiological differentiation, the determination of the differences in endocrine activity, is of vital importance for the differentiation of sexual behavior.

Therefore, before dealing with behavior, we shall briefly discuss anatomical and physiological differentiation of the reproductive system.

ANATOMICAL DIFFERENTIATION OF THE REPRODUCTIVE APPARATUS

The reproductive system of the embryo develops in three rather discrete stages. First the gonad is formed, then the internal reproductive organs (accessory sex glands, vas deferens, etc., in the male and uterus and fallopian tubes in the female), and finally the external genitalia differentiate. Before each of these three phases there is a stage in which it is impossible to determine to which sex the structure belongs. However, in the first two stages the *anlagen* or precursors of male and female structures are both present, then one regresses and the other develops. In the case of the external genitalia, a common *anlage* exists that is capable of development in either the masculine or feminine direction. How does this come about? In the following description the human will be used as an example, but the same principles are believed to apply to mammals in general, the only differences being those of detail.

Under the influence of the sex chromosomes, presumably exerted via some undefined organizer substances referred to as "cortexin" and "medullarin," either the cortex of the indeterminate gonad develops into an ovary, or the medulla develops into a testis (see Figure 7.2). In the male the cortex regresses, and the medulla regresses in the female. This occurs during the seventh and eighth weeks of intrauterine life.

The anlagen of the male and female internal reproductive duct systems are the Müllerian duct for the female (develops into uterus and tubes), and the Wolffian duct for the male (develops into epididymis, vas deferens, seminal vesicle, etc.). In the third and fourth month of embryonic life, one or other of these develops and the other regresses (see Figure 7.3). The stimulus for the differentiation of these organs is a secretion from the testes. This organizing substance acts locally by diffusion through the tissues, rather than via the bloodstream, at least in some species. If a male gonad is present, the secretion of some still undetermined substance(s) by it causes development of the Wolffian and regression of the Müllerian ducts. Experimentally, testosterone administration reproduces the former but not the latter effect of the developing testis. There are probably two "duct-organizing" substances: a male activator akin to testosterone, and a female suppressor of unknown constitution (Jost, 1969). If the animal is castrated, the female pattern develops; if unilateral castration is performed, differentiation occurs only on the opposite side. It appears then that the female pattern is basic; in the absence of any gonadal influence a uterus and fallopian tubes will develop. The superimposition of the influence of a male gonad is required for male differentiation.

The external genitalia are differentiated also in the third and fourth months and here again the female pattern is basic. There are several dif-

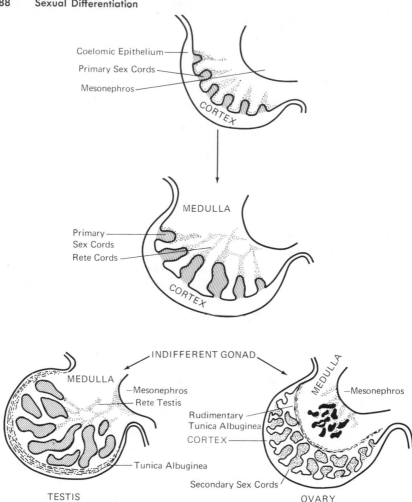

FIGURE 7.2 Diagrammatic representation of the development of an ovary from the cortex or a testis from the medulla of a bipotential primordial gonad in man. From Grumbach, M. M. (1960) Some considerations of the pathogenesis and classification of anomalies of sex in man. In Astwood, E. B. (ed.), *Clinical Endocrinology I.*, pp. 407–436. New York: Grune & Stratton. Used by permission.

ferences from the process of reproductive tract differentiation. First, there is only one set of anlagen or precursor structures. The genital tubercle and groove and the urethrolabial fold and labioscrotal swelling have the bipotential ability to develop either into a vulva, clitoris, and so on in the female, or a penis, scrotum, and so on in the male (see Figure 7.4). The other differences are that, although the stimulus for differentiation comes from the developing gonad, in this case testosterone appears to be involved, and the effect is not local but apparently bloodborne. Thus unilateral differentiation cannot be brought about by unilat-

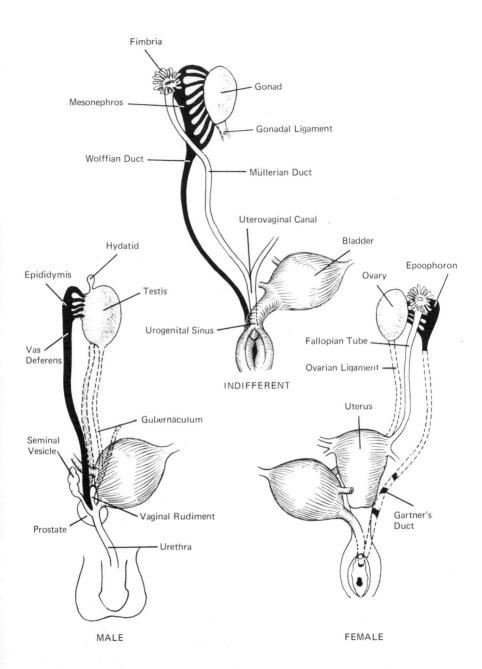

FIGURE 7.3 Embryonic differentiation of male and female internal genitalia (genital ducts) from Wolffian (male) and Müllerian (female) primordia. From Van Wyk, J. J., and Grumbach, M. M. (1968) Disorders of sex differentiation. In Williams, R. H. (ed.), *Textbook of Endocrinology*, pp. 537–612. Philadelphia: Saunders.

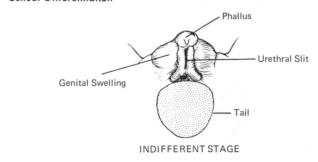

Phallus

Urethral Slit

Genital Swelling

Tail

INDIFFERENT STAGE

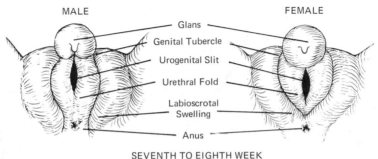

MALE   FEMALE

Glans
Genital Tubercle
Urogenital Slit
Urethral Fold
Labioscrotal
Swelling
Anus

SEVENTH TO EIGHTH WEEK

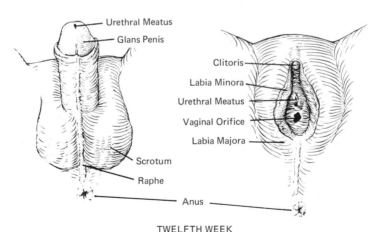

Urethral Meatus
Glans Penis

Clitoris
Labia Minora
Urethral Meatus
Vaginal Orifice
Labia Majora

Scrotum
Raphe
Anus

TWELFTH WEEK

FIGURE 7.4 Differentiation of male and female external genitalia from indifferent primordial structures in the embryo. In the absence of androgenic stimulation, female development will always occur. From Van Wyk, J. J., and Grumbach, M. M. (1968) Disorders of sex differentiation. In Williams, R. H. (ed.), *Textbook of Endocrinology*, pp. 537–612. Philadelphia: Saunders.

eral castration. Disturbances of these processes may occur at any of the three stages. Some of the drastic results that may follow such aberrations of differentiation in the case of the human are discussed in the last chapter.

PHYSIOLOGICAL DIFFERENTIATION OF THE REPRODUCTIVE SYSTEM

The hierarchy of mechanisms controlling reproductive function were discussed in Chapter 6. The differences in reproductive function between the sexes stem from gonadal differentiation; they have to do with the different patterns of gametogenesis and steroidogenesis in the ovary or testis. Because the gonadotropic hormones, which control both of these functions of the gonads, are identical in both sexes, it would seem that all the sex differences in reproductive physiology are the direct result of anatomical differentiation of the gonads. In considering whether this is so, it will be useful to speculate as to the possible outcome of a hypothetical experiment.

We remove the whole reproductive system—gonads, reproductive tract, and accessory sexual structures—from a male and a female mammal of the same species. We then reverse these organs, transplant the male reproductive structure into the female, and vice versa. Assuming this difficult experiment to be successful, with all the grafted tissues growing normally in their new sites, could we expect a complete reversal of reproductive function in the two animals?

Although transplantation experiments of such a wholesale nature have almost certainly never been attempted, enough work along these lines has been done to allow an excellent guess as to the probable outcome. When ovaries are transplanted into various sites in the bodies of male animals, they rapidly become vascularized, follicles develop, and estrogen is secreted. The one great defect in these ovaries is that they show none of the cyclic changes found in normal females.

The cyclic function of the ovary is of course the direct consequence of the cyclic secretion of gonadotropic hormones by the pituitary. The basic differences between physiological functioning of the male and female reproductive systems is that in females gonadotropic hormones are secreted in cyclic fashion, whereas in males they are secreted, so far as we know, in a relatively steady, tonic manner. All other differences flow from this and from the differences in the substrate of gonads and other reproductive organs on which the hormones act.

Some confusion might arise out of the classification of male animals as acyclic. Although domesticated mammals generally do not show marked seasonal fluctuations in reproductive function, breeding seasons are the rule among wild animals (see Chapter 3). This is not, however, to be confused with female reproductive cyclicity. In males of nondomesticated species there is an alternation between function and nonfunction —a *quantitative* change from an inactive state in which the gonad may be completely atrophied, to a state of full functioning. This type of seasonal cyclicity is also found in females of many wild and domesticated species. However, the female also shows a different type of *qualitative* alternation within the active breeding season, between two very different anatomical-physiological reproductive conditions.

In the one condition the predominant ovarian secretion is estrogen, the animal is behaviorally receptive to the male, and the reproductive tract is physiologically prepared for receiving the spermatozoa and for fertilization. The full development of this stage is seen at estrus, and in primates, at the midpoint between two periods of menstruation. Following ovulation and mating, the second, progesterone-dominated phase commences. At this time the female is unreceptive to the male, and the reproductive system is geared to receiving the fertilized ovum and development of the embryo.

This form of sexual cyclicity, which is essential for the implementation of reproduction, is of course completely absent in the male. It is achieved basically by a cyclic alternation in the secretion of gonadotropic hormones, as we have seen in Chapter 5. The reader will remember that these hormones are identical in males and females, so that differences in the control of gonadal function between the sexes have to be understood as resulting from differences in the quantitative pattern of secretion of the gonadotropins. It appears, therefore, that the problem of physiological sexual differentiation can be approached by asking what determines whether an organism will secrete gonadotropic hormones in a cyclic or an acyclic fashion. The obvious answer is to postulate a basic difference in the pituitary glands of males and females. That this is false was clearly demonstrated by Harris and Jacobsohn (1952) in experiments in which pituitary glands were actually exchanged between male and female rats. When males were hypophysectomized and their pituitaries replaced with transplants taken from female rats, normal male reproductive function continued. Similarly, female rats that were hypophysectomized and transplanted with pituitaries from males showed normal estrus cycles and the other manifestations of female sexuality. The conclusion from these studies, which since have been confirmed by other investigations, was that the pituitary gland itself is a completely passive agent that functions in the manner appropriate to the environment in which it is placed. The pituitary is in fact controlled by hypothalamic neurohumors carried to it via the pituitary portal vessel system, as described in Chapter 5.

The relevant question about physiological sexual differentiation then is, what determines whether a hypothalamus will develop into the male acyclic type or the female cyclic type? We shall see that the basic processes here bear a striking resemblance to those we encountered in considering the control of anatomical differentiation.

In 1936 Pfeiffer performed a series of experiments whose real significance was not recognized for about 20 years. He found that if male or female rats were castrated at birth and an ovary transplanted into these animals when they reached adulthood, cyclic ovarian function with corpus luteum formation ensued in both sexes. If castration was performed later in life, the male never developed cyclic gonadotropic function. The presence of the testes in the neonatal period thus seemed to be the essential factor suppressing the potentiality for cyclicity in male rats. Pfeiffer thought that the pituitary itself became differentiated, and this was an understandable mistake because his work was done before the role of the brain in controlling the pituitary was understood. In the late 1950s other investigators reopened this question, working now with a knowledge of the newer neuroendocrine concepts. Barraclough found that a single dose of testosterone administered to neonatal female rats resulted in permanent sterility and a "masculinization" of the reproductive system to the extent that the rats never ovulated, but rather showed a steady acyclic level of ovarian function. These animals, which are referred to as "androgen-sterilized," show vaginal smears that are constantly cornified, suggesting continuous secretion of estrogen. The ovaries have well-developed follicles, but corpora lutea are absent. If the injection of testosterone is delayed beyond the first few days of life, its effect is minimal or absent, and the animal manifests normal female cyclic ovarian function in adulthood (Barraclough, 1966). These experiments have been repeated with many variations by different investigators in recent years, and the main conclusions with regard to the physiological control of cyclicity are generally agreed to be as follows.

The female (cyclic) pattern of gonadotropin secretion is the basic one, just as in the case of anatomical sexual differentiation. Differentiation occurs postnatally in the rat, and it involves basically a suppression of the inherent cyclic pattern in the male rats by exposure to testosterone in the immediate postnatal period. The presence or absence of the ovaries is immaterial at this stage. It appears (see Figure 7.5) that the mechanism whereby this differentiation is achieved involves the action of neonatal androgen on the anterior hypothalamic-preoptic region of the brain to prevent its becoming organized as the initiator of cyclic gonadotropin release in the future, postpuberal period (Barraclough, 1966; Gorski, 1966).

BEHAVIORAL DIFFERENTIATION

Because the work of Barraclough and others showed that the neuroendocrine regulation of the female reproductive system could be permanently changed by a single injection of testosterone in the immediate postnatal period, it was of considerable interest to find out whether similar manipulations could affect the establishment of patterns of reproductive behavior. Much work has been done on this question in the past few years, and the results have far-reaching implications for understanding the physiological control of sexual behavior.

THE CRITICAL PERIOD IN RATS
If 0.1 mg testosterone propionate is administered in a single injection to a new-born (neonatal) female rat, the development of behavioral patterns is

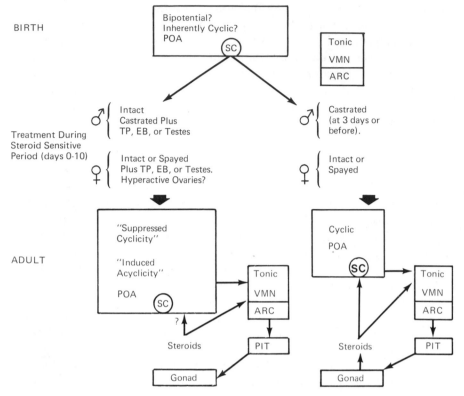

FIGURE 7.5 Theoretical concept of the influence of perinatal hormone levels on sexual differentiation of the hypothalamic regulation of LH secretion. ARC: arcuate nucleus; EB: estradiol benzoate; PIT: pituitary; POA: preoptic area; SC: suprachiasmatic nucleus; TP: testosterone propionate; VMN: ventromedial nucleus. "Tonic": relatively continuous. From Gorski, R. A., and Wagner, J. W. (1965) Gonadal activity and sexual differentiation of the hypothalamus. *Endocrinology* 76:226–239.

profoundly altered for the rest of the animal's life. In adulthood, normal lordosis and other components of estrus behavior cannot be elicited even after treatment with large quantities of estrogen and progesterone. The single exposure to testosterone at a critical period in the animal's life results in a complete and permanent suppression of the capacity to display feminine[1] sexual behavior. The critical period for the behavioral effect, as in the case of the neuroendocrine effect, is approximately the first five days of life in rats, injections at ten days or later being without significant effect. Similar administration of testosterone to neonatal males, however, appears to have no significant effect on subsequent sexual behavior. What can these experimental findings teach us about the differentiation of sex behavior under normal conditions? Obviously they suggest that testosterone can have an "organizational" effect on behav-

---

[1]We will use "male" and "female" to refer to the genetic sex of an animal and "masculine" and "feminine" to refer to the behavior typically shown by that sex.

ioral patterns,[2] determining future behavioral responses for the duration of the individual's life. Before general hypotheses of behavioral differentiation could be constructed, however, an additional body of experimental evidence was required relating to the effects of neonatal removal of the gonads. As might be expected, the results of these experiments were opposite to those of neonatal testosterone administration, the experimental treatment being effective in males but not in females.

In 1945 it was noted that rats born with congenital absence of the ovaries showed normal female sex behavior when injected with estrogen and progesterone in adulthood (Beach, 1945b). Subsequent experiments have demonstrated that experimental removal of the ovaries on the first day of life likewise has no apparent effect (Adams-Smith, 1967). On the other hand, when the testes are removed from male rats on the first day of life, a profound "feminization" of the behavior results. When these animals are injected with ovarian hormones in adulthood they respond to mating attempts by males in a manner essentially similar to that of normal females (Harris and Levine, 1965).

The theory of behavioral differentiation that evolved from these investigations closely resembles that advanced to explain differentiation of the gonadotropin control mechanism (Harris, 1964; Levine and Mullins, 1964). It is presumed that the rat is born with a potential capacity to manifest both male and female behavioral patterns in adulthood, on exposure to the appropriate hormones. Testicular hormone secreted in the first few days of life[3] suppresses the capacity to manifest the female pattern and develops the capacity for male behavior. As in the case of physiological differentiation, estrogen is presumed not to play any role, and the female pattern is "basic."

In concrete terms, it is assumed that the brain of neonatal rats of both sexes contains mechanisms capable of responding, after puberty, to male or female hormones by initiating male or female sexual behavior respectively. If exposed to testosterone in the first five days of life, that part of the cerebral mechanism that would have subsequently become sensitive to the female hormones becomes permanently inactivated, and its sensitivity declines.

DETERMINATION OF MALE BEHAVIOR

So far we have emphasized the evidence relating to female sexual behavior. This was done to simplify matters, but the problem of interpreting the available data on early differentiation of male sex behavior must now

[2]The first investigator to appreciate the significance of early exposure to androgen in terms of the organization of brain mechanism for behavior was W. C. Young (e.g., 1961), who based his conclusions on the effects of androgen treatment of pregnant guinea pigs.

[3]It has not been conclusively proven that the hormone that is active in the neonatal male under physiological conditions is testosterone. However, there is evidence that the testes can secrete testosterone at this time (Niemi and Ikonen, 1963; Resko et al., 1968), and no other testicular hormone that could fulfill this role has been demonstrated.

be considered. This deserves detailed attention, if only because it exemplifies rather clearly the conceptual and experimental handicaps to research in developmental psychobiology.

The main difficulty is that in none of the animal "preparations" used to study these problems—neonatally castrated males or females, or androgen-treated females—does a normal penis develop in adulthood. Hence intromission and ejaculatory behaviors are either impossible or severely impaired. But what of mounting behavior? As mentioned earlier, mounting can hardly be regarded as a uniquely male behavior pattern, being found, often to a considerable degree, in intact or castrated animals of both sexes. It is not necessarily correlated with other elements of the male pattern because long-term castrates that have lost the ability of intromission may actually show high frequencies of mounting (Davidson, 1966b). These theoretical considerations suggest that it will not be easy to arrive at substantive conclusions about the mechanisms involved in the differentiation of male sex behavior.

Unfortunately the results of experimental investigations on this point do not dispel the difficulty. In guinea pigs, early exposure to testosterone enhances the mounting behavior of the female offspring (Phoenix et al., 1959). In rats, however, the reports on the effects of neonatal testosterone on "malelike" behavior in females are quite contradictory. This treatment in female rats has been variously reported to stimulate intromission-like behavior (Whalen and Edwards, 1967), to leave mounting behavior unchanged (Harris and Levine, 1965; Whalen and Edwards, 1967), and to decrease mounting (Mullins and Levine, 1968a). Undoubtedly differences in experimental design and methods of behavioral scoring are responsible for the divergencies in these and other reports.

Neonatally castrated male rats show more mounting behavior as adults than do normal females. This could be taken as evidence that behavioral mechanisms are partially differentiated before birth in the male rat. Whalen and Edwards (1967) reject this interpretation and favor the view that mounting behavior, in both sexes, is genetically "built in," rather than differentiated by hormonal influences during development, with males having a higher potential to manifest this behavior than females. It would not be rewarding to delve deeper into the inconsistencies of the experimental data on differentiation of mounting behavior. Even if the picture were clear, information on mounting could not form an adequate basis for a theory of differentiation of male sexual behavior because, as we have seen, it is much less sexually dimorphic than other elements of the mating pattern.

We fall back, therefore, on the other components of male copulation—intromission and ejaculatory behavior. Can the dilemma created by the dependence of this behavior on penile structure be resolved? We need to be able to manipulate the hormonal milieu during early development without preventing the development of a viable penis during adulthood. Unfortunately, the development of the penis in neonatally castrated male rats is permanently impaired; testosterone treatment in adulthood cannot restore its normal structure (Mullins and Levine, 1969c; Beach et al., 1969). Therefore the same kind of critical period effect seems to apply to development of penile structure as to behavioral development. Thus we cannot distinguish whether the failure of ejaculatory behavior following neonatal castration is due to absence of androgen at the level of the external genitalia or of the brain-behavior mechanism.

To shed further light on this problem we can consider two other experimental situations. Differentiation of the penis is the result of "organizational" effects of androgen exerted both in the prenatal and in the postnatal period in the rat. By administering testosterone to female rats both in the last days of pregnancy and the first 40 days after birth, Ward (1969) was able to produce females possessing a hypertrophied (overdeveloped) clitoris almost indistinguishable from a penis (see Table 7.3). Of course these females were incapable of lordosis, but when given further androgen in adulthood, they manifested the full pattern of male sexual behavior. Seminal vesicles, prostates, and other male sexual accessory glands were also present. In similar experiments, Whalen and Robertson (1968) demonstrated that females so treated would show similar numbers and spacings of intromission-like and ejaculation-like behavior at intervals similar to those shown by normal males. Curiously enough, neonatally estrogen-treated females may also show occasional "ejaculatory" responses following androgen treatment in adulthood (Levine and Mullins, 1964), but considerable phallic (clitoral) development is also found in these animals. Again in these experiments one cannot distinguish between organizational effects on the brain and on the genitalia.

Another interesting finding relates to the behavior of neonatally castrated male rats treated for several days with androstenedione, the second most important androgen secreted by the testes. This secondary androgen does not suppress the development of lordosis behavior in the neonatal castrate, but the severe impairment of male sexual behavior and of genital development normally found in neonatal castrates is prevented (Goldfoot et al., 1969; Stern, 1969). As a result one produces animals that

TABLE 7.3. Summary of the Mean Anogenital Distance, Body Weight, and Penile Length of the Experimental Groups at About 100 Days of Age[1]

Groups	Anogenital Distance (cm)	Body wt (g)	Penile or Clitoral Length (cm)
Normal males	4.3	292	3.4
Pre- and postnatal females	3.6	298	3.2
Prenatal TP females	2.8	215	2.2
Postnatal TP females	2.0	280	2.1
Normal females	2.1	252	0.4

(From Ward, 1969.)
[1]Daily 1 mg TP injections for all female groups began at 60 days of age.

when confronted with receptive females will show normal male intromission and ejaculatory patterns, but when paired with vigorous males will show normal female mating patterns. The mechanism of these effects has not been elucidated yet but what is clear is that again the development of male genitalia is correlated with the presence of male intromission and ejaculation patterns.

Thus in all the cases we have mentioned and in other experimental situations such as when antiandrogenic drugs are administered prenatally, full male behavior patterns are found only when a well-developed phallus is present. This leads us to one of two alternative conclusions. The first is that there is no organizational effect of androgen on the development of brain-behavioral mechanisms for male sexual behavior. This implies that all that is needed for male sexual behavior is possession of a normal male genital apparatus and that the necessary cerebral mechanisms are inherent. The alternative is that developmental organizational effects occur both at the level of the brain and of the penis, but these two have so far not been distinguished because they coincide in time.

Why, in the androstenedione experiment, was the capacity for lordosis not suppressed by this androgen treatment although the treatment was effective in development of the capacity for male sexual behavior? The most plausible explanation seems to be that this is a question of relative thresholds. We can assume that it takes more androgen during the early critical period of development to suppress the potential for showing feminine mating behavior than it does to stimulate the male potential (either of the phallus or of the brain, or of both). Androstenedione, being a weak androgen, may suffice for the one but not for the other. At any rate, there is an additional interesting implication of this experiment. It demonstrates clearly that maleness and femaleness of behavior is not an either-or matter. The neural basis underlying sexual behavior is not like the *anlage* of the genital tubercle in which the degree of male organization is inversely proportional to the extent of female organization, resulting in a continuum of development with a clitoris at one end and a fully developed penis at the other.

The demonstration that the potentiality to manifest both male and female behavior patterns can coexist in the same animal at the same time has implications that should be emphasized at this point. This bisexuality is not merely an artefact of the peculiar experimental circumstances described. As alluded to on pp. 180 and 182, spontaneous lordosis is sometimes observed in males of several species without evidence of endocrinologic or other abnormality. Whether the situation in these cases is the result of a genetic endowment at the extreme end of the normal spectrum, or whether the observed bisexuality has more to do with unusual stimulus configurations, we cannot judge. But apart from the lesson these cases teach us about the measure of bipotentiality that can be retained in the adult brain, they also provide us with a background in mammalian sexual behavior with which to view the phenomenon of bisexuality in human beings (see Chapter 8).

To summarize and reiterate our main conclusions from this section, we must take the position that the main demonstrated effect of androgen during the developmental period is to suppress the future potentiality for the manifestation of female sexual behavior and to allow for the development of full-fledged male external genitalia. Whether there is an additional action of enhancing the development of male sexual behavior patterns independent of the latter effect remains to be demonstrated.

OTHER SUBHUMAN SPECIES

It may seem surprising that the critical period for behavioral differentiation of rats should occur after birth, rather than in the more protected embryonic environment. However, the newborn rat is at a relatively early stage of somatic maturation; it is instructive to compare this with the situation in other species. In 1937, Dantchakoff found that the internal and external genitalia of female guinea pigs born to androgen-treated mothers were incompletely differentiated. The clitoral hypertrophy and masculinization of reproductive tract structures justified the designation of the female offspring as "pseudohermaphroditic" by Young and his co-workers (see Chapter 8), who made a detailed study of the effects of prenatal androgen some 20 years later (Phoenix et al., 1959). These females showed an increase in testosterone-induced mounting behavior, as well as decreases in the percentage of tests positive for estrus and a shorter duration of estrus. There was little effect on the male fetuses, except for some precocity in the development of their sexual behavior. The optimal time for injections of testosterone to produce this effect in guinea pigs was between the thirtieth and thirty-fifth day of gestation. The work of Price et al. (1963) has shown that the fetal guinea pig testis is capable of producing androgen at this stage.

Interestingly enough, there is little or no effect on the mother when testosterone is administered prenatally, even in large quantities. Although such injections into nonpregnant guinea pigs do produce a temporary masculinization of genitalia and behavior, the effect is not seen in the pregnant animal. This is apparently because of the protective effect of progesterone, which is present in large quantities during pregnancy (Diamond and Young, 1963). It is not clear why the young are not similarly protected from injections of androgen, but perhaps this steroid does not adequately cross the placental "barrier" to reach the embryo's circulation. In the neonatal female rat it has been found that large quantities of progesterone will protect the young animal from the masculinizing effect of testosterone administration (Kincl and Maqueo, 1965). This antagonistic effect of progesterone is that of a "competitive inhibitor," which means that efficacy depends on the relative levels of androgen and progesterone in the circulation. It is not unlikely that under natural conditions maternal progesterone helps to protect the behavioral and reproductive system development of female twins from the otherwise masculinizing effects of androgen produced by the male fetus (Diamond, 1967).

If the time of the critical period is compared in guinea pigs and rats by computing the age from time of fertilization rather than the time of birth, it will be seen that the two periods are relatively close in these two rodent species. However, the temporal relationship between the time of reproductive tract and behavioral differentiation is not the same in the two species. In the guinea pig these two processes virtually coincide, whereas in rats differentiation of the tract occurs mostly before birth, and behavioral differentiation extends into the postnatal period. As noted, however, the male *external* genitalia are not completely differentiated before this time.

In both rats and guinea pigs the period of behavioral differentiation appears to coincide fairly closely with that for differentiation of gonadotropic cyclicity. Despite this, and the fact that both presumably reflect the action of androgen on the hypothalamic region of the brain, the two processes can be separated experimentally. Thus in the rat, low doses of neonatal testosterone (approximately 10 μg) that prevent estrus cycles and ovulation are insufficient to suppress the development of feminine sexual behavior (Barraclough and Gorski, 1962).[4] It appears that the threshold of the mechanism for behavioral differentiation is higher than that for neuroendocrine differentiation. Strangely enough, the situation is apparently reversed in the guinea pig, so that behavioral suppression is achieved with less prenatal testosterone than is inhibition of cyclicity (Brown-Grant and Sherwood, 1971).

In mice, too, similar effects on physiological and behavioral differentiation as occur in the rat are found to follow various postnatal hormonal manipulations (Edwards and Burge, 1971). An interesting rodent species to study in this respect is the hamster. As mentioned before, it is exceptionally easy to induce lordosis in male hamsters; furthermore, females of this species do not show mounting behavior. Thus the hamster appears to be "more differentiated" in the female direction than other animals, and this makes it of interest to consider the effects of early hormone manipulations.

Hamsters respond in the same direction as rats and mice to neonatal androgenization of the female and neonatal castration of the male, but there are interesting differences of detail (Eaton, 1970). Neonatal androgen eliminates the development of the lordosis response in females (Crossley and Swanson, 1968) as well as in neonatally castrated males (Eaton, 1970). Thus the single injection of testosterone propionate in a sense "overdefeminizes" the newborn hamster, which as an adult will show less of the female behavior than a normal male. Furthermore, the same treatment seems to "overmasculinize" the neonatal castrate, which will respond to androgen with considerably more male behavior than the rat (including some intromissions and ejaculatory behavior). These findings in the hamster point up rather strikingly that the experimental treat-

---

[4]In fact these animals manifest a state of persistent sexual receptivity in the absence of hormone administration, suggesting a steady output of estrogen from their ovaries. According to Mullins and Levine (1968a), however, the behavior of these animals is "aberrant" in that they show exaggerated lordotic postures and also manifest general passivity and inactivity.

ment with androgen may fail to duplicate the natural events by which endogenous androgen determines the process of behavioral differentiation. Among possible causes for this failure are that the experimentalist is using the "wrong" androgen, or more likely, the "wrong" regimen of hormone administration (Eaton, 1970).

Recent work on the dog by Beach and Kuehn (1970) gives us an indication of the pattern of differentiation of reproductive behavior in a species with more complex courtship and mating patterns. Perinatal androgen prevented development of overt sexual receptivity in the adult female offspring ("presentation" of the vulva to the male, etc.), and a variety of "positive social responses" of the female to prospective male suitors were decreased (prancing and barking at the male, etc.). In addition, the attractiveness of the female to the male was impaired, as indicated by the decline in "positive social responses" and mounting by the male. Although genital differentiation in dogs is mostly prenatal, androgen-induced insensitivity to the behavioral effects of ovarian hormones appears to develop more slowly, over the intrauterine and postnatal phases.

Initial attempts to define the critical period for neuroendocrine and behavioral differentiation in purebred rabbits (Campbell, 1965) failed because prenatally injected mothers cannibalized their young. Postnatal injections of testosterone were ineffective in this species. Working with different strains of rabbits, Anderson et al. (1970) avoided the problem of the failure of the mothers to rear their young and made interesting observations on the effects of prenatal testosterone injections on behavioral differentiation. They did not study mating but rather maternal behavior in the female offspring as measured by the building of nests. This sexually dimorphic behavior is activated by gonadal hormones in the female but not in males. Only 11 percent of the female offspring born to mothers treated with androgen built nests, as compared to 90 percent of the control animals. The important point that this experiment exemplifies is that the "organizational" effect of early androgen on behavioral differentiation is not limited to sexual behavior but in this case extends to another sexually dimorphic behavior connected with reproduction.

Experiments in other species have demonstrated that other sexually dimorphic activities not directly related to reproduction are similarly "organized" by androgen during early development. Male mammals are generally more aggressive than females; and this is well documented by studies on other primate and subprimate species. This difference is androgen-dependent; fighting behavior in male rats and mice is inhibited by castration and restored by androgen injections (see Gray, 1971). The propensity to fight when treated with androgen is markedly reduced in males that were neonatally castrated, and increased by administration of testosterone to neonatal females; aggressive behavior follows the same basic pattern for early differentiation as reproductive behavior (Edwards, 1969; Conner and Levine, 1969).

An important aspect of these investigations, to which we shall return later, is that this is a case of early androgenic induction of a masculine behavior that is totally independent of the external genitalia. Similar conclusions can be drawn from observations on fear-related behavior (Gray, 1971). When placed in novel environments, males of various rodent species explore less and/or defecate more, both regarded as indications of fearfulness. Neonatal steroid treatment tends to eliminate the sex differences in fear-related behavior.

An additional important point is illustrated by the work on rhesus monkeys, a species in which analogies to the human are much easier to make. As in the guinea pig, prenatal androgen treatment in these primates results in the birth of genetic females possessing a well-developed phallus and scrotum with absence of the vagina. These offspring have been studied extensively before puberty. Infant rhesus show a variety of behavior patterns that are clearly sexually dimorphic, including some elements of the adult mating patterns, such as mounting by the males. The infant males initiate play much more often than the females, and it is much rougher ("rough-and-tumble play"). In addition, the frequency of chasing and threatening behavior is considerably higher in the infant male. The important point for us is that these elements of aggressive play behavior are not dependent on testicular hormones. Androgen is virtually absent from the blood in the first year of life, and neonatal castration does not eliminate the sex differences. The play of the female pseudohermaphrodites was intermediate between that of normal males and females, or closer to that of males. Thus, exposure to androgen during the "critical" prenatal period does more than alter the future manifestation of those adult behaviors that are dependent for their activation on gonadal hormones. In fact it can markedly affect behavior that is not so dependent (Goy, 1968). It seems that in this respect the fear-related behavior of rodents is also partially independent of the presence of hormones *at the time of testing* (Gray, 1971). This point is significant in relation to the mechanism of the organizational effects of hormones on behavior, which we will now consider.

THE MECHANISM OF BEHAVIORAL DIFFERENTIATION

What is the nature of the change produced by androgen during early development that results in the far-reaching effects on the development of behavior? The customary use of the term *organizational* to describe this change carries with it implications that must now be questioned. The term, borrowed from embryologists studying the organization of tissues in the developing fetus, implies a structural change, analogous to the differentiation of the reproductive system by androgen and/or other "organizers."

However, as emphasized by Beach in a recent incisive review (1971), we know of no anatomical difference between the brains of males and females that might be correlated with the behavioral differences between the sexes. It is therefore pointless to regard the term *organizational* as retaining its original morphological connotation in the present context. Promising research is being conducted in several laboratories in the attempt to demonstrate sexual dimorphisms of brain chemistry. In particular, recent research suggests that differences in protein synthesis in certain brain areas may be correlated with the developmental effects of early androgen (Clayton et al., 1970). The significance of these observations is not yet clear. We have pointed out that the differentiating effect of androgen on male sexual behavior might conceivably be mediated by effects on the genitalia. We must now stress that the differentiation of other male behavior patterns such as those related to aggression, infant play, and fearfulness cannot be due to such peripheral effects of androgen. For these male behavior patterns and for female sexual behavior the differentiating change must be thought of as occurring in the brain, and we must view the effect of androgen as a change in some unspecified *mechanism* that may or may not be reflected in anatomical change.

The most direct demonstration that early androgen acts on the brain comes from experiments of Nadler (1968). This investigator showed that behavioral and physiological effects similar to those obtainable by subcutaneous neonatal injections of testosterone resulted from implantation into the hypothalamus of smaller quantities of the hormone. However, precise localization within the brain of this effect was not obtained by Nadler, presumably because the methods available were not sufficiently refined for such localization in the newborn rat.

Insofar as development of sexual behavior is concerned, the effect on brain function is best characterized as a change in the development of future thresholds to the activational effects of gonadal hormones. As such it may be thought of more as a quantitative than a qualitative effect. Thus neonatally androgenized female rats require greater amounts of ovarian hormones in adulthood to produce receptive behavior than do normal females (see Figure 7.6). The suppression of responsiveness to progesterone seems to be more extreme than that to estrogen. Thus male rats (see Table 7.2) and neonatally androgenized female rats (Clemens et al., 1969) do not show facilitation by progesterone of the lordosis-activating effect of estrogen. Furthermore, unlike normal males, neonatally castrated males primed with estrogen do respond to subsequent progesterone with facilitation (see Figure 7.7).

The concept of early androgen acting via a change in brain threshold to postnatal activation by hormones is in line with our earlier discussion on the differences in sexual behavior between males and females. These were seen to be largely the result of differential sensitivity to the activational effects of homotypical and heterotypical gonadal steroids. However, this concept breaks down when applied to certain sexually dimorphic behaviors that are not directly related to reproductive function. Thus, as we have seen, the typically masculine play behavior of infant rhesus monkeys is differentiated by androgen during the fetal stage, but is not dependent on postnatal androgen for its activation. A slight modification

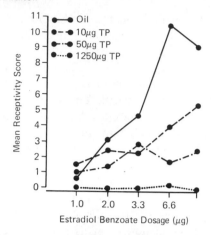

FIGURE 7.6 With increasing neonatal doses of testosterone propionate (TP), the sexual sensitivity of adult female rats to injections of estradiol progressively declines. Oil = control animals receiving oil injections neonatally. From Gerall, A. A., and Kenney, A.M. (1970) Neonatally androgenized females' responsiveness to estrogen and progesterone. *Endocrinology* 87:560–566.

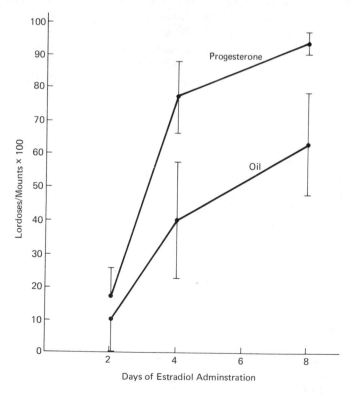

FIGURE 7.7 Lordosis quotients (means ± S.E.) in rats castrated neonatally and treated postpuberally with estrogen or estrogen and progesterone. From Davidson, J. M., and Levine, S. (1969) Progesterone and heterotypical sexual behavior in male rats. *Journal of Endocrinology* 44:129–130.

of the concept will, however, bring these facts into line. We can regard the differentiation of these sexually dimorphic nonsexual behavior patterns as a change in the threshold to environmental stimuli. Of course we must not forget that mating behavior is no less dependent on environmental stimuli (usually coming from members of the opposite sex) than other behaviors. The interchangeability of environmental and hormonal stimuli was well exemplified in the study of Clemens et al. (1969), who found that lordosis could be easily elicited in neonatally androgenized female rats (low testosterone dose) if they were exposed to prolonged adaptation to the testing situation or were mounted many times by vigorous males.

The concept that emerges is that of a series of sexually dimorphic behavior patterns presumably all built into the central nervous system of the developing individual. The levels of circulating androgens during critical periods of early development determine the probability that each of these behavior patterns will appear in later life. The higher the androgen level at these times, the higher the threshold to environmental and/or hormonal stimuli for each and every one of the sexually dimorphic patterns. Threshold changes in response to gonadal hormones are a most prominent component of the process for those behaviors most directly related to reproduction, that is, courtship and mating behavior.

Chapter 8
Approaches to the Study of Human Sexual Behavior

The purpose of this final chapter is to acquaint the student with four different approaches to the study of human sexual behavior. The recent huge increases in the amount of literature available on the topic, reflecting rapidly changing sexual attitudes and practices in this country, make a complete scientific or scholarly review impossible here. We also ought to make the following perhaps obvious observation: the material in this chapter may increase the reader's knowledge about sexual behavior, but it is neither intended nor likely to change his competence in it. And finally, we hope that it becomes clear as the chapter unfolds, that what one takes as an example of human sexual behavior depends in some measure on his approach in studying it. Like all important human activity, sexual behavior is not easily brought under a single definition that is at once rigorous and satisfying. No completely general definition will be attempted here.

What are the four approaches to sexual behavior that we will present? First we will consider results of studies involving direct observation and experimentation. This straightforward approach to sexual behavior, particularly as exemplified in the works of Masters and Johnson (1966), has been particularly helpful in elucidating some of the fundamental anatomical changes that occur in the genitalia during sexual stimulation. A slightly modified version of this approach, in which individuals are exposed to still or motion pictures of human sexual behavior, has been used in attempts to determine genital and general autonomic responsiveness to this form of erotic stimulation.

A second approach has relied on indirect observation and on answers to questions concerning personal sexual behavior. Field anthropologists, with the help of local informants, have attempted to determine the sexual practices of numerous societies in this way. A pioneering study in this area was performed by Bronislaw Malinowski on the Trobriand Islands, near New Guinea; we will consider his findings in some detail. Within our own society, intensive questionnaire and interview methods have been used to accumulate data on various physiological and social aspects of sexual responding; the work of the late Alfred Kinsey and his colleagues (1948; 1953) are the best known of these researches. In our treatment we will consider Kinsey's data and information from other cultures together, in discussions of masturbation and homosexual behavior.

A third major approach has developed within the fields of clinical medicine, in particular clinical endocrinology. We will discuss the results of studies in this area with an eye toward understanding the activational and organizational roles of hormones in the determination of sexual behavior and sex-typed behaviors (gender roles).

Fourth and finally, no research has had more profound effects on popular attitudes about human sexual behavior than the work of Sigmund Freud and the subsequent psychoanalytic tradition. The terminology of Freudian and other "depth" psychologies has so permeated everyday vocabulary that much material that has at best the status of hypothesis or theory appears to be accepted as fact or well-known principle. This is particularly true regarding the concepts of sexual need and sexual instinct, and the roles that these notions play in theories of personality and its development. Psychoanalytic theory originated in psychiatric observations of individuals with marked behavior disorders; as such it can be classified as a fourth major approach to the study of human sexual behavior.

DIRECT OBSERVATION AND EXPERIMENTATION: ANATOMY AND PHYSIOLOGY OF HUMAN SEXUAL RESPONSE

In many mammalian species, including our own, the male's ejaculation is a relatively well-defined response. Data on some mammalian penile receptors, and the spinal innervation of ejaculation and emission, have been described in earlier chapters. Experiments in Sweden by Kollberg and his colleagues (1962) on the human male permit some tentative conclusions to be drawn as to the relations between contractions of the muscles serving the penis and the sensation of orgasm.

Small recording electrodes were inserted into the deep transverse perineal muscle and the membranous urethral sphincter of male volunteers. The men then masturbated to ejaculation while electromyographic recordings were made. Approximately three seconds prior to the first seminal expulsion the transverse perineal muscle began and maintained a strong tonic contraction. The urethral sphincter showed phasic contractions whose peak amplitudes correlated with spurts of ejaculate. The subjective report of orgasm coincided with the tonic contraction of the transverse perineal muscle. A sample electromyogram is reproduced in Figure 8.1. It appears that at least some of the sensation of orgasm is produced by afferent impulses from the transverse perineal muscle.

A general picture of changes in male genital anatomy during copulation may be presented as follows: penile erection is produced by marked vasodilation of the corpora cavernosa. The rapidity of erection varies with many factors, not the least of which is age. During stimulation prior to ejaculation the corona glandis (ridge of the glans), swells slightly, the skin of the scrotum thickens, and the testes swell and move upward to rest closely against the body wall. The ejaculatory reaction is always preceded by at least some elevation of the testes; it is not clear at present whether this is a necessary condition for ejaculation or is only correlated with it.

Accompanying these local changes are variations in circulatory and respiratory functions. Pulse and respiration rates and blood pressure all increase during active copulation, reaching their peak at approximately the time of orgasm. A rosy flush may appear on the head and neck. These changes are all indicants of increased activity of the sympathetic branch of the autonomic nervous system. The role of the ANS in the control of copulatory reflexes in other species was discussed in Chapter 4.

The distinction between ejaculation and orgasm needs to be emphasized. In addition to electromyographic data, there are other reasons for separating the two events, of which the most obvious is the orgasm without seminal expulsion experienced by prepuberal boys. Some investigators have claimed the existence of orgasm in male infants.

Loss of erection and sexual desire occur very shortly after ejaculation. The duration of the postejaculatory recovery period varies with many factors, including age and general physical condition. Changes in male external genital anatomy following orgasm are obvious; similar changes in the female are not. This apparent lack of gross change, and the difficulties associated with obtaining adequate reports from females on the occurrence of orgasm, has in the past prompted much speculation but little specific and reliable information about the nature, or even the existence, of the phenomenon. In the last ten years, however, our knowledge of female sexual response has been greatly enlarged. The following anatomical and physiological information, as well as much of the foregoing description of male activity, stems from work done at the Reproductive Biology Research Foundation, St. Louis, Missouri, directed by Dr. William H. Masters (Masters and Johnson, 1966).

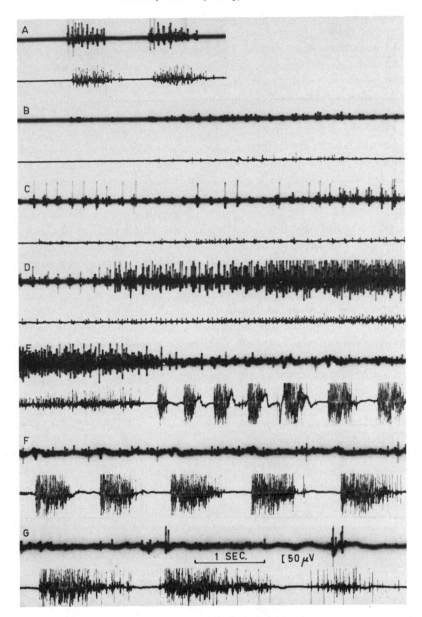

FIGURE 8.1 Electromyographic recording during orgasm and ejaculation. The upper part of each section is a recording from a muscle in the vicinity of the striated urethral sphincter, probably the deep transverse perineal muscle. The lower parts are recordings from the membranous urethral sphincter. (A) Two voluntary contractions as on attempting to interrupt micturition. (B–G) Continuous recording. Orgasm was experienced simultaneously with the steep rise in activity during D and at the start of E. Ejaculation occurred during E through G. From Kollberg, S., Petersen, I., and Stener, I. (1962) Preliminary results of an electromyographic study of ejaculation. *Acta Chirurgica Scandinavica* 123: 478–483.

Changes in female anatomy during erotic behavior may conveniently be grouped under the four phases of what Masters and Johnson have called the *sexual response cycle*: the excitement phase, the plateau phase, the orgasmic phase, and the resolution phase. Perhaps it should be emphasized that these names are only descriptive labels and should not be thought to explain the reactions.

Commencement of stimulation by direct (e.g., a caress) or indirect (e.g., exposure to erotic literature or photographs) methods may start the *excitement phase* of the response. One of the early reactions is a contraction of fibers in the nipples of the breasts, causing nipple erection. Somewhat later in the phase the veins of the breasts become congested. The vascular reaction produces a noticeable increase in total breast size toward the end of the excitement phase. Beginning during the late excitement phase is the rosy flush of the skin described earlier for the male. In the female this reaction of the skin vasculature starts on the area over the stomach and grows during the course of the response to cover the surface of the breasts and other body areas.

The changes in external genitalia during the excitement phase depend on whether the woman has borne children. The outer lips of the vulva (labia majora) may move up and apart in the childless female, whereas in women who have had several children they tend to become engorged and somewhat pendulous. The inner lips (labia minora) become bright pink during the excitement phase, and they begin to swell.

During the excitement phase the veins in the head (glans) of the clitoris become congested, producing a microscopically visible tumescence. There is also an increase in the shaft diameter of the organ and, in a small percentage of women, an increase in its length. This reaction does not constitute an "erection" of the clitoris, as has sometimes been claimed.

The earliest reaction of the excitement phase occurs on the walls of the vagina. It is the beginning of vaginal lubrication. Within several seconds after stimulation has begun, small liquid droplets appear on the vaginal walls. As stimulation continues the quantity of material increases so that eventually the vaginal barrel is covered by a thin liquid film. The mechanism behind this "sweating" reaction is not completely known, but it appears to be related to the congestion of the veins supplying the vagina. The discovery of this vaginal reaction, with the additional finding that the sexual accessory glands are not functionally significant in vaginal lubrication, is a considerable advance in our understanding of female reproductive physiology. As the excitement phase progresses, the inner portions of the vagina elongate and widen, and the entire passage becomes easily dilatable.

During the *plateau phase* the area immediately surrounding the protrusion of the nipple continues to be engorged. The color of the labia minora changes from pink to a bright or dark red, and the lips continue to swell to twice their original size or greater. The clitoris is retracted beneath the swollen labia and reduced in length by approximately half.

There is marked congestion of blood in the veins of the vagina, particularly in the outer third, which swells to an extent that produces an appreciable narrowing of the canal. The extent of vaginal lubrication does not change significantly.

The duration and intensity of these reactions is quite variable among different women. In order for satisfactory orgasm to be achieved, however, the plateau-phase reactions must occur. Of particular importance is the congestion of the labia minora and the outer third of the vagina. In their swollen state these structures allow the muscular contractions of orgasm to be fully expressed; they have been labeled the "orgasmic platform."

The female *orgasmic phase* consists of strong phasic contractions of the muscles supplying the pelvic area. The transverse perineal and the bulbo- and ischiocavernosus muscles, mentioned earlier for the human male and for other species, are all active in the response. The muscular reactions produce large contractions in the congested outer third of the vagina at approximately one-second intervals. There is considerable variability within and between individuals in the intensity and duration of the orgasmic phase, and in the capacity to achieve multiple orgasms before passing to resolution.

The *resolution phase* marks a retreat from the vasocongestion produced in the earlier stages of the sexual response. The highly distended orgasmic platform shrinks at a rate approximately equal to that of the loss of male erection; accompanying this reaction is a movement of the clitoris to its usual, unstimulated position, and a change in the color of the labia minora to the pink of the plateau phase. Reduction of vasocongestion in the inner two-thirds of the vagina proceeds more slowly. The flush on the breasts recedes, followed by shrinking of the areolae surrounding the nipples' protrusions. The marked blue-purple venous patterning on the breasts then subsides. The resolution phase may last for 30 minutes or more. A graphic review of the changes in the vagina and associated structures occurring during the phases of the response cycle is presented in Figure 8.2.

In addition to the appearance of marked vascular flush, other autonomic changes occur in the female as orgasm approaches. Various studies have demonstrated increases in heart rate, blood pressure, and respiration rate.

The identification of the vaginal walls themselves as the source of lubricating material was mentioned as one major contribution of the recent research on the feminine response cycle. Another finding of considerable importance concerns the relation between orgasm achieved through vaginal stimulation (as in normal intercourse) and that achieved by clitoral stimulation alone. Evidence accumulated to date suggests that the anatomic changes leading to the orgasms produced by these two kinds of stimulation are extremely similar if not identical. An earlier view, based on Freudian theory, held that clitoral and vaginal orgasms

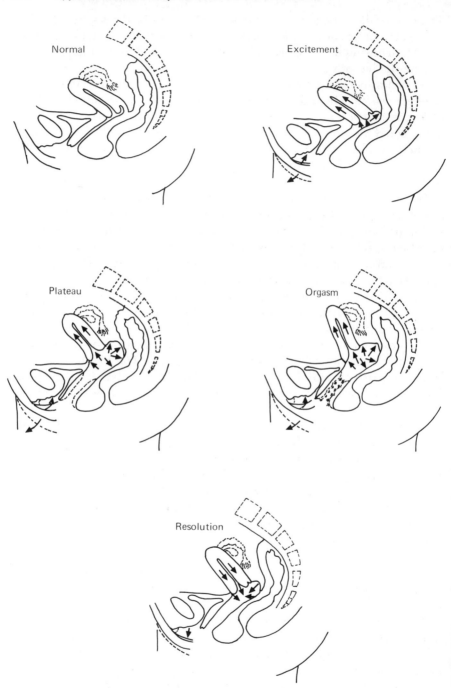

FIGURE 8.2 Anatomical changes in the clitoris, vagina, and uterus during the female sexual response cycle. See text for explanation. From Masters, W. H. (1960). The sexual response cycle of the human female: I. Gross anatomic considerations. Western Journal of Surgery, Obstetrics and Gynecology 68:57–72.

were separate entities and that in the course of adequate sexual development women passed from maximal response to the first to maximal response to the second. There appears to be at present no *anatomical* basis for this argument.

PHYSIOLOGICAL CORRELATES OF INDIRECT EROTIC STIMULATION

One obvious and fundamental fact about human sexual behavior is that many of us can become sexually aroused by the sight, verbal representation, or imagining of sexual activity. The practical, even commercial, significance of this fact of human psychobiology is great; indeed it is so great that its commercial exploitation, the various forms of pornography, is a subject of some social concern. There is no doubt that more pornographic material is available, at a lower cost, than ever before. In most major cities in the United States one need pay only 3 to 5 dollars to watch several hours of film depicting frankly all forms of sexual behavior. In some areas, the traffic in this and related trades has become so heavy that civic officials have felt strong pressures to reduce it.

There are a number of interesting questions that we might raise and discuss about pornography; for example, for what reasons do people use or possess it? But, in keeping with our general plan, we will present only a brief discussion of the experimental evidence related to the direct physiological consequences of visual exposure to pornographic or less frankly sexual material; readers are referred to the *Report of the President's Commission on Obscenity and Pornography* (1970) for a presentation of other facets of the issue.

Zuckerman (1971) has provided a thorough review of approximately 35 experiments that measured one or more physiological responses to the visual presentation of erotic material. Most of the response measures were the ones typically used to assess the functions of the autonomic nervous system, including changes in electrical conductance of the skin (electrodermal response or EDR), cardiovascular changes (blood pressure, heart rate, finger volume, etc.) and changes in breathing rate and amplitude; other physical measures included penile erections and uterine contractions, changes in body temperature, changes in the diameter of the pupil of the eye, and changes in averaged evoked electrical potential recorded from the scalp. Some experimenters also measured chemical changes, including levels of catecholamines (e.g., adrenaline and noradrenaline), 17-hydroxycorticoids, and urinary acid phosphotase.

Few if any unambigous generalizations can be drawn from these experiments, particularly those that utilized the rather gross EDR or other general autonomic nervous system indicants. It is quite clear that frank visual presentation of copulation, fellatio, and so on can produce changes in these measures as well as in verbal reports of sexual arousal; but it is equally clear that other "emotionally loaded" visual displays (e.g., concentration camp scenes or scenes depicting aboriginal penile subincision rites) can produce essentially the same pattern of changes; it appears not to have been possible, so far, to describe a unique autonomic profile of response to indirect sexual stimulation. The measurement of penile erection gives less ambiguous results; at least some males can voluntarily control tumescence and detumescence while watching erotic material; this technique appears to have potential value in both experimental and clinical settings (Freund, 1967; Laws and Rubin, 1969).

Finally, we note that physiological response to visual displays of sexual behavior may be in part a function of the general and specific social contexts in which the material is presented. If one is nervous, anxious, or otherwise emotionally involved with the mere fact of being in this *kind* of experiment, then this will be reflected in the typical physiological measures; and although the experimenter may work to control for differences within and between individuals of these background levels of response, individual variations may nevertheless be larger than the effects produced by the experimental conditions. In these experiments, as in all studies of human sexual behavior, it is probably unwise to assume that there is a strict separability between the sexual component of the situation and other emotional and environmental variables.

THE SURVEY APPROACH:
ANTHROPOLOGICAL AND SOCIOLOGICAL PERSPECTIVES

In the process of education each of us becomes used to the influence of cultural expectations and norms on our behavior. Acceptable modes of dress, speech, and general social conduct are taught in direct and indirect ways from infancy onward. So ingrained do these standards become that the individual, when faced with making a decision concerning the normality or propriety of his own or another's behavior, may assume that his standard of judgment is based on some fact about "human nature" that holds true for all persons in all phases at all times. The possibility that a given standard judgment reflects more the idiosyncrasies in the internalized rules of a particular culture or subgroup than it does the biological nature of man is not always clear. When the behavior involved has strong emotional components and consequences, this problem is particularly likely to arise. In order to objectify our understanding, therefore, we need a frame of reference that places the biological and cultural determinants of sexual behavior in proper relation to each other. One way to gain this frame of reference is to examine the sexual practices and attitudes of various cultures and to record observed similarities and differences.

One of the earliest important descriptions of the sexual practices and attitudes of a foreign culture was made by the anthropologist Bronislaw Malinowski (1884–1942), who lived for several years at the time of World War I in native villages on the island of Boyawa, the largest of the Trobriand Islands, northeast of New Guinea. As an early member of functionalist tradition in anthropology, Malinowski wanted to relate the sexual behavior of the Trobrianders to the total cultural framework within which it developed: "Sex, in its widest meaning . . . is rather a sociological and cultural force than a mere bodily relation of two individuals." He also insisted that neither of the two earlier popular views of the Oceanic peoples, that they were either "noble savages" untouched by the artificial restraints of civilization, or that they existed in a state of sin from which only the missionaries could save them, was correct. Rather, he pointed out, the islanders had a highly developed set of moral codes that they obeyed with the same approximate fidelity that can be found in Western society. The key to understanding their sexual practices came not from a direct comparison with our culture but from an appreciation of how these practices interacted with other aspects of behavior to create and maintain social stability. Some examples taken from Malinowski's classic ethnography, *The Sexual Life of Savages in Northwestern Melanesia* (1929) will help to illustrate these points.

Trobriand children of both sexes were relatively free from direct parental constraints on their behavior. The children roamed freely about the village and neighborhood engaged in pastimes of their own devising and obedient primarily to their own natural leaders; in this way they formed what Malinowski called the "children's republic." Some of the games that they played were frankly sexual, and imitated the courtship and copulatory practices of their elders. As was the custom among adults, a boy would present a small gift of nuts, fruits, or beads to a girl. The girl accepted the gift, and the pair then retreated to the bush surrounding the village for a period of sex play that might or might not include penetration. All of this was performed discreetly and without a great deal of attendant comment by other children; but it was not a secret from adults that the behavior occurred with regularity. On the contrary, the adults viewed this and most other heterosexual practices of children with amused complacency, as a natural feature in childhood development.

The major exception to this indulgent attitude toward general and sexual familiarity among children concerned the relation between brother and sister. In this case familiarity or warm friendship of any sort was strongly prohibited from the very earliest age onward. An older brother had to maintain total reserve and restraint in the presence of his sister. At adulthood her behavior toward him became formally obeisant: she had to bow to him whenever they met. The extreme derference to her brother shown by a Trobriand woman related to his eventual role in the upbringing of her children, for the Trobrianders practiced a matrilineal kinship system: all descent was figured on the mother's side. In addition there was no concept of physiological paternity in the culture. According to native belief, the baby was placed on the hair of the prospective mother by a special spirit, and then transported to the womb on a tide of the mother's blood. The blood helped to nourish the baby, and that is why the pregnant woman stopped menstruating. These two factors, the matrilineal kinship system and unrecognized physiological paternity, strengthened the role of the mother's brother as a provider and figure of authority for her children. And his high position in his sister's house was made obvious by her low bow whenever they met. As we shall point out in more detail later on, the relation of the mother's brother's role in Trobriand society to the role of the father in Anglo-European society has definite implications for the general validity of one of the major sex-related concepts in Freudian theory, the Oedipus complex.

Premarital intercourse among Trobriand adolescents and young adults was a totally expected and approved occurrence provided only that the members of the pair did not bear certain kinship relations to each other. Trobriand attitudes toward adultery, however, had a much more familiar ring: "Any breach of marital fidelity is as severely condemned in the Trobriands as it is in Christian principle and European law; indeed the most puritanical public opinion among ourselves is not more strict. Needless to say, however, the rules are as often and as easily broken, circumvented, and condoned as in our society." In this passage and in many others, Malinowski emphasized that sexual proscriptions and sexual practices were often at variance among Trobrianders, as indeed they are among ourselves.

Trobriand copulatory positions and sexual foreplay practices were governed by definite rules of propriety. Two major components of the Western mode of stimulation, the protracted, relatively motionless kiss, and face-to-face copulation with the man prone over the woman, were considered by the natives to be either slightly silly or sexually unstimulating. Copulation usually took place with the woman supine and the man squatted between her legs. In the performance of the act as well as in its initiation, women responded as freely and openly as men. Female orgasm was well known and an expected feature of copulation; the men apparently were able to forestall ejaculation for extended periods. During foreplay and copulation, partners often scratched at each other with some severity; the resulting insignia of copulation were worn proudly by young men and women alike.

Anogenital and oral-genital variations of intercourse were considered unpleasant. Although no specific social sanctions were placed against them, they were the subject of considerable ridicule. Homosexual intercourse and masturbation were treated in the same way: not as frightening, unhealthy, or taboo practices, but as a subject for ribald jokes and social ostracism.

Opportunities for premarital heterosexual copulation were abundant in the Trobriand culture. In addition to the individual pair-formations, there were also occasional trips late at night by groups of young men to neighboring villages for amorous adventure (called *ulatile*), and similar expeditions made by girls (*katuyasi*). Certain dangers were inherent in participating in the *ulatile*, however, for if the boys of the visited village discovered the intruders in the presence of the village girls, violence was likely to result. The role of the *ulatile* in Trobriand culture was summarized by Malinowski: "The childish erotic experiences with which the sexual life history of an individual begins always takes place within the community; the *ulatile* is one among the customs which carry erotic interest and those transitory affairs, which are the next stage in development, beyond the village. Such intrigues may become permanent and thus the *ulatile* is one of the ways in which matrimonial choice is extended beyond a single village."

The copulation that occurred at these occasions, and in virtually all premarital occasions in native life, was performed in private. There was little direct evidence for mass public copulation or orgiastic festivals that led to open intercourse without prior retreat to a dwelling or the bush. Natives in the northern sections of Boyawa, where Malinowski spent most of his time, attributed these and other violent sexual practices to the inhabitants on the extreme southern portion of the island. The natives in the south were also considered by the northerners to be generally socially and morally inferior to themselves.

Strict discretion regarding the observation of copulation did not carry over into the family unit. A child who happened to observe his parents in intercourse was apparently scolded without too much force and instructed to look away. Complete verbal discussion of sexual matters was also carried on in the children's presence. As mentioned above, children began early to imitate the well-known practices and vocabulary of their elders.

Postpuberal boys had no occasion to observe their parents' sexual behavior, for they spent their nights at a "bachelor's house" (*bukumatula*), a separate dwelling typically owned by the eldest of a group of young bachelors. Love affairs were actively pursued within the *bukumatula*, and thus attractive adolescent girls of the village were also unlikely to sleep at home.

The copulatory position adopted for use in the *bukumatula* was the only frequently practiced variant on the female supine, male squatting posture. In this case the members of the pair faced each other while lying on their sides, with the woman's upper leg placed over the man. This position was favored for intercourse in the bachelor's house because it was less likely to awaken or attract the attention of the other people who were present. The integrity of boy-girl pairs was maintained, and there were no reports of exhibitionistic or voyeuristic tendencies surrounding sexual behavior in the *bukumatula*.

It should not be imagined that what Malinowski found to be true for the Trobriand Islanders also applies generally to other Oceanic or even other Melanesian groups. The behavior of the Trobrianders may profitably be contrasted to that of another Melanesian group, the "East Bay Society," studied by William Davenport (1965). In this Melanesian society social separation of boys from girls is emphasized as soon as the children have begun to walk. Little girls typically accompany their mothers throughout the day, while boys play in gangs within the confines of the village. From the earliest ages onward, there is little opportunity in childhood for the free heterosexual relationships practiced by the Trobrianders. After the age of five all discovered physical contact between boys and girls is met with at least some punishment; however, these restrictions are in some ways less strict for brothers and sisters than for unrelated children. Although brother and sister must avoid each other in public, the avoidance relation is relaxed within the home and at the family garden. In fact, occasional brother-sister sex play or copulation is an acknowledged occurrence that occasions strong reproof but does not scandalize the parents or other adults.

Formal restrictions on heterosexual activity in East Bay are continued until the time of marriage, that is, until a woman is 18 or 19 and a man is at least 20. Between the onset of puberty and the time of marriage, young men and women are encouraged by their elders to control their sexual desire by solitary masturbation: "It is impossible to overstress the fact that in this society masturbation to orgasm is not only accepted, but approved and encouraged as the proper sexual behavior for young, unmarried adults of both sexes. It is regarded as a safe and normal sexual outlet for young unmarried persons." Unmarried men typically masturbate at night in the men's quarters in which they live. No particular effort is made to hide masturbatory practice from the other men in the dwelling, as it is in no way considered shameful. Young women masturbate alone at night while in the sleeping quarters.

Homosexual contacts among men are considered to be a socially sound practice both before and after marriage. A prepuberal boy may form a sexual liaison with an older man; in return for taking the passive role in anal intercourse the boy receives gifts of tobacco or other valued goods. The child's motive is not entirely economic, however, because his compliance with the adult's request is a feature of general etiquette. Homosexual relations including mutual masturbation and anal intercourse are also practiced by pairs of young unmarried men, sometimes between brothers. Davenport emphasizes that "men who behave thus are not regarded as homosexual lovers. They are simply friends or relatives, who, understanding each other's needs and desires, accommodate one another thus fulfilling some of the obligations of kinship and friendship." Apparently in neither type of homosexual interaction is fellatio practiced.

At least some homosexual intercourse is practiced by many men after marriage. East Bay men consider this continuation of premarital practice to be an expression of normal sexuality and find our Western concepts of total homosexuality and total heterosexuality difficult to comprehend. The continuation of masturbation after marriage is claimed by the natives to be unnecessary. As far as could be determined, there is no pre- or postmarital homosexual behavior among the women of East Bay.

Marriage in this society is the culmination of a set of complicated economic arrangements by the parents and other kin of the couple. Bethrothals are often made while the prospective spouses are still in infancy or childhood; in either case the separation of the boy and girl following engagement is strictly maintained. In general, young people have little to say in the initial stages of mate selection. However, because the East Bay society believes strongly that mutual sexual attraction is a necessary feature of a good marriage, parents will usually inquire whether the members of the bethrothed pair find each other physically appealing. Strong objection to the union by either partner may result in termination of the bethrothal. Elopement provides the only other means of expressing individual choice of a marriage partner; it is a serious crime, however, tantamount to murder, that in earlier times could lead to open violence between the families of the eloped couple. A boy and girl who would like to get married but have difficulty winning parental approval may threaten to elope as a means of obtaining permission.

Marriage in East Bay demands a rapid adjustment to unfamiliar ways of living. Prior to marriage there is little opportunity for prolonged encounter with age-mates of the opposite sex. Naturally there is a period immediately after the ceremony, sometimes lasting for several months, during which the bride and groom experience considerable apprehension and embarrassment in each other's presence. This period of awkward avoidance and uncertainty is considered to be a sign of good upbringing, and the parents of the groom, with whom the couple initially lives, act as confidants and advisors until the marriage achieves stability.

Marital copulatory behavior develops within this initial period of adjustment. Even before harmony has been achieved, intercourse with orgasm twice in 24 hours is given by the natives as the norm. During the day intercourse is accomplished in the private garden given to the couple by the groom's parents, and again at night in the private home that is eventually acquired. This rate of sexual contact is expected to continue throughout the first several years of marriage and then decline gradually. Sexual contact within the garden or home begins with an extended period of foreplay in which the partners stimulate each other manually. Coitus occurs with the woman supine and the man fully prone over her or resting above her on his hands. Pelvic thrusting is vigorous immediately after insertion, and no premium is placed on the prevention of early orgasm. On the basis of reports from informants, Davenport estimated that the interval from initial insertion to ejaculation was approximately 15 to 30 seconds. Single female orgasm is an expected result of this sexual sequence, and multiple orgasms are known to occur. Fellatio and cunnilingus apparently are not practiced.

There are two major variants on this course of events. East Bay couples will adopt a lateral position, as described for the Trobrianders, when the female is in late pregnancy or whenever they desire to be particularly quiet. Secondly, coitus interruptus is typically practiced for as long as two years following the birth of a child. Unlike the Trobrianders that Malinowski studied, the East Bay people are aware that seminal fluid is involved in reproduction, so they use this method to space their families. Penetration without seminal deposit is also practiced during adulterous affairs.

East Bay views on adultery are conditioned by the complex economic arrangements that go into making up a marriage. Numerous members of the four families involved in an affair between two married people become parties to the dispute that follows its discovery. At the time of marriage these kinsmen obtained certain proprietary rights on the sexual availability of the couple, and extramarital intercourse constitutes an abridgement of these rights. Compensation for this form of larceny is made by the payment of money; working out the details may be a very complicated undertaking, and as Davenport points out, "there is always the lingering suspicion that a person who has once committed a sexual offense is likely to repeat it."

In earlier days the men in East Bay bought young girls from the islands to serve as concubines. The possession of a concubine was often shared among several men. The girls held a unique position within the society; to own a share of a concubine was to have high social status. The wives of the owners were proud of their husband's possession and apparently felt no sexual jealousy toward her. Sexual privileges with the concubine were rotated among the owners. Other interested men could buy her services on a temporary basis, and thus the concubine acted in part as a prostitute. The practice of prostitution offended the colonial government that came to control the East Bay people. To remedy the situation, a law was passed and strictly enforced that made the entire concubine system illegal. This forced change in an important part of the sexual system in East Bay has, in the opinion of the natives, produced undesirable social effects. The men claim that their sexual vigor declines too rapidly with age because of the lack of stimulation and excitement produced by the presence and availability of the concubines; this in turn makes continued marital sexual compatibility more difficult, because the men lose interest in their wives. In addition, fathers may no longer satisfy the sexual desires of their unmarried sons by providing them with the temporary services of a concubine. As we saw, there is no other socially approved heterosexual outlet for the youth of East Bay.

These examples taken from the Trobriand and East Bay societies serve to introduce some of the basic concepts that arise from a cross-cultural analysis of sexual behavior. First, it is essential to distinguish between the sexual acts that a society approves, condones, or forbids, and the acceptable partners, if any, with whom these acts can be practiced. Similarly, an act that is considered appropriate at one age may be scorned or condemned at another. The same may be true for the time of day and the place at which a sexual act is performed. In short, the sexual behavior of individuals is thoroughly understood only when it is viewed with respect to the general social framework in which it occurs.

In 1951, C. S. Ford and F. A. Beach collaborated to produce a review of the sexual practices of 191 different societies, including our own. The broad perspective gained from this extensive cross-cultural survey allowed the authors to reach sound conclusions about the degree of generality of various aspects of sexual behavior observed in peoples throughout the world and to make some comparisons of human behavior with the sexual behavior of other animal species. We will combine their findings with the results of other studies to gain some insight into two forms of sexual activity that are often totally or partially misunderstood by members of our own society: masturbation and homosexuality.

MASTURBATION

What is typically meant by masturbation is self-masturbation: stimulation of one's own genitals to produce a condition that may and in many cases eventually does lead to orgasm. In males this stimulation is typically provided by rhythmic manual friction; in females direct or indirect manual stimulation of the clitoris seems to be the rule, although in some societies women employ penis substitutes for direct vaginal stimulation. As was pointed out previously, the anatomical and physiological changes in female genitalia under these two conditions of stimulation are the same. Manual stimulation of another's genitalia may also occur as a part or the entirety of a sexual episode between two people of the same or opposite sex. If the action is reciprocal, the term *mutual masturbation* is often employed. In the rest of this discussion we will consider only self-masturbation.

The techniques used in masturbation vary but slightly across cultures, but popular attitudes toward the practice vary widely. One end of the continuum is represented by the East Bay people, who encourage premarital automanipulation for both sexes as a substitute for illicit heterosexual intercourse. Female masturbation, accomplished by sitting on the heel, is a frequent and openly discussed practice among the women of Lesu, another Oceanic society; it is practiced when heterosexual stimulation is not available. In most of the societies for which evidence is available, masturbation by adults of either sex meets with at least some public disapproval, although attitudes about adolescent masturbation may be somewhat more lenient. The disapproval of masturbation can take more than one form. As among the Trobrianders, it may be expressed as scorn for the person's inability to achieve sexual satisfaction through heterosexual contact. Or, as in some segments of our own society, it may be expressed either as a concern for mental or physical defects allegedly caused by masturbation, or as outrage about the breaking of a moral code against masturbation.

If there are feelings of shame, guilt, or embarrassment concerning masturbation in a society, reliable information concerning its incidence and frequency will be relatively hard for a foreigner to obtain. In spite of this difficulty, the available ethnographic data indicate that masturbation does occur in at least some societies in which it is strongly disapproved or forbidden. The most complete data for any society are those collected in the United States by Kinsey and his colleagues, who found that 92 percent of the men and 58 percent of the women in their samples had masturbated to orgasm at least once by the time they were interviewed. Men who had not begun to masturbate by age 20 were very unlikely to begin after that time, but women were approximately equally likely to begin at any age up to 35.

As might be expected, several social variables influence the frequency of masturbation in adulthood, of which the most obvious is marital status. Many men and women who masturbated prior to marriage continue the practice after marriage, typically with reduced frequency. For both sexes the average percentage of total sexual stimulation provided by masturbation is much smaller after marriage than before.

There are more or less specific sanctions against masturbation in all of the major religions in the United States. Devout men and women of any of the denominations are less likely to masturbate, or masturbate less frequently, than less devout or religiously uncommitted people. It should also be pointed out that there is less sexual activity of all forms in strongly religious groups than in mildly religious or nonreligious groups; for devout single people of any denomination, therefore, the percentage of total sexual activity accounted for by masturbation may be as high or higher than it is for less strongly affiliated individuals.

Social factors such as marital status, religious affiliation, family background and income, educational level, and so on, may be expected to interact in complicated ways to influence the types and frequencies of sexual behavior, including masturbation, practiced by an individual. For example, men and women who enter graduate or professional training tend to get married at a later age than those whose education stops after high school or college. In addition, there is, in general, a positive relation between a person's final educational level and that of his parents. Either directly or indirectly, these and other social variables determine in part the availability and subjective desirability of opportunity for heterosexual intercourse and the subjective necessity for masturbation.

The widespread occurrence of masturbation in human societies and its occurrence in nonhuman primate and other mammalian species indicates that it is not per se a biologically "unnatural" practice or a perversion. There are no reliable data to support the claim that masturbation causes physical ailment or debilitation. The short-term changes in muscle and in the circulatory and nervous systems accompanying masturbation activity appear to be very similar and in some cases identical to those accompanying coital activity. Mental defects or emotional disturbances are not necessarily produced by masturbation any more than they are by heterosexual coitus.

It is quite clear, however, that many individuals in this country experience inner conflict about masturbation, often because they feel that the practice violates a moral code, and yet they are not totally able to abstain. Prohibitions against masturbation are very long standing in the Judeo-Christian tradition. In this as in other cases of behavior forbidden by tradition, when the rule has been internalized by an individual, he is likely to feel conflict when he breaks it.

HOMOSEXUALITY

Homosexuality is a very hard word to define, and considerable controversy exists today about whether it is scientifically meaningful to label some people as "homosexuals" and other people as "heterosexuals." It will be recalled that in the discussion of the Trobriand and East Bay peoples, reference was made to homosexual *practices* but not to homosexual *individuals*. If two men or women practice mutual masturbation or oral-genital stimulation, they are clearly engaging in a homosexual act. If they are found to engage in this activity to the partial or total exclusion of heterosexual activity even in the presence of apparently ample opportunity, then we may safely label their sexual behavior as partially or totally homosexual, as the case may be. In this case the reference point for the definition is the sex of the partner with whom an individual practices genital stimulation that may lead to orgasm. In their study of sexual behavior in the United States, Kinsey et al. relied partially on this approach to devise a 7-point ranking scale of homosexuality in which a score of 0 indicated that an individual neither engaged in direct homosexual activity nor experienced a desire to do so, and a score of 6 was given to a person who desired and practiced sexual behavior only with members of his own sex. Intermediate degrees of overt behavior and desire were given intermediate numbers. This scoring system was useful in codifying the vast amount of data collected from more than 11,000 men and women; however, as the authors emphasized, there is an essentially unlimited number of gradations between "total heterosexuality" and "total homosexuality," even under this relatively explicit definition. It would be unwise to place much emphasis on the quantitative nature of the procedure; in terms of measurement, it is at best an ordinal scale. Nevertheless, this method of defining homosexuality is relatively free from ambiguity.

In our society there is a considerable folklore concerned with the personal and social characteristics of people, particularly men, who engage primarily in homosexual behavior. For example, it is often believed to be true that all men who practice homosexual acts must have accepted a "feminine" role in many aspects of their lives. The classic stereotype of the male homosexual includes references to exaggerated posturing and gait, affected speech mannerisms, conspicuous style in dress, and a general tendency toward emotional overreactivity. It is not assumed that these characteristics apply normally to females, but rather that they occur whenever males attempt to express themselves in a feminine fashion. Accompanying this stereotype is the assumption that male homosexuals are ineffective and unproductive in the conduct of commerce or government, although they may possess artistic sensitivity and creativity. In regard specifically to sexual practices, the assumption is that the homosexual prefers the passive, "feminine" posture or role.

Our stereotyped concept of the male homosexual appears to have become institutionalized in the behavior of law enforcement agencies when dealing with homosexuality. Gebhard et al. (1965) have put the problem this way:

It is, in essence, our cultural concept that sexual techniques indicate who is a "real" homosexual and who is not. This concept has become a part of psychology and psychiatry, giving rise to the distinction between "active" and "passive" homosexuals. If a policeman discovers individual A masturbating individual B or with B's penis in his mouth, individual A is certain to be arrested and charged; however, there is a fair possibility that B will not be arrested, but released after a scathing reprimand. The police attitude seems to be that individual A is a proven "fairy" or "queen" because he was stimulating B, whereas B is simply an ordinary person with a regrettable but understandable tendency toward sexual opportunism. Yet, illogically, if the policeman finds B inserting his penis in A's rectum, this attitude is unshaken and A is still considered the homosexual. Apparently the crux of the matter is this: the man who is bringing the other man to orgasm is the "real" homosexual; the man being brought to orgasm is not. [pp. 324–325]

The results of recent research (Bieber et al., 1962; Gebhard et al., 1965; Hooker, 1965a, b; Marmor, 1965) show clearly that this stereotype of male homosexuals is grossly oversimplified and in many cases totally inaccurate. We now understand that there can be no simple characterization of the sexual or general social behavior of all American homosexual men. The complexity of the problem can be illustrated with several examples.

To begin with, there are marked individual differences in preferred methods of sexual stimulation. In one sample of 30 men (Hooker, 1965b), 13 expressed preference for fellatio. Anal intercourse was less preferred. Mutual masturbation was apparently employed sometimes in foreplay and sometimes continued to orgasm. In another sample of 106 male homosexuals undergoing psychoanalysis (Bieber et al., 1965), two-thirds of the men expressed a preference for one or another form of sexual stimulation. These preferences can be considered along two dimensions: oral vs. anal contact, and "insertor" vs. "insertee" role in the contact. Of the 38 men who stated preference for the inserter role, only 10 declared preference for the orifice of insertion. But of the 33 men who usually chose the insertee role, all stated a preference for the orifice of insertion.

These data on preferences should not be confused with a rigorous classification of types of homosexual responders. Bieber et al., Gebhard et al., and Hooker all emphasize that many homosexual men occasionally or regularly engage in forms of activity other than those they claim to prefer. This point is of considerable importance because it relates to the problems involved in applying the dimensions of "active-passive" and "masculine-feminine" to the sexual behavior and personality characteristics of male homosexuals.

An examination of male homosexual practices does not permit the conclusion that these men find sexual gratification by accepting only an active or only a passive role during sexual behavior. It was for this reason that Bieber and his colleagues used the terms *insertor* and *insertee* to describe operationally the details of homosexual interactions. It is not the case that one who partially or totally prefers or takes the insertee role is being passive in his acceptance or execution of this role. Moreover, these roles are often reversed for the reciprocal satisfaction of the partners. And in the case of mutual fellatio, a practice performed by at least 37 percent of Bieber's sample and at least 30 percent of Hooker's, an active vs. passive distinction becomes meaningless.

Our traditional sexual mores have confounded the concepts of masculine and feminine with positions on the dimension of activity-passivity. We tend to suspect that within any homosexual pair there must be a "male" member and a "female" member. This is clearly false if all masculinity means is taking the active role in sexual behavior and femininity the passive role. Thus, neither the dimension of activity-passivity nor a simple division into masculine and feminine role taking is useful for a classification of the details of homosexual practices. We shall also inquire about the usefulness of these labels as they might apply to the general social behavior or personality characteristics of male homosexuals.

In her research into this problem, Hooker has attempted an analysis of the subjective response of male homosexuals to the question of their masculinity or femininity. The question of what constitutes "maleness" and "femaleness" (i.e., the question of *gender identity*) is an extremely complicated one, both in regard to our own society and within the context of cross-cultural considerations. We return to this point in the next section. But there is no doubt that considerable social pressure is brought to bear upon individuals who transgress the limits of traditionally defined gender roles. Therefore, men who engage exclusively in homosexual practices, and thereby break the most obvious rule for being a man in American society, are forced to resolve in one way or another their anomalous situation. Each must find a solution or "working hypothesis" for defining his own gender identity within the confines of a generally hostile community.

Hooker has outlined three general strategies that male homosexuals may take in resolving their own sexual preferences with traditional prescriptions. The first involves an acceptance of the premise that "masculine" and "feminine" are dichotomous (nonoverlapping) categories and an assignment of oneself to one or the other. If self-assignment is feminine, then the individual is likely to possess some or all of the characteristics of the classic stereotype, and may in extreme cases wear female clothing. Such individuals may strongly prefer or insist on the insertee role in sexual interactions.

Men who accept the gender dichotomy and identify themselves as masculine typically engage in activities commonly associated with the pursuits of the American male. These may include active interest and participation in competitive business practice and sports. According to Hooker these men have little difficulty in social interactions with non-homosexual men and give no indication of their sexual preferences to the general community. Within this class there are some men who carry the masculine identification to the extent of insisting on the insertor role, whereas others distinguish between their general social masculinity and their homosexual practices, which they permit to vary as the occasion demands.

The second strategy for resolving the homosexual's dilemma, and the one most commonly adopted by the 30 men in Hooker's sample, involves a rejection of the masculine-feminine dichotomy. These men deny that the definition of masculinity need include preference for a female sexual partner. According to this view a man's preference for a male sexual partner and the details of his participation in homosexual activity (whether insertor or insertee or both) are simply not valid determinants or criteria of masculinity. A continuum of masculinity-femininity is assumed to be based on psychological attributes other than the sex of the preferred partner for erotic activity.

The third strategy, adopted by one member of Hooker's sample, postulates a "third sex": a category that is separate from and intermediate to psychologically defined masculine and feminine genders, and contains characteristics of both. Under this assumption the individual is apparently able to cope with the socially forced issue of gender identity by creating a unique class that accounts for both his masculine and feminine traits as defined by the standard set of expectations.

These three strategies represent different modes of adjustment to the pressures the society places on individuals whose sexual preferences and practices are considered repugnant or immoral. As Hooker has clearly documented, there is also a social means of dealing with the prevalent hostility toward homosexuals. This is the formation of a "homosexual subculture." In a large city this subculture includes favored meeting places ("gay bars"), preferred residence districts, rules specifying appropriate conduct in social situations, and a generally efficient communication network for information of common interest, particularly regarding the activities of local law enforcement agencies. There has also been increasing politicalization of some homosexual groups, leading to the formation of the Gay Liberation Movement, which attempts to bring about changes in laws prohibiting homosexual conduct.

Cross-Cultural Data on Homosexuality

Ford and Beach (1951) were able to obtain information concerning the presence or absence of homosexual behavior for 76 societies within their total sample of 191. In 64 percent of these societies at least one form of homosexual activity for at least one age group was socially condoned or encouraged. Among the societies for which information was available, social reaction to discovered homosexual behavior ranged from ridicule to penalty of death (the Rwala Bedouins).

The foregoing description of male homosexuality in the East Bay society exemplifies a system in which male homosexual relations are encouraged between age peers prior to marriage and sanctioned between older men and young boys. Among adolescent boys, insertor and insertee roles in anal intercourse are freely interchanged.

Widespread anal intercourse between unmarried men also occurs in the Aranda of Australia, the Siwans of northeast Africa, and the Keraki and Kiwai of New Guinea. In all of these cases it appears that participation as an insertee in this activity does not necessarily imply femininity as judged by local standards.

In other societies approved homosexual activities occur between one man who maintains full masculine status within the community and another who has adopted, more or less completely, the prevailing feminine gender role, including feminine dress. The position attributed to the latter man (who is known technically as a *berdache*) in the community varies in different societies from high prestige and religious status, as among the Chuckchee of Siberia and the Koniag of northern Alaska, through relatively neutral status, as among the Tanala of Madagascar, to relative inferiority, as among the Lango of east central Africa. Among some American Indian tribes the role of *berdache* was imposed on a young boy whose physique or apparent interests suggested to his elders that he would not succeed in the competitive, aggressive activities of men.

Our discussion to this point has been concerned only with male homosexuality. There is a simple reason for this: relatively little systematic information on female homosexual behavior exists in the scientific literature. The opinion of those who have made careful searches for the existence and frequency of female homosexual activity in other cultures (Ford and Beach, 1951; Kinsey et al., 1953) is that this form of erotic expression is much less frequent among women than among men. Within our own society the same appears to be true (Kinsey et al., 1948, 1953). Much of the discussion of female homosexuality has been within a clinical, primarily psychoanalytic, framework (e.g., Romm, 1965; Wilbur, 1965).

Causes of Homosexuality

At present there is no theory that can account rigorously for why some people engage primarily in homosexual behavior and others do not. Moreover, experts in the field disagree about the validity of describing homosexuality as a result of physical or mental pathology. Those who have insisted that homosexual behavior should be viewed from the standpoint of "disease" have searched for causes within the genetic structure, endocrine physiology, and early psychological experiences of the "afflicted" individuals. We may briefly review some of their findings.

Most of the data concerning a possible genetic basis for a homosexual preference have come from the investigation of its transmission from generation to generation within single families. Of particular interest has been the occurrence of homosexual behavior in identical and fraternal twins. Kallman (1952) presented evidence to indicate that homosexuality in both members of a pair of twins was much more likely if the twins were identical rather than fraternal. This finding appears to support the conclusion that there are genetic factors directly linked to the expression of the behavior. However, as Kallman himself and others (Bieber et al., 1965; Ellis, 1963) have pointed out, this conclusion is not warranted by the data, for several reasons. One of the more striking reasons was the generally low level of social-psychological adjustment in the sample of 85 pairs of twins that Kallman had available for investigation. In addition to expressed homosexuality, many subjects had also been diagnosed as schizophrenic or schizoid. This leaves open the definite possibility that homosexual behavior was but a single, secondary characteristic of a more general deterioration or pathology.

We will pause for a moment to investigate some basic problems underlying the hypothesis that there is an inherited tendency toward homosexual behavior. First, as we have seen, homosexuality takes diverse behavior forms. Thus it would be difficult to argue that what is inherited is the tendency to *behave* in a particular way, that is, to perform certain postures or gestures. In other words, it is a mistake to conceive of *homosexuality* as some sort of unitary behavioral entity that is directly predisposed by a particular genotype. Therefore the argument must be shifted to claim that what is inherited is a *preference* for a male partner during sexual behavior. This argument appears to conceive of the selection of a human sex partner as a process akin to the "release" of an instinctive act by a species-typical "sign-stimulus," as these terms are used by ethologists in discussing the unlearned behavior of nonhuman species. The argument quickly runs into numerous difficulties, including the obvious one that at least some homosexuals respond primarily to men who have gone to great lengths to appear female. Furthermore, as several writers (e.g., Ford and Beach, 1951) have pointed out, the capacity to respond with the species-specific copulatory postures of both sexes is inherent in many mammalian species. This is true, if only in a trivial way, for humans as well. From this point of view, there is no question that all men (and women) *can* respond in a homosexual fashion; the question is why some do and some don't.

During the past several years there has been a reawakened interest in the search for endocrine correlates of homosexual preference. This new interest was stimulated by the report of Margolese (1970) that the relative amounts of two urinary metabolites of androgen (androsterone and etiocholanolone) differed in groups of homosexual *vs.* heterosexual men. Specifically, the ratio of androsterone to etiocholanolone in the urine of 10 homosexual men was less than the ratio in the urine of 10 heterosexual men. Also in 1970 Loraine et al. reported that estimates of testosterone levels (again based on urine samples) were lower in two exclusively homosexual men than in a group of heterosexual men. A lowered androsterone/etiocholanolone ratio has also been reported by Evans (1973), along with several other biochemical and morphological differences between a group of 44 homosexual men and 111 heterosexual men. Finally, Kolodny and colleagues (1971, 1972) have measured testosterone and other sex hormones directly in the blood plasma of 30 homosexual and 50 heterosexual men. They reported significantly lower average plasma testosterone levels in men who were exclusively or predominantly homosexual (5 and 6 on the Kinsey scale) than in other homosexual men or heterosexual men. On the other hand, Tourney and Hatfield (1973) reported no differences either in plasma testosterone or the urinary androgen metabolites, between a group of 13 homosexual men and 11 heterosexual men.

How are these various results to be interpreted? In the first place, the failure of Tourney and Hatfield to replicate the results of the other studies suggests a conservative position be taken on the generality of the phenomenon of lowered androgen levels in homosexual men. Second, the interpretation of a relatively low androsterone/etiocholanolone ratio needs to account for the finding of Rose et al. (1969) that similarly low ratios of these metabolites were found in army recruits during the early portion of basic training and in Green Berets just before a combat mission in Vietnam. Given these findings, one could not reasonably conclude that low ratios are indicative of physiological circumstances that cause (that is, serve as a sufficient condition) for homosexual preference. A more likely interpretation might be that low ratios are indicative of a physiological stress response that some homosexual men exhibit in response to societal pressures on them as a result of their deviant status. This interpretation is somewhat strengthened by the fact that the urinary metabolites in question reflect the activity of androgen secreted from the adrenal gland (which is very responsive to stress) as well as that secreted from the testes. Finally, given the reasonable assumption that the foundations of homosexual preference are established prepubertally, the levels of sex hormone found circulating after puberty could not be taken as more than an index, as opposed to a cause, of the preference. This consideration led Money and Ehrhardt (1972) to stress the importance of prenatal hormonal dysfunction as a more likely physiological determinant of homosexual preference than postpubertal hormonal dysfunction. But after reviewing the relevant evidence, they concluded that environmental variables operating during the years of late infancy and early childhood are so powerful in the determination of gender identity and sexual preference that endocrine conditions are unlikely to have other than relatively indirect consequences for eventual sexual preference. However, as they and other specialists have emphasized, at present the best course is to keep an open mind and await the presentation of further data.

The "disease-oriented" accounts of the development of homosexuality that stress the importance of early psychological experience have been based primarily on psychoanalytic theory. Many psychoanalytic theorists have expressed themselves on this issue (Bieber et al., 1962). On the basis of their intensive analysis of 106 male homosexuals, Bieber and his colleagues supported an earlier contention of Rado's (1949) that homosexual behavior stems from "hidden but incapacitating fears of the opposite sex." Bieber et al. concluded that this fear develops out of experiences with a mother who is, for various reasons, overly "close-binding-intimate," and a father who is hostile to or detached from his son. The seductive behavior of the mother toward her son creates anxiety in him that may generalize to an aversion for all heterosexual contact. The overprotectiveness of the mother also leads to inadequate opportunities for heterosexual adjustment in play and other peer groups. The effective "demasculinization" of the boy makes all future attempts at heterosexual adjustment difficult and subjectively threatening. Repeated failures during childhood and early adolescence "pave the way" for entrance into homosexual episodes or the homosexual community.[1]

As Bieber et al. have pointed out, all psychoanalytic theories of homosexuality assume that the behavior is pathological. We need to consider the problems involved in making that assumption. To begin with, there is a substantial problem involved in the definition of the term *pathological behavior*. It is safe to assert that no firm criteria have been devised for deciding what behavior is to be considered pathological and what behavior is not.

Second, it has not been possible to show that homosexuals are less "well adjusted" than heterosexuals, as measured by standard psychological testing (Hooker, 1965a). It must be recalled that the subjects of Bieber's study were men who were voluntarily undergoing psychoanalytic treatment. Sixty-four of the 106 men were unhappy enough about

[1]Since this chapter was written, a number of developments have occurred related to homosexuality that date the treatment given here. For example, in December 1973 the American Psychiatric Association removed "homosexuality" from its official list of mental disorders. The Association instead classified homosexual preference as a "sexual orientation disturbance." The lobbying efforts of organized homosexual groups were in part responsible for the change in psychiatric nomenclature. This development emphasizes the subtle but important relationship between the apparently "objective" basis of medicine and political process.

their homosexual behavior to want to change it. Presumably these men were considerably distressed by their behavior and its social consequences. They viewed their homosexual activities as deleterious to their achievement of personal and social stability, and in that sense the behavior could be classified as pathological. But it does not follow that all homosexual men find themselves in such precarious psychological positions. Hooker's data (1965a, b) indicate that many male homosexuals lead substantially productive lives. Within our society, behavior disorders associated with homosexuality may well arise as a result of the intense social pressures on the deviating individual, particularly if the social norms are in part internalized.

Another problem with the assumption of pathology is presented by the evidence for socially sanctioned homosexual behavior in societies other than our own. It appears that in the East Bay society, for example, refusal to engage in anal intercourse at the request of a friend would be a breach of etiquette at least, and might be considered an unfriendly, antisocial act, perhaps even symptomatic of a "character disorder." In evaluating the nature of homosexual behavior, we must be careful not to describe behavior disapproved in the United States as pathological wherever it occurs.

The cross-cultural data also present a problem to psychoanalytic theories of the etiology of homosexual behavior. If Bieber and his colleagues are correct in their assessment of the roles of mother and father in the development of homosexual preferences, are we to expect similar parent-child dynamics to be present in all societies where many or most men regularly engage in homosexual acts? Are all East Bay mothers "close-binding-intimate" and East Bay fathers hostile or detached? Answers to these kinds of questions await more detailed ethnological investigation. But it would be a test of the generality of any developmental theory of homosexuality to determine whether the presumedly relevant patterns of family interactions are frequent where there is widespread occurrence of the behavior and infrequent where the behavior is rare.[2]

We conclude that no single theory presently available can account for the occurrence of homosexual behavior. It is clear, however, that a basic requirement for the establishment of a successful theory includes the collection of much more data without prejudgment of the behavior as *necessarily* pathological.

[2]Note that the fundamental premise of Bieber's theory is an "incapacitating" fear of the opposite sex. Within the psychoanalytic theory, this fear results from particular kinds of interaction with parents. We can speculate that this fear might, in other societies, be transmitted by the parents in other ways. If, as in the East Bay society, there are penalties for any heterosexual contact prior to marriage, "fear" of girls might develop as a natural consequence of normal socialization. It could be argued that socially sanctioned homosexual behavior (and masturbation) is present as a means of insuring the established ban on heterosexual activity. Therefore, even if a positive correlation were established cross-culturally between attitudes toward girls and the prevalence of homosexuality, it would not necessarily follow that the psychodynamics postulated by Bieber et al. would thereby be shown to account for the behavior.

THE CLINICAL MEDICAL APPROACH
TO SEX AND SEXUAL BEHAVIOR

It is clear from the preceding discussion that at least part of the problem involved in understanding homosexual behavior stems from a definite lack of precision in the uses of the words *male* and *female*. For example, we saw that there is an (incorrect) tendency to relate maleness or masculinity to "activity" and femaleness or femininity to "passivity." This and similar misconceptions result from the presumption that our cultural norms and traditions reflect physiological reality. Perhaps now, when so many fundamental questions about the roles of women in American society are being raised, we are all more sensitive than we were to the arbitrary nature of some of our sex-role prescriptions. Nevertheless it may still be true, as Hampson and Hampson put it in 1961, that "all too often we have wrongly identified that which is current and prevalent in our own culture as the natural and normal state of affairs."

Just what, then, *is* the normal state of affairs? What are we to mean when we call a person "male" or "female"? In this section we will consider the insights gained from studies of the roles of male and female hormones as they influence human sexual behavior. For obvious reasons this information has been collected within the context of clinical medical practice; the kinds of experimental information that we collect from non-human animals we may not collect from human beings. Clinical data seldom meet the criteria for rigorous evaluation that are met by experimental data; hence our answers are not likely to be totally clear-cut. Nevertheless, the general outlines of the argument should be clear.

As we have seen, hormones play two roles in determining behavioral outcomes: they play an organizational role before and just after birth, and they play an activational role in puberty and adulthood. Here we will consider the activational role of hormones first; we will examine the available evidence on the effects of hormone loss and replacement on human sexual behavior. Then we will turn to the organizing function of gonadal hormones on human sexuality. In this discussion we will emphasize the research that has been done with the condition of human hermaphroditism. One result of this research has been the realization that an adequate concept of sex, in the sense of masculinity or femininity, must take into account a number of different biological dimensions.

ACTIVATIONAL ROLES OF GONADAL HORMONES IN MEN
Castration of the male is one of the earliest surgical techniques known. It has been used to attain a variety of objectives in both men and animals. Men have been castrated ever since ancient times for various reasons, for example to supply "safe" companions for Roman ladies, to provide eunuchs for harems, to subdue captives, for religious reasons (as in the case of the Russian Skoptsi sect, which practices either castration or complete emasculation), to prolong the usefulness of choir boys in the

medieval church, and to punish enemies or criminals. In modern times orchidectomy has been performed in treatment of such conditions as genital tuberculosis, malignancies or traumatic injury, and also as a legal measure for eugenic reasons or to deter sex offenders. In view of the countless times this operation has been performed, it is amazing that there is little precise and scientifically valid information on its effects and that there is still controversy as to whether it has important effects on libido and/or potency.

In the writing of antiquity there are a variety of references to the ability of postpuberal castrates to cohabit with women, and it is not clear to what extent these comments exemplify the special attention accorded to extreme curiosities. However, it is said that Roman ladies demanded that their male companions be castrated after puberty so that the capacity for cohabitation be spared. On occasion, when the need to prevent potency was felt strongly, additional penectomy or prepuberal castration was performed. Although in both these cases there is still the possibility of obtaining some stimulation from the remaining stump or undeveloped penile tissue, the castraters could feel secure that their subjects would not be capable of intercourse. On the other hand, there are simply too many reports through the years, albeit mostly anecdotal, for there to be much doubt that postpuberally castrated males can at least occasionally perform coitus. In one of the better documented cases a man of 43, castrated 18 years previously was reported to have intercourse with his wife one to four times weekly (Hamilton, 1943).

Of course even in the cases where coitus occurs, a question still remains as to the "quality" of this behavior. One rather dramatic example of this question (cited in Tauber, 1940) is that of a 42-year-old Russian Skoptsi who had been castrated at 21 years of age. He claimed that erections were short-lived, that orgasm occurred very early, and that there was a very scanty and watery ejaculate. Tauber comments that this kind of information "represents more or less a medical curiosity rather than a source of useful information." However, this report points up the need for more reliable information on the details of sexual behavior in castrates.

The most extensive investigation of the effects of castration in man was a follow-up study by Bremer (1959) on several hundred men who were legally castrated in Norway for a variety of sexual offenses and "psychiatric indications." The generalizability of the findings is somewhat questionable because of the high percentage of psychopaths, mental defectives, and schizophrenics among the subjects, and the behavioral data are far from impressive. Nevertheless because of the scope of this survey its results merit our attention.

Of 157 individuals who were sexually active preoperatively, 74 were asexualized shortly after castration. The term *asexualized* was used to imply a cessation of activities leading to orgasm, and a loss of sexual interest. Twenty-nine cases were asexualized in the course of the first year after castration. In 54 cases "a certain sexual interest and reactivity, possibly activity or pronounced sexual coloring of psychotic syndromes, persisted for more than one year." Despite the difficulties of accepting at face value reports from psychotic individuals, as well as the fact that the operation itself can have profound psychological effects, the results of this study do confirm the general conclusions of many other authors. Apparently the effects of postpuberal castration are quite variable in different individuals. Sexual intercourse with orgasm may continue for one or more years postoperatively, but eventually the ability to copulate is lost in most cases along with a decline in the desire to copulate. In the cases studied by Bremer, the legal objective of the operative procedure was generally attained—to prevent the recurrence of a variety of forms of sexual behavior regarded by society as criminal. This was so whether the direction of the behavior involved was heterosexual or homosexual.

Unfortunately we can say very little of the probability of retention of different components of male sexual behavior after castration. It appears that one of the components of the decline in potency is a partial or complete failure of erection (Miller, Hubert, and Hamilton, 1938; Yamamoto and Seeman, 1960). We do not know, however, whether this is the indirect result of decreased libido. The psychological factor of the knowledge of their deficiency must play a role for some if not all men in the effects of castration. However, we simply do not have reliable information on the factors that affect the tremendously variable rate of decline in sexual activity following castration. To define the specifically biological, hormonal role is difficult without such information.

One additional general comment is in order before leaving this question. A very great variety of types of pleasurable sensations can be derived by men from their genitalia and from sexual contacts. That many of these may continue, and new ways of deriving pleasure be found after castration, does not contradict the important role of the testes in controlling sexual behavior under "normal" conditions. What gives satisfaction in a sexual relationship is defined by the needs of partners, and human beings are the most flexible of organisms. In the light of this consideration one should evaluate the reports of satisfactory marital relations following castration. It should also be borne in mind by medical practitioners in setting aims for therapy in cases of castration or hypogonadism.

Clinical Disorders and Effects of Testosterone Treatment

It is widely accepted in the practice of clinical endocrinology that testosterone treatment in hypogonadal men increases abilities for sexual performance. It is generally accepted, too, that testosterone stimulates sexual drive in castrated men and restores the ability of erection (Miller et al., 1938).

Of course the information we have on the effects of testosterone treatment in men comes mostly from the treatment not of castrates but of a variety of forms of impotence. It should be stated at the outset that in most cases impotence is not due to any demonstrable endocrinologic abnormality but mainly to psychological causes. Among the less important causes are a variety of drugs, including some narcotics, tranquilizers, alcohol (see the Porter's speech in *Macbeth*), and malformation of the genitalia. Parenthetically, probably the most celebrated case of the last-mentioned (or any) cause of impotence was that of Louis XVI of France, who married Marie Antoinette when he was 16 and was impotent for the first seven years of his marriage. Finally, after extensive consultation with the medical authorities of the time, it was determined that the young king's problem was a tight foreskin, and circumcision resulted in an apparently complete cure (Hastings, 1963).

Once the variety of other causes is removed, one remains with a number of conditions in which impotence may be related to deficient production of testicular androgen. Among the more important of these conditions are secondary hypogonadism due to insufficiency of pituitary gonadotropic function, and lack of testicular development resulting from chromosomal abnormalities. The most common example of the latter is Klinefelter's syndrome, which is caused by an extra chromosome (sex chromosome complement XXY instead of XY). Testicular function is quite variable in these patients, but impotence is a common finding. Some seem to have normal testicular function and sexual behavior until after puberty, and loss of potency can be an especially acute problem for those individuals who are sexually experienced. Although the sterility cannot be reversed, treatment with testosterone often does cure the impotence. This is a fairly good example of a medical equivalent of the castration-replacement therapy experiments in animals.

Hypogonadotropic eunuchoidism, or failure of testicular development due to pituitary failure, is one of the most common forms of hypogonadism apart from Klinefelter's syndrome. This may be a hereditary disorder. Although testosterone is helpful in treating the impotence found in these men, gonadotropic hormone is more effective. Gonadotropin secretion is of course defective (by definition), unlike in Klinefelter's syndrome, where excess production of gonadotropin secretion is found. In this fact may lie the explanation for the otherwise puzzling discrepancy between the effectiveness of testosterone and gonadotropin treatment. That is to say, it is conceivable (though unlikely) that gonadotropin may have a direct effect on behavior, independent of androgen, or more likely that some other testicular product that affects behavior may be stimulated by the gonadotropin.

The final form of endocrinologic impotence to be discussed is that found in some aging men. Unlike women, men very seldom undergo a climacteric (i.e., a rapid cessation of gonadal function). There is, however, a gradual, albeit variable, decline in testicular function and potency beginning in middle age, and in some cases a true male climacteric occurs. This involves the hot flashes, psychologic depression, and other classical signs of the withdrawal of gonadal hormones during menopause.

These men may become impotent, although libido (as loosely defined earlier in this chapter) may be less affected, creating a severe psychologic problem. In some cases testosterone treatment can improve potency, but such treatment should be used with care because "too much" sexual activity in the aging can also create problems of a social or medical nature. However, androgen therapy is frequently ineffective. This fact does not seriously compromise the hypothesis of the androgen dependence of male sexual behavior because the male climacteric is a poorly understood syndrome. Furthermore, many gerontologic factors other than a decline in androgen level must surely be of importance for sexual behavior. Masters and Johnson (1970) have argued that age-related male impotence is very often produced by male *expectation* of impotence rather than physiological alterations.

Comparative Considerations

What then can we infer from our knowledge of the effects of castration, testosterone treatment, and clinical disorders about the role of androgen in controlling sexual behavior in normal men? There can be little doubt that androgen (mainly testosterone) plays an important role. However, that role is not absolute. As in carnivores (and to a lesser extent in rodents), some individuals may show only minor impairments for prolonged periods after castration or pathologic changes in testicular function. The deficits, which do occur, are correctable with androgen treatment. Unfortunately there is little beyond this that can be said. Dose-response relationships have not been worked out, and in fact there are no definitive, carefully controlled studies on the effects of testosterone on erectile capacity and various other specific elements of male sexuality.

The paucity of quantitative data on human males is sufficient reason to restrain us from making too many correlations with events occurring in animal species. However, a cautious attempt to find such analogies can be useful, if only in suggesting directions for future research. For example, we may take the problem, examined in the animal literature, of whether precastration experience influences postcastration sexual behavior. Although Rosenblatt and Aronson (1958) found that cats with preoperative coital experience maintain sexual behavior for longer periods following castration than inexperienced animals, no effects of preoperative experience were found in dogs (Hart, 1968c) or rats (Bloch and Davidson, 1968; Larsson, 1966). This could have relevance to the expected severity of the effects of castration in humans and the efficacy of testosterone treatment in hypogonadal males. Possibly relevant, too, is the report by Money (1965) of a 21-year-old sexually infantile eunochoid in whom testosterone increased the incidence of erection and also resulted in his first experience of ejaculation. Lloyd (1968) mentions that variations in plasma testosterone levels of as much as 100 percent among men were consistent with "normal sexual relationships." This is reminiscent of the conclusions arising from the rodent studies that individual variations in male sexual behavior are not related to differences in androgen level. Properly conducted studies of men could lead to the conclusion that a considerable "safety factor" exists in relation to the androgenic control of sexual behavior in both men and animals.

Critique of Kinsey's Conclusions

The view of the role of hormones in human sexuality presented here is somewhat different from that expressed by Kinsey et al. (1953). Because of the extraordinarily influential nature of their work, we shall devote some detailed attention to their conclusions, which were based mostly on a review of the literature relating to hormones and male sexuality. The major points are best presented by full quotation:

> While hormonal levels may affect the levels of sexual response—the intensity of response, the frequency of response, the frequency of overt sexual activity— there is no demonstrated relationship between any of the hormones and an individual's response to particular sorts of psychologic stimuli, an individual's interest in partners of a particular sex, or an individual's utilization of particular techniques in his or her sexual activity. . . . There seems to be no reason for believing that the patterns of sexual behavior may be modified by hormonal therapy. [p. 761]

> Various hormones may affect the levels of sexual responsiveness in the human female and male because of their effects on the levels of all physiologic activities, including those capacities of the nervous system, on which sexual behavior depends. [p. 760]

What these authors are saying is, first, that although hormones affect the level of sexual response, they do not affect its quality. Second, there are no specific effects of sexual hormones on sexual function, but rather general metabolic effects of all hormones, including gonadal ones, any of which may indirectly affect sexual function. With the first of these conclusions one cannot quarrel, at least insofar as it applies to the treatment of homosexuals. There is no positive (and a great deal of negative) evidence that treatment of homosexuals of either sex will change the direction or pattern of their sexuality. Applied to other qualitative aspects of sexual activity, there is little evidence with which to either substantiate or deny the conclusion, but it does seem in accord with the animal data (see Chapter 5).

The bases for the second argument, that hormonal effects on sexual behavior are basically nonspecific, have to be considered in some detail.

1. Kinsey et al. point out the inadequacies in studies by investigators who have purported to show an involvement of sexual hormones. However, to the extent that these studies are generally inadequate (as cannot be denied) it is difficult to draw any conclusions from them, let alone a sweeping generalization about the minor and nonspecific role of sexual hormones.

2. It is argued that the ontogenetic changes in secretion of sex hormones in humans do not parallel the age-related changes in sexual behavior. Insofar as puberty is concerned, Kinsey et al. point out that their data on the onset of sexual activity leading to orgasm show a more sudden increase in the adolescent than do data (taken from an early investigation in the literature) on excretion of urinary androgen. Many questions could be raised about the validity of this comparison, but we will raise only two. First, how do we know that sexual behavior does not show a "threshold" type of onset when a certain level of androgen production is reached? In animal species where the androgen dependence of male sex behavior is unquestioned, different species show different degrees of prepuberal sexual activity. Second, because of the considerable variability in puberal onset of both hormone secretion and behavior phenomena, comparisons should be made on the same individuals rather than on different populations. Of course it is true that the decline in sexual behavior with old age is not necessarily correlated with the decline in hormone production (except for certain cases of male climacteric). This is likely no more than one manifestation of psychologic senility; it may be due to the aging of neural structures involved in sexual behavior (Reichlin, 1968, p. 1010).

3. A review of the earlier studies of castration in men showed that the results were inconsistent. This problem has been discussed earlier. Kinsey's conclusion that the decline was due to old age simply is not valid when one considers the many reports of rapid loss of potency and libido.

4. "The effects of androgens are presumed to be due to their ability to step up the general level of metabolic activity in an animal's body, including the level of its nervous function and therefore of its sexual activity" (Kinsey, p. 748). It is true that sex hormones do have effects on metabolism, but there is no reason to believe that any of the reported metabolic effects bear any relationship to stimulation of sexual behavior. As far as general metabolic rate is concerned, thyroid hormones have a far greater effect than do gonadal hormones. By Kinsey's own admission (p. 753), the evidence for a stimulatory effect of thyroid hormones on human sexual behavior is extremely poor. In a surprisingly weak argument, cytological changes in the pituitary of fowls are correlated with age-dependent changes in human sexual behavior. The changes in the fowl pituitary might mean that gonadotropic hormone production, and therefore gonadal function, was correlated with sexual behavior *in the fowl*, a conclusion supportive of the concept that sex hormones *do* affect behavior. However the extrapolation from fowl pituitary anatomy to human sexual behavior is so remote as to be almost incredible. Indeed it is surprising that such careful workers should have drawn such unwarranted conclusions from such poor data. In all fairness, it must be added that the level of scholarship and objectivity that characterizes this part of the study is far from typical of the study as a whole.

ACTIVATIONAL ROLES OF GONADAL HORMONES IN WOMEN

Effects of Ovariectomy and Menopause

In women, as in men, there is a severe paucity of reliable objective scientific data on the role of hormones in sexual behavior. There is, however, an "experiment of nature" that occurs in all women and that provides an opportunity to observe the extent of involvement of ovarian hormones. At the menopause the ovary effectively ceases to function as a producer of hormones. Yet there is no reason to believe that sexual behavior is impaired as a result of this functional castration except for a possible problem of decreased vaginal size and lubrication, which is unlikely to arise in women who maintain regular intercourse (Masters and Johnson, 1966). In fact, women may find increased pleasure in sexual activity, presumably due to emancipation from the fear of pregnancy.

Ovariectomy is not infrequently performed on relatively young women for conditions that are not debilitating; thus valid pre- and postoperative comparisons on sexual behavior can be made. Although some authors have claimed a decrease in sexual urge following this operation, the general opinion is that the only major effect is an occasional increase in libido, as in the case of menopause.

Kinsey's report on 123 castrated women "does not provide evidence that females deprived of their normal supplies of gonadal hormones have their levels of sexual responsiveness or their frequency of overt activity lowered by ovarian operations" (p. 735). Bremer's study on legal castration included 27 women who were ovariectomized for promiscuity or psychiatric disorders. No significant changes in sexual behavior or psychiatric condition in general were found. The conclusion appears clear: gonadal hormones are certainly not essential for the continuation of normal female sexual behavior; their complete removal appears to have no important effects thereon. In this, women differ from men, and apparently from females of all other animal species.

Comparative Considerations

Such a qualitative evolutionary leap in the control of sexual behavior seems surprising in comparison at least to what we found in males. It is necessary at this point, therefore, to examine whether nonhuman primates show trends that would link the human female to subprimate species.

The patterns of sexual behavior are more complex in the rhesus monkey, the most studied primate, than in lower species. To detect relevant changes one cannot rely on a single measure of receptivity, as often suffices in rodents. Frequency of presentations to the male and of "refusals" to his mounts are the common measures of receptivity, but these measures cannot be interpreted without consideration of factors of sexual preference, aggressiveness, and other social interactions that complicate the analysis (Herbert, 1970).

Nevertheless, early studies (Ball, 1936; Young and Orbison, 1944) and more recent ones (Freedman and Rosvold, 1962; Michael, Herbert, and Welegalla, 1967) show that the sexual behavior of female monkeys is adversely affected by ovariectomy, so that the cyclic variations in both copulatory and social (e.g., grooming) behavior disappear. However, the degree of dependence of female primate behavior on the ovaries appears to be considerably less than that of subprimate species. Thus Michael et al. (1967) found cyclic variations in the level of several parameters of receptivity in less than 50 percent of the spayed monkeys he studied.

The administration of estrogen can restore various components of the behavior to the level seen at the midcycle peak. However, it is not clear to what extent estrogen directly affects the female's behavior, as opposed to increasing her attractiveness to the male, thus indirectly influencing her behavior in response to his advances. Thus the cyclicity of copulatory activity in monkey societies may not be a function of the females' inability to mate outside the midcycle period, but rather of their greater attractiveness for the males at that time. After that phase is completed, they may simply lose the male's attention in favor of other more attractive females (see Herbert, 1970).

It seems, therefore, that the phenomenon of absolute estrogen-dependence of female receptivity does not disappear suddenly with the evolution of the human species; the ground has been prepared for this change in nonhuman primates.

The Question of Adrenal Androgens

The nonessentiality of gonadal hormones for sexual behavior in women does not by any means close the discussion on this topic. For one thing, there is another source of steroids—the adrenal cortex—that produces small quantities of estrogen and larger quantities of biologically weak androgens. In contrast to the well-known ineffectiveness of estrogen[3] in women, it has often been claimed that libido is stimulated by administration of testosterone. Unfortunately there is a dearth of well-controlled and behaviorally sophisticated studies of the effects of androgen on sexual activity in women. Salmon and Geist (1943) reported that testosterone treatment of patients suffering from "primary frigidity with somatic gynecological disorders as well as secondary (endocrinopathic) frigidity" had the following effects: increased susceptibility to psychic sexual stimuli; increased sensitivity of the external genitalia, particularly the clitoris; and a greater intensity of sexual gratification. Other clinical investigators have reported similar findings. Schon and Sutherland (1960) found that

---

[3]Estrogen has some usefulness in enhancing vascularity in the vaginal walls and increasing lubrication during coitus, although Masters and Johnson (1966) reported that the "sweating" type of lubrication they discovered was quite marked in the ovariectomized women they studied.

hypophysectomy, unlike ovariectomy, decreased sexual desire, activity, and gratification (including orgasm) in women, and interpreted this as due to adrenal hypofunction. In a widely quoted study, Waxenberg et al. (1959) found that a decrease in libido followed adrenalectomy in ovariectomized women suffering from advanced cancer.

Because of these data, the view is fairly widely held that adrenocortical androgen is "the" hormone involved in the activation of sexual behavior in women (e.g., Money, 1961). There are two main problems with this conclusion. First, it is extremely difficult to draw general conclusions from the results of such severe operations as adrenalectomy and hypophysectomy on patients with advanced metastatic cancer. Surgery of this kind may be almost a last-ditch procedure and life expectancy is no more than a few years. It seems neither surprising nor particularly meaningful, therefore, that libido would be decreased postoperatively. Second, testosterone in the doses administered in the foregoing studies causes clitoral hypertrophy. It seems quite likely that the reported increases in libido may be due to increased sensitivity of the clitoris with the result that orgasm is facilitated. Such a peripheral effect would be quite different from the manner in which we believe androgens and estrogens act on sexual behavior in both male and female animals (see Chapter 7).

The crucial question, however, is whether the doses of testosterone administered in these studies result in circulating testosterone levels that resemble those present in women under normal circumstances. Although many of the clinical reports are not explicit as to the precise dose regimen, it appears very probable that the doses were indeed supraphysiological whenever they were effective, as seen for instance by the presence of clitoral hypertrophy. The total output of androgen in normal women is but a fraction of that produced by males. The conclusion we must arrive at, in the absence of more definitive data, is that there is no convincing evidence that the effects of androgen on female sexual behavior represent an important physiological function under normal circumstances.

Of course, more research is required on this, as on all other aspects of the endocrinology of human sexual behavior. It may be that such research might provide the reliable evidence needed to prove involvement of the adrenals in women. It was pointed out earlier that situations exist in which the adrenals have a facilitatory influence in female rats. But here the adrenal hormone involved was almost certainly progesterone rather than an androgen. Of greater relevance are recent findings by Everitt and Herbert (1969) in the rhesus monkey. These investigators found that dexamethasone, a potent synthetic corticosteroid, markedly reduced sexual receptivity in female monkeys as measured by decreases in the number of sexual "presentations" and increases in "refusals" to the male's advances. By suppressing ACTH secretion, dexamethasone would reduce both adrenal androgens and progesterone. However, because progesterone has been shown to have only inhibitory effects in female monkeys, it is likely that changes in adrenal androgen are involved in this effect. This is supported by the finding of Everitt and Herbert that testosterone administration reversed the effects of dexamethasone in these experiments. Because corticosteroid treatment is an extremely widespread clinical procedure, research to determine whether similar effects might occur in the human would be both of practical and theoretical interest.

The Menstrual Cycle

None of the previous discussions exclude the possibility that ovarian hormones may be involved, in a more subtle way, in modifying the intensity and quality of sexual behavior in women. As a result of recent advances in chemical methodology, we now have very reliable measurements of hormone levels in the blood at different stages of the menstrual cycle. It is clear that estrogen reaches a peak shortly before ovulation at midcycle, and progesterone peaks several days before menstruation, after which it declines rapidly (see Figure 5.5, p. 140). With these facts in mind we can evaluate the results of studies on the relationship of cycle stage to female sexuality.

The well-known investigations of Benedek and Rubinstein (1942) suggested that a number of interesting changes in the direction of the sexual impulse occur at different phases of the menstrual cycle, presumably due to the hormonal changes occurring at those times. On the basis of psychoanalytic interviews Benedek and Rubinstein concluded that women manifest an active, extroverted heterosexual tendency in the first (estrogen-dominated) half of the cycle. Shortly after ovulation a decrease was reported in "heterosexual tension." In the second (progesterone-dominated) half of the cycle there were changes in mood in the direction of introversion, and the dream material was weighted with themes of pregnancy and mother-child relations. Waxenberg et al. (1959) disagreed with these conclusions on the basis of their failure to find changes in libido in relation to the vaginal smear in women. However, their techniques were quite different and their findings are not really relevant to the type of subtle psychological changes proposed by Benedek and Rubinstein.

There have been a variety of studies attempting to correlate frequency of coitus and stage of the menstrual cycle. Kinsey et al. (1953) and Hampson and Hampson (1961) conclude that peaks in sexual activity and responsiveness have most often been found (in the earlier studies) before and just after menstruation. But aside from the problem of obtaining reliable data when retrospective data gathering is used, these peaks could be explained by conscious or subconscious desire to avoid the period of presumptive ovulation during midcycle, as well as abstention during the period of menstrual flow resulting in increased desire immediately thereafter. In a recent study, the best of its kind, care was taken to collect data each day throughout the cycle (Udry and Morris, 1968).

The results clearly showed that the incidence of coitus as well as orgasm was highest around the middle of the cycle and that there was a trough in these events some days before menstruation, at approximately the time when progesterone is at its highest point in the cycle. The midcycle peak may represent an atavistic remnant of the estrus period so ubiquitous in animals. Similarities between data from human and rhesus monkey females are depicted in Figure 8.3.

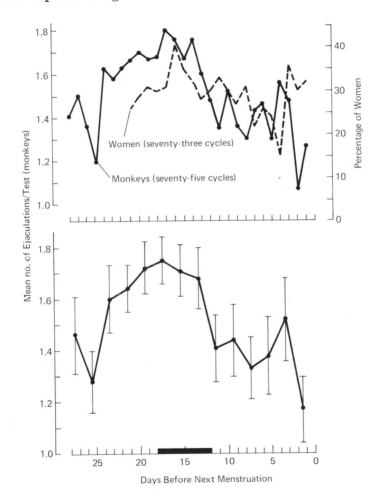

FIGURE 8.3 Upper part: comparison of the copulatory activity of rhesus monkey and man in relation to the menstrual cycle. ●———●: mean number of ejaculations per test (32 pairs of rhesus monkeys); -----: percentage of women reporting sexual intercourse (40 women) Udry and Morris, 1968). Lower part: rhesus monkey data smoother by plotting means of two consecutive days. Vertical bars give standard error of means. Horizontal bar gives expected time of ovulation (Hartman, 1932). From Michael, R. P., and Zumpe, D. (1970) Rhythmic changes in the copulatory frequency of rhesus monkeys (*Macaca mulatta*) in relation to the menstrual cycle and a comparison with the human cycle. *Journal of Reproduction and Fertility* 21:119–201.

Possible Effects of Oral Contraceptives

The introduction of oral antifertility steroids has resulted in the fact that millions of women are now artificially raising the levels of estrogenic and progestational steroids in their blood. In the newer sequential method the two steroids are being taken separately. This could provide an excellent opportunity, as yet insufficiently exploited, for studying the effects of experimentally increased levels of ovarian hormones in normal women on a completely unprecedented scale. Because of great interindividual variability, only large-scale investigations are meaningful. In the original studies on "the pill" only superficial observations were made. The conclusion was that the only effect of oral contraceptives is a marked increase in the frequency of coitus, due simply to the decrease in anxiety about becoming pregnant (Pincus, 1965). Increased libido has been reported in various later studies (e.g., Glick, 1966).

All of the various side effects of oral contraceptive agents are characterized by tremendous variability in their incidence among different women (Salhanick et al., 1969). Nevertheless, several studies show that a far from negligible proportion of women report decline in libido when using these agents, and presumptive evidence suggests that it is the progestational component that may be responsible for this (Grant and Mears, 1967; Kane, 1968). Psychiatric explanations have been offered to explain oral-contraceptive-induced decline in libido. These include the distaste reportedly felt by one woman when pills were substituted for condoms as a means of contraception; this patient felt disgust that her husband was leaving his semen inside her. There is also the possibility that some women may have a masochistic need to run the risk of pregnancy and may derive less pleasure when taking the pill because of the absence of this danger. Other women find that the pill activates "prostitution fantasies" of uncontrolled promiscuous sexual activity. In general the pill does seem to have unique and interesting psychological significance (Orchard, 1969). Nevertheless, the reported effects of progesterone in monkeys raises a serious question about possible *biological* effects. That other women (probably a smaller percentage when highly progestational agents are used) seem to show an increase in libido is not surprising in view of the factors we have discussed.

To summarize, the evidence that there may be an inhibitory effect of progestins on elements of sexual behavior in women, although by no means conclusive, is steadily mounting. It comes from the following sources: (1) the data showing that both progesterone and synthetic progestational agents depress sexual receptivity in the rhesus monkey. From the information on ovariectomy and on behavior in relation to menstrual cycles, it appears quite likely that these animals occupy an intermediate position between subprimates and humans with respect to the hormonal control of female sex behavior. (2) The finding of a trough in coitus and orgasms as well as reported decreases in sexual interest at the time during the luteal phase of the menstrual cycle when progesterone levels are highest. (3) The mounting evidence on effects of oral contraceptive "pills" in some women.

To settle the question whether progesterone indeed has an important effect of the kind suggested would not be difficult. It requires carefully controlled double-blind studies of the effects of progesterone and oral progestational agents on various aspects of psychosexuality and actual sexual behavior in volunteer women. It should seem that the potential importance of this information should outweigh the objections to performing research of this kind.

THE ORGANIZATIONAL ROLE OF HORMONES AND ANOMALIES OF HUMAN SEXUAL DIFFERENTIATION

In the last section we moved within the general context of the activational role of gonadal hormones to a more detailed consideration of possible effects of the exogenous progestins present in oral contraceptives. We need now to regain our broader perspective, which centers on the question of the biological determinants of the human sexes and sexuality, and to examine the evidence that relates sex to the organizational effect of hormones. A general discussion of this topic, based on experimental treatment of nonhuman species, has already been provided in the chapter on sexual differentiation. Here we will consider the clinical research on the behavior and biology of human hermaphrodites, individuals whose anatomic status as male or female is singular or mixed according to one or more criteria (Money, 1961a, b, 1965; Money and Ehrhardt, 1972; Hampson and Hampson, 1961). Before proceeding, the reader may wish to review the earlier treatment and also the discussion of functional hermaphroditism in Chapter 2.

The Need for Multiple Definitions of Sex

This research into hermaphroditism and other anomalous conditions has revealed the need for a multidimensional definition of sex. Five separate biological categories have been distinguished: nuclear or chromosomal sex, gonadal sex, hormonal sex, sex of the internal accessory organs, and sex as determined by the appearance of the external genital organs. A sixth category is social-cultural in nature; it is the sex of assignment and rearing. The seventh and final category is a social-psychological one; it is the sex that an individual accepts for himself, the gender role that he finds suitable. This seventh category is a dependent variable that is determined by the variables in the other six categories. The goal of research has been to discover the relative importance and details of interaction of the first six categories in the determination of the seventh. We will now review briefly the five biological characteristics of sex.

Chromosomal Sex

This is sex as defined by the characteristics of the twenty-third chromosome pair. The presence of two X chromosomes defines a "genetic female"; if there is one X and one Y chromosome, the individual is a "genetic male." The most common test of chromosomal sex is based on the presence of "sex chromatin," a darkly staining mass of approximately 1 micron diameter attached to the inner nuclear membrane of the cells of females in many species (Barr, 1959). This material is indicative of the presence of paired X chromosomes, for it seldom appears in the cells of normal males. The presence or absence of sex chromatin in human cells can be determined routinely from a smear of oral mucosa. Other, technically more difficult, procedures have been developed for the counting and identification of the entire set of human chromosomes (Gowen, 1961).

Under normal circumstances the "sex" of the chromosomes agrees with all other sexual criteria. However, in several clinically recognized conditions there is incongruence between the results of the chromatin test and one or more of the other categories. In some of these conditions the sex chromosome complement is abnormally large (e.g., XXY, as in Klinefelter's syndrome) or small (e.g., a single X, as in Turner's syndrome). In these cases the definition of chromosomal sex becomes ambiguous.

Gonadal Sex

Under normal circumstances a male can be defined by the presence of testes and a female by the presence of ovaries. The first systematic classification of hermaphroditic conditions, made by Klebs in 1873, included as *true* hermaphrodites only those individuals who possessed both ovarian and testicular tissue (Money, 1961b). *Pseudo*hermaphrodites, on the other hand, were individuals with an external appearance at odds with their gonadal structure and/or internal accessories. Both male and female pseudohermaphrodites were identified. As Money (1961b) has pointed out, the discovery of sex chromatin has shown this three-way classification to be oversimplified. Moreover, as we pointed out in the section on hermaphroditism in Chapter 2, true functional hermaphroditism is not found in any mammalian species. All forms of human hermaphroditism are pathological.

Hormonal Sex

A distinction between hormonal sex and gonadal sex arises whenever the gonads produce insufficient levels of hormone. This occurs, for example, in genetic females suffering from congenital hyperadrenocorticism (abnormally large output of androgens and other steroids from the adrenal cortex). Although ovaries are anatomically present in these patients, their output of female gonadal hormones is suppressed and insufficient to counteract the progressive masculinizing effects of the adrenal androgens (Hampson and Hampson, 1961).

Sex of the Internal Accessory Organs

As we described in Chapter 7, internal sex accessory structure stems from the development of one of the two systems present at an early stage in the embryo: the Müllerian system, which, if fully developed, results in typical female structure; and the Wolffian system, from which typical masculine structure evolves. The presence of the testes is required for the development of masculine structure and the inhibition of feminine development. In the absence of testicular secretion, the Müllerian system continues to differentiate, and the Wolffian system degenerates. Normally, of course, the complete differentiation of one or the other system is congruent with the sex-chromatin pattern, gonadal and hormonal sex, and the appearance of the external genitalia. In some varieties of hermaphroditism, however, the differentiation of the internal system is incongruent with one or more of the other characteristics. For example, if the prenatal testicular function of a genetic male is abnormally low, Müllerian development may continue, and Wolffian development is more or less stunted.

Sex of External Genital Appearance

The appearance of the external genitalia is the primary factor in determining the sex to which an infant will be assigned by the hospital staff and subsequently by the parents. Normally this appearance is perfectly predictive of all the internal dimensions of sex. But cases do arise in which the external appearance is anomalous and classification therefore difficult. For example a genetically and gonadally female infant may be born with a hypertrophied clitoris and labia that have a definite scrotal appearance (Hampson and Hampson, 1961; see Figure 8.4). Without examination of internal characteristics, the correct assignment of sex is very difficult. At puberty, individuals of this type undergo feminine secondary sexual development. Similarly, there is a striking divergence between gonadal sex and sex of external appearance in the case of testicular feminization (see Figure 8.5). In this syndrome, an individual with male genotype possesses androgen-secreting testes, but the normal target tissues of the hormones are incapable of utilizing the hormone. At birth these individuals appear as normal females, but at puberty they do not menstruate or become fertile.

BEHAVIORAL SIGNIFICANCE OF HERMAPHRODITIC CONDITIONS

Given these five biological dimensions of sex, a hermaphrodite can be defined as an individual whose position along one or more dimensions is at variance with his position on the others (Money, 1961b). This means

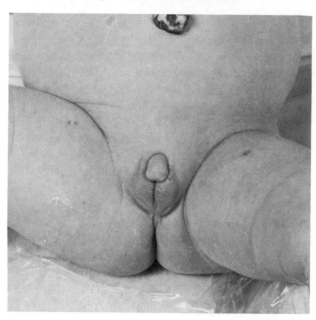

FIGURE 8.4 Female infant with masculinized external genitalia. The internal reproductive structures and ovaries are normal. From Hampson, J. L., and Hampson, J. G. (1961) The ontogenesis of sexual behavior in man. In Young, W. C. (ed.), *Sex and Internal Secretions*, vol. 2. Baltimore: Williams & Wilkins. Photograph courtesy of Dr. John Hampson.

that in all cases of hermaphroditism, the sex of assignment and rearing will be contradictory to the sex of at least one of the biological variables. Therefore, the stage is set for a potential conflict between the physical and social-cultural determinants of the individual's gender role. By examining the case histories of hermaphrodites with respect to the direction and success of their gender-role identification, the relative importance of these two general classes of gender-role determinants should be made clear.

The results of research reported by Hampson and Hampson (1961) were unequivocal in affirming the strength of social-cultural factors (sex of assignment and rearing) in the eventual determination of gender role. Individuals raised as boys came to see themselves as boys and to identify with masculine pursuits. Socially appropriate identifications were also made by individuals raised as girls. The acceptance of socially prescribed sex often occurred in spite of definitely anomalous or contradictory external genital appearance. However, the discrepancy between sex of assignment and external genital appearance led to substantial psychological distress that was not always successfully resolved. More than any other single biological dimension of sex, an external genital appearance at variance with the sex of rearing was likely to produce conflict in the hermaphrodite's gender-role adaptations.

How are these clinical results to be evaluated? First, it is possible that the various hermaphroditic conditions may obscure or prevent the expression of innate sex differences present in normal children. The plasticity of hermaphroditic children in accepting the sex of assignment as their gender role might be due in some part to the absence of normally present behavioral predispositions. An experimental approach to this problem, which for obvious reasons is unlikely to be pursued, would be to make random assignments of sex to a large number of physiologically normal infants. Systematic observations of these children over a number of years, in stable social environments, would reveal whether individuals raised under the label ("boy" or "girl") that was congruent with the label applied to their chromosomal, hormonal, and genital structures ("masculine" or "feminine") adapted to the role demands better than individuals raised under an incongruent label.

There is in fact some evidence that girls who were exposed during

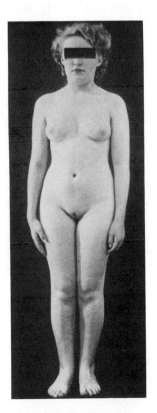
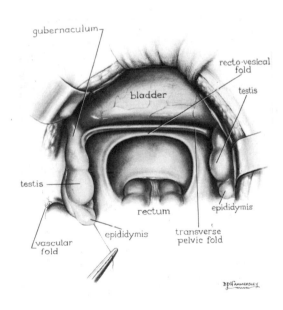

FIGURE 8.5 A patient, 17 years old, with "testicular feminization." Well-developed breasts and external feminine characteristics (in fact, won a beauty competition) but no pubic or auxiliary hair and infantile vulva. At laparotomy, no internal female genitalia, but undescended testes found. From Winton, F. R., and Bayliss, L. E. (1968) *Human Physiology*, 6th ed. London: Churchill Livingstone. Photograph courtesy of C. N. Armstrong.

gestation to relatively large amounts of androgenic activity showed a tendency to tomboyishness, as reflected in strong interests in outdoor physical-athletic activities and boyish clothes and toys (Money and Ehrhardt, 1968). This finding suggests the interpretation that the prenatal androgen acted on the developing nervous system so as partially to "masculinize" it. And indeed, given traditional definitions of presumably typical pursuits and interests of boys and girls, this interpretation seems reasonable. However, underlying the interpretation is the assumption that the reason boys tend to prefer "masculine" activities more than girls do is the natural constitutional difference between them. In other words, we have here the assumption that what we normally expect to see in our own cultural setting is a true reflection of innate biological processes.

This assumption is always questionable, as we pointed out when we introduced this section. Its utility is particularly suspect in our current society because of changing attitudes about adult gender roles. There can be no doubt that the position of women in American society is an issue of heightened public consciousness. Many social and legal practices that clearly discriminate between the sexes are being examined and, in some cases, abolished. Among young people there appears to be a lessening concern about highly sex-related styles in clothing, hair length, and so on (Winick, 1968). A large number of writers, who today are collectively described as representing the Women's Liberation Movement, have documented the arbitrary channeling of girls into particular forms of life (e.g., Firestone, 1970; Friedan, 1963; Greer, 1971; Herschberger, 1948; Morgan 1970). Much of this material is admittedly political and polemical rather than scientific, but that is not the important point. The important point is that American society is in flux regarding its modal judgments of what it will expect from boys and girls, men and women. The eventual outcome of these changes, should we again reach a time of stability, cannot be known today. But at the very least, we should refrain from using current social labeling rules when we attempt to relate biological sex differences to socially shaped behavior, simply because these labeling rules are changing.

THE PSYCHOANALYTIC APPROACH: THE CONCEPT OF INSTINCT (DRIVE), THE INCEST TABOO, AND THE OEDIPUS COMPLEX

When we consider some of the fundamental ideas about human sexual behavior that arose in the work of Freud, we are faced with kinds of theory and data, and relations between them, that are in some ways significantly different from those so far encountered in this book. To begin with, the scope of the theory, the extent of its intended generality, is far greater than that of any theory so far developed from other approaches. Freud did not conceive of sexuality as simply one aspect of human behavior and experience that could be treated more or less independently.

Rather, he placed the concept of sex, in his own terms of sexual instincts or drives, right at the core of human development. The unfolding expression of the personality was conceived as an interaction between these inborn sexual factors and the supports and demands of the environment. Many varieties of adult behavior, particularly those considered deviant or pathological, were explained by reference to events occurring earlier in life, wherein the normal processes of sexual development were thwarted or otherwise modified.

Second, Freud did not encapsulate his ideas into a single systematic theory. He wrote almost continuously on human sexual behavior and related topics for more than 40 years; his collected works fill more than 20 volumes. During his career Freud revised some of his most fundamental ideas several times. This means that the student who wishes to obtain a deep appreciation of the work must approach it with a firm historical perspective. In recent years attempts have been made to place Freud's major views into a systematic framework (Pumpian-Mindlin, 1952; Fletcher, 1957; Hartmann, 1959; Rapaport, 1959, 1960).

Finally, the information on which Freud based his conclusions was drawn originally from his private interviews with individuals with undiagnosed illness, his extensive analysis of his own personality, and his wide and deep appreciation of classic and contemporary literature. These sources are not typical of the kind of data on which other scientific theories are based. Originally, Freud developed his ideas as attempts to account for the illness of patients whose symptoms did not respond to more orthodox medical treatment. Only over time did psychoanalytic theory come to be seen as a general psychological theory. Calvin Hall (1954) put the matter simply: "The room in which he treated his patients became his laboratory, the couch his only piece of equipment, and the ramblings of his patients his scientific data. Add to these the restless, penetrating mind of Freud, and one has named all of the ingredients that went into the creation of a dynamic psychology."

Basic to Freud's theories of sexual development and behavior is the concept of sexual instincts, or drives.[4] Freud's first major presentation of his theory of instincts appeared in the 1915 paper "Instincts and Their Vicissitudes." We will first consider his treatment of the general nature of human instincts, and then have a closer look at his conception of the sexual instincts.

[4]Students familiar with the concept of instinct as used by ethologists (Hess, 1962; Lorenz, 1965; Tinbergen, 1951) and the concept of drive in American psychology may be puzzled by the apposition of these words in the context of Freudian theory. The problem is in the translation of Freud's German word *trieb*. In the standard edition of Freud's work in English, *trieb* has been translated as *instinct*. Fletcher (1957) and Jones (1955, vol. 3, p. 317) have argued that this is the proper translation. However, Rapaport (1959b) pointed out that in some ways *drive* catches Freud's meaning more clearly; he advocated the use of the term *instinctual drives* to avoid confusion among the ethological, psychoanalytic, and American behavioristic conceptions. In the following discussion we will develop Freud's ideas using the term *instinct*, and make occasional comparisons with ethological usage of this term.

According to Freud, a human instinct involves a relation between an internal, physiological event or state and its representation in the mind. In part, an instinct is a "stimulus to the mind." What distinguishes instinctual stimulation from other kinds of stimulation is that it comes from an internal source and that its termination is dependent on a particular resolution of the conditions that led to its occurrence. This resolution is the "satisfaction" of the "need" that the stimulation represents. Thus, Freud conceived of instincts as inborn generators of demands for satisfaction. Satisfaction is achieved simply by the reduction of stimulation to the lowest possible level. He described this process as the "purpose" of the nervous system. "The nervous system is an apparatus having the function of abolishing stimuli which reach it, or of reducing excitation to the lowest possible level: an apparatus which would even, if this were feasible, maintain itself in an altogether unstimulated condition.[5] However, he goes on to say, this quiescent condition is not permitted in the nervous system because of the barrage of stimulation emanating from the somatic sources of the instincts; just because these sources reside within him, the individual cannot escape from them. He must instead develop ways of coping, methods of manipulating his environment that promote maximal satisfaction of the instinctual demands without exceeding the limits on satisfaction imposed by parents, peers, and society in general. The inevitable tensions arising among instinctual demands, the "natural" tendency of the nervous system, and external restraints, produce the dynamic matrix in which human development occurs.

Thus every instinct can be considered as having four components: source, impetus or pressure, aim, and object. The source is the anatomical structure or physiological system from which the stimulation originally arises. Although Freud specified certain anatomical areas (oral, anal, genital) as being directly related to sexual instincts, he did not believe that detailed anatomical or physiological knowledge was required for the development of his theory.

Freud conceived that all instincts impelled the individual to some form of activity. The amount of "impulsion" present at any time he called the impetus (pressure) of the instinct. In some cases, for example hunger, the pressure may be felt consciously and identified correctly. But this is not always true. Much of the "work" of the instincts is done at an unconscious level and has its effect on consciousness in indirect and sometimes symbolic ways. It was in fact Freud's contention that much of the pressure arising from the sexual instincts is "rechanneled" (sublimated) to motivate various forms of highly socially approved behavior.

[5]The assertion that the primary function of the nervous system is the abolition of stimulation, which Freud adopted in part from the nineteenth-century psychophysicist Gustav Fechner, will undoubtedly seem naive to the student of modern biological psychology. However, the student should notice that this general idea, in various guises, has been present in much of the motivational theory developed in ethology and behavioristic psychology. Consider, for example, the build-up and discharge of action-specific energy, and the concepts of need-reduction, drive-reduction, and drive stimulus-reduction. See Fletcher (1956) and Hinde (1960) for analyses of the issues involved.

We believe that civilization has been built up, under the pressure of the struggle for existence, by sacrifices in gratification of the primitive impulses, and that it is to a great extent forever being recreated, as each individual, successively joining the community, repeats the sacrifice of his instinctive pleasures for the common good. The sexual are amongst the most important of the instinctive forces thus utilized: they are in this way sublimated, that is to say, their energy is turned aside from its sexual goal and directed towards other ends, no longer sexual and socially more valuable. . . . These sexual impulses have contributed invaluably to the highest cultural, artistic, and social achievements of the human mind. [Freud, 1924]

The aim of an instinct is the "satisfaction" produced by reducing accumulated pressure. Satisfaction is gained through direct action in the environment; for example, hunger is satisfied by eating. But pressure reduction may also be achieved in the purely internal activities of dreams, fantasies, and delusions. Freud termed this form of activity *wish fulfillment*. The general principle that accumulated pressure is discharged he called *the pleasure principle* (Rapaport, 1960).

The object of an instinct is any object that is effective in producing the required satisfaction. In Freud's view, the object is an instinct's most variable property. It may be an inanimate object, a part of one's own body, or another person. During an individual's lifetime, the object associated with an instinct will change several times, and different people, particularly in adulthood, will not necessarily respond to the same objects. This does not mean that there can be no way of specifying the *typical* objects associated with achieving the aim of instincts. For example, in the case of the normally functioning genital stage of the sexual instinct, the typical object may be specified as a member of the opposite sex. However, several factors may enter into an individual's development in ways that force the utilization of an atypical object. Psychological trauma may lead to the substitution of one object for another. One example is sexual fetishes. And, as we have seen, homosexual behavior may be explained in similar psychoanalytic terms.

The undetermined nature of the objects of instincts is at once a major strength and a major weakness of Freudian theory. The strength lies in the theory's ability to allow for the development of the tremendous variety of human behavior, particularly in regard to the objects used for sexual gratification. The clinical psychiatric literature has made hundreds of references to the diverse and bizarre practices that some people engage in for the achievement of orgasm (see, for example, Krafft-Ebing, 1892). Any theory that attempts to account for this variability must, at the very least, allow some behavioral plasticity (conditioning or learning) to be present as a factor in the determination of the object used in obtaining satisfaction.[6] As Fletcher (1957) pointed out, Freud's analysis of

[6]The flexibility in the object of an instinct in Freudian theory is in strong contrast to the fixity conceived for the analagous concept in ethological theory. Although the parallel is not exact, "object" in psychoanalysis corresponds to "sign-stimulus" or "releaser" in ethology. However, the sign-stimulus that releases a particular consummatory response is the same for all members of the species, or more precisely, all

the possible rechannelings and transformations of instinctual pressures and objects provides a sort of "learning theory" to meet this requirement.

There is a logical weakness inherent in allowing theoretically for the total flexibility between the aim of an instinct and its objects: it permits the theory to explain everything, and therefore to be in danger of explaining nothing. Almost all of the data that might be used to test the validity of the psychoanalytic concepts "begin at the end," that is, consist of the case histories of individuals undergoing psychoanalytic therapy.[7] Presumably these case histories provide the information that shows how the aims of particular instincts are being achieved through particular objects. But what objects are "attached" to which instincts? There are no testable rules for making the proper assignments. Therefore, although the explanatory psychoanalytic accounts of rechannellings of instinctual gratification through diverse objects often have dramatic appeal and a certain surface validity, they may be as much a product of the analyst's imagination as they are of the patient's condition. Consider the following brief case history as an example.

[A woman] suffered from an imperative craving to eat bread. When a little girl, she, like many children, gained the impression that the fertilising substance entered the body by way of the mouth, that pregnancy was brought about by the mother's swallowing something—a seed or some medicine. Bread is very familiarly the "Staff of Life" and hence could be a symbol of the penis. Her compulsion came on when, after a not very happy wedded life, her husband died, leaving her childless. The unhappiness of her married life had caused the patient to resolve not to make a second venture. But her natural cravings for sex satisfaction and for motherhood, thus denied their natural outlet, were displaced, to find expression in a substitute action. The symbolism of the substitute action was, as is easily seen, compounded partly from infantile and partly from adult conceptions of the initial act of reproduction. [Frink, 1918, pp. 170–171]

THE SEXUAL INSTINCT[8]

Basic to Freud's conception of human sexual behavior is the idea of *infantile sexuality*. We might be accustomed to assume that human sexual urges or desires originate with the onset of puberty and therefore are related to gonadal maturation and the secretion of sex hormones. But Freud insisted that this view was totally incorrect. He maintained that the sexual life of an individual begins essentially at birth, and that sex-

appropriate members of the species, for example, adult males during the breeding season. That the stimulus may be specified as applying to the species, and not just to a single individual, gives the ethological theory *precision* and verifiability lacking in psychoanalysis. However, ethological theory is derived from the study of birds, fish, and the lower mammals, while the psychoanalytic theory of instinct deals only with human behavior. There is no good reason at present to believe that the details of the one theory should apply to the data on which the other theory is based.

[7]See Cofer and Apley (1964, Chapter 12) for a recent review of the types of evidence relating to psychoanalysis.

[8]Unless otherwise noted, the quotations in this section are from Freud's major exposition of the sexual instinct, *Three Essays on the Theory of Sexuality* (1905).

related events in infancy and childhood are crucial determiners of adult sexual practices. The occurrence of "normal" sexual behavior (heterosexual copulation) in adulthood was not to be considered as the natural result of an innate tendency, but rather as the typical outcome of a series of dynamic sexual interactions beginning with contact at the mother's breast. "A thorough study of the sexual manifestations of childhood would probably reveal the essential characters of the sexual instinct and would show us the course of its development and the way in which it is put together from various sources." Through the data collected in psychoanalytic interviews, Freud attempted to describe and account for these sexual manifestations. What are the manifestations, and how did Freud deal with them in terms of instinctual aims, objects, pressures, and sources?

A sucking response is one of the earliest behaviors a newborn child exhibits. The response is effective in obtaining nourishment. But babies also suck on objects that do not provide nourishment; most obviously, they suck their thumbs. Freud argued that sucking is the earliest infantile sexual response. The aim of the sexual instinct in this case is the satisfaction produced by stimulation of the lips and mouth; it is related to the incorporation of the object being sucked. Freud did not contend, however, that the satisfaction gained from this activity is sexual at the time of its first appearance. "To begin with, sexual activity attaches itself to functions serving the purpose of self-preservation (in this case, feeding) and does not become independent of them until later." Originally, hunger provides the pressure that produces the response. Sucking and the ingestion of milk satisfies hunger; and satisfaction produced by hunger reduction creates a need for the repetition of sucking. This need [the pressure of the instinct] is felt as a "peculiar . . . tension, possessing rather the character of unpleasure, and by a sensation of itching or stimulation which is centrally conditioned and projected onto the [lips and mouth]." The pressure is not directly related to hunger. It is the pressure of the sexual instinct as manifested in the mouth and lips. This area is thus an *erotogenic zone*. Stimulation of the zone in a rhythmic manner, as occurs in sucking, produces satisfaction by removing the stimulation, felt peripherally, that is the instinctual pressure. Normally, predominantly oral sexuality is a characteristic of infancy only; Freud believed that individuals who carry their orality into adulthood are likely to become "epicures in kissing, will be inclined to perverse kissing, or, if males, will have a powerful motive for drinking and smoking."

Freud also considered the anus to be an infantile erotogenic zone. Anal sexual satisfaction is achieved by the sensation accompanying the muscular contractions involved in retaining feces and the final movement of the feces over the anal mucous membranes. Freud maintained that infants retain feces intentionally; the accumulated material becomes a "masturbatory stimulus" in the anus. But voluntary bowel control also serves another, social function. By gaining and utilizing this control, the infant may obey or disobey the commands of the adults who are interested in his toilet training. The feces become in Freud's word, the infant's first "gift," which he may bestow on or withhold from his environment. A particularly strong anal-erotic disposition in childhood, Freud believed, could lead, through the process of sublimation, to an adult personality characterized by high degrees of orderliness, thrift, and obstinacy (Freud, 1908).

Oral and anal sexuality are the initial stages of sexual development. In the normal individual their manifestations are neither obvious nor prolonged; they have disappeared by approximately five years of age. They are both *autoerotic* stages of the sexual instinct because their objects are portions of one's own body. Moreover, for an obvious reason, Freud called them *pregenital* stages of sexual development. Presumably, therefore, there is little if any difference between boys and girls during these periods of sexual development. They become important in the adult only when some accident in the course of normal development causes their repressed or sublimated manifestation to predominate in the determination of adult behavior and attitudes.

The initial occurrences of pregenital sexuality are supported by their relation to nonsexual functions: feeding and defecating. This is also true in the beginnings of genital sexuality. The earliest stimulation results in part from the attention given to the infant's genitals by the parents during bathing and diaper changing. These forms of stimulation "make it inevitable that the pleasurable feeling which this part of the body is capable of producing should be noticed by children even during their earliest infancy, and should give rise to a need for its repetition." Following this very early introduction to genital stimulation there occurs what Freud called the first phase of infantile masturbation, which he assumed was of very short duration. Second and third stages of masturbation were presumed to develop before approximately the age of five.

The major occurrences that Freud wanted to account for in this part of his theory were the eventual predominance of the genitals in the expression of sexuality and the choice of the appropriate sexual object in adulthood. These were difficult problems to solve because, in Freud's view, sexual development enters a state of dormancy (the *latency period*) from about age five until puberty. Yet the sexual manifestations at puberty appear to be, typically, genital and heterosexual. What events occur during the latest stages of infantile sexual development to set the stage for the development of typical adolescent sexual behavior?

In 1923, 18 years after the initial publication of his concepts of infantile sexuality, Freud supplemented the theory in an attempt to account for the predominantly genital sexuality observed in puberty. In his earlier work he had argued that the primacy of genital function does not occur until puberty; throughout the latency period the components of the sexual instinct (oral, anal, genital) remain imperfectly coalesced. But in 1923 he stated that there is a final phase of infantile development, the *phallic phase*, in which a sort of genital predominance is attained. The distinction between phallic sexuality and mature genital sexuality is seen most clearly by considering the characteristics of the latter phase first.

The adult man or woman understands and is comfortable with the reproductive and pleasure functions of his genitalia. The woman is presumed to achieve sexual satisfaction via vaginal stimulation.[9] Most important, individuals of either sex understand the genitalia of the other to be natural anatomic structures well suited to their roles in reproduction. Concepts such as feminine "envy" for the masculine genitalia or masculine "fear" of being reduced to a condition of female genital appearance are inappropriate in describing the attitudes of the nonneurotic adult.

But it is just these concepts that Freud applied to the emotions of boys and girls during the phallic phase of development. During this phase children of both sexes believe that there is but one genital organ: the penis. Moreover, when a boy discovers that some people, girls, do not have a penis, he will at first deny its absence or imagine that it is in fact there but has not grown yet. Eventually, the child forms a more complicated and dramatic hypothesis: any person lacking a penis has had that organ cut off as punishment for misbehavior. From this premise the child concludes that if he were to behave very badly he, too, would have his penis removed. For little boys, the fear of losing the penis (known typically as the *castration complex*, which is a misnomer) plays an important role in the resolution of the Oedipus complex. Little girls, on the other hand, upon discovering that a male possesses a penis, become envious of the organ and wish that they themselves possessed one. "Penis envy" during the girl's phallic phase is also of critical importance in the theory of the Oedipus complex. One function of the phallic phase is thus to produce one form of genital primacy in both sexes prior to the onset of the latency period during the sixth year. We turn now to our other problem: the choice of a heterosexual object at puberty. In Freud's view, this choice is understood by reference to the final event in the period of infantile sexuality: the successful resolution of the Oedipus complex. What is this "complex," and how is it resolved?

Freud made his initial major exposition of the Oedipus complex in the first edition of *The Interpretation of Dreams* (1900). His clinical experience had convinced him that unresolved feelings of erotic love for one parent and hate for the other are at the root of adult neurosis. But it is the manner of resolution of the feelings, and not the feelings themselves, that produces later difficulties. The feelings are experienced by most children, but in the neurotic individual they are greatly magnified.

Freud believed that the intense appeal of the Oedipus legend, dramatized in Sophocles' *Oedipus Rex*, results from the genuine psychological universality of its theme: a series of fateful circumstances causes Oedipus

---

[9]Recall the results of Masters and Johnson described previously.

to murder his father and marry his mother, Jocasta. Inevitably he discovers what he has done, and in his shocked and guilt-ridden state he stabs himself blind with the brooches from the dead Jocasta's robe, and leaves his kingdom for exile. Freud says of Oedipus: "His fate moves us only because it might have been our own, because of the oracle laid upon us before our birth the very curse which rested upon him. It may be that we were all destined to direct our first sexual impulses toward our mothers, and our first impulses of hatred and violence toward our fathers; our dreams convince us that we were."[10]

It would be impossible to overemphasize the importance of the Oedipus complex as a central concept in psychoanalytic theory. In the 1920 revision of *Three Essays on the Theory of Sexuality* Freud claimed the complex arises in everyone, with no exceptions, and that failure to master its tensions inevitably results in neurosis. He devoted several essays to the description of the ways in which this mastery is achieved (Freud, 1924, 1925, 1931). For both boys and girls the process of resolution occurs during the phallic phase of sexual development. The details of the process are different for the two sexes. We will consider the masculine situation first.

A little boy's erotic attachment to his mother develops naturally out of his intimate contact with her during the first two stages of infantile sexual development. He comes eventually to desire her exclusive attentions, and he feels that his father is an obstacle in the way of achieving his goal. However, with the onset of the phallic stage, the boy is faced with a profound threat: if he continues to compete with his father for the attention of his mother, his father may very well defeat him by castrating him. The boy recoils from the threat and represses his incestuous wish. The act of repression, if completely successful, is equivalent to the destruction of the Oedipus complex. It also is the culmination of infantile sexuality, and marks the beginning of the latency period. During this period the direct sexual demands of infancy are transformed into the desire to give and receive affection. A successful adjustment at puberty requires the integration of the "affectionate current" of the latency period with the newly developed sexual capacities.

The development of the Oedipus complex in girls is more complicated, and its resolution less distinct. To begin with, girls are necessarily as attached to their mothers during infancy as boys are. How do they become attached strongly to their fathers and hostile to their mothers? Freud (1931) described several possible motives for resentment of the mother, including the girl's opinion that her mother has withheld a penis from her. During the phallic stage the girl may feel cheated because she lacks a penis, and holding her mother responsible, withdraw from her

[10]Freud's use of "the oracle" in the first sentence is obviously not to be taken as a literal expression of his view of causation. The Oracle at Delphi is an important element of Sophocles' play. However, the quotation does illustrate Freud's easy transitions between "literary" and "scientific" styles in presenting his ideas.

and turn to her father.The transition is accomplished by way of a symbolic analogy or equation between penis and baby. The girl substitutes the desire to have a baby for her desire to have a penis. In this sense she wants to assume her mother's role with her father. But her wish remains unfulfilled, and the sexual attachment to her father wanes gradually. During the latency period the direct sexual manifestations of the Oedipal stage become blunted, as it were, into feelings of tenderness.

Thus for the girl, the effect of the castration complex is to *lead* to the Oedipal attachment, whereas for the boy the effect is to *end* it. Resolution of the conflict is less crucial and traumatic for the girl. Freud (1931) believed that this was a fundamentally important determinant of later differences in behavior: "We should probably not err in saying that it is this difference in the interrelation of the Oedipus and the castration-complexes which gives its special stamp to the character of woman as a member of society."

MALINOWSKI ON THE OEDIPUS COMPLEX

We will complete this brief discussion of psychoanalytic theory by presenting the views of Malinowski on the cross-cultural generality of Freud's formulations of infantile sexuality and the Oedipus complex. Unlike some anthropologists of Freud's time and since, Malinowski was not hostile to psychoanalytic theory. He believed Freud had done a great service by insisting that human behavior should be studied "without irrelevant trappings, and even without the fig leaf." However, he was concerned that Freud had paid inadequate attention to the importance of culture as a determinant of sexual expression. Freud maintained that the Oedipal conflict, just as he had described it, was a universal human occurrence. But Malinowski contended that, to the extent the theory was valid at all, it was a product of the patriarchal family structure found in Europe; he evidenced his argument by comparing the child-rearing practices and family structure of Europe with those found among the Trobriand Islanders.

For the purposes of his description, Malinowski (1927) divided the years from birth to adulthood into four periods: infancy, babyhood, childhood, and adolescence. Infancy ended when the child was weaned; in Western society weaning occurred at approximately age one, whereas among the Trobrianders it might not be enforced until the child was three. Babyhood referred to the period from weaning until about age six. Thus, the end of babyhood was approximately equivalent to the end of the Freudian period of infantile sexuality. Childhood lasted until puberty, and therefore corresponded chronologically with the Freudian latency period. And adolescence ended when the individual achieved "full social maturity." In Western society this typically corresponded to the beginning of higher education; in more primitive societies adolescence was considered ended by the act of marriage.

In comparing the treatment of infant children in European and Trobriand societies, Malinowski noted that the major difference was in the role played by the father. In the Europe that Malinowski described, the father was usually a secondary, albeit helpful, agent in caring for the infant's needs.[11] The Trobriand father, on the other hand, shared fully in caring for the newborn. His duties included cleaning, feeding, and carrying the infant. He was, in Malinowski's words, a hard-working and conscientious nurse. Yet, in contrast to the European father, he took no pride in the child as a biological extension of himself. Being ignorant of physiological paternity, he accepted his nursemaid's role as part of his obligation as "mother's husband."

By the time a Trobriand baby was fully weaned he was already able to eat a regular diet, and had achieved substantially more independence from his mother than his European counterpart. The mother's breast was not denied the Trobriand baby until he had lost his need for it; no trauma was produced by the eventual separation. In Western society, by contrast, weaning often took place as soon as possible, primarily to suit maternal convenience.

The Trobriand mother's attitude toward weaning was paralleled by her indulgent treatment of the baby in other ways. She was not responsible for "training" her baby, in the sense that that term was understood in Europe. Moreover, because the Trobriand father had no legal or social status as the head of the household, he was never in a position to deal severely with his children. Throughout the period of babyhood, he continued in his role as an affectionate nurse. In his own and his society's view, he did not deserve automatically the reciprocal affection of the baby—he had to earn it.

During the period of childhood, children in both cultures began to emerge from the shelter of the nuclear family. In Western society the transition was abrupt; affluent children started school, and working-class children were apprenticed into a trade. The child had no choice in this, for all arrangements were made by his parents. Among the Trobrianders, this was the time when the child joined one of the autonomous play groups that roamed the village. The child's entrance into the groups was as gradual as he cared to make it. His parents neither coaxed him to join nor dissuaded him from doing so. During the child's transition, his father remained warm and supporting, acting as protector, companion, and instructor in basic skills and crafts.

It was also during the early part of childhood that a child became fully aware of the importance of the *kada,* his maternal uncle. The uncle instructed the child in the history and tradition of his clan (his mother's blood relatives) and in the laws and prohibitions of the tribe. He encouraged his nephew's ambitions for material success and status; part of a

---

[11]Fifty years have elapsed since Malinowski made these comparisons. Although some of his descriptions of Western patterns of child rearing are perhaps still valid today, others probably are not. In addition, it is highly likely that there have been substantial changes in Trobriand culture during the past half-century. Therefore, we use the past tense in our descriptions of both cultures.

boy's time was spent working on the land that belonged partly to him and his uncle by virtue of their mutual clan membership. Thus, although the child naturally acquired some knowledge of moral standards and traditions from his parents, it was the *kada* who made the rules explicit and had the authority to enforce them (the reader will not be incorrect if he thinks of the *kada* as the Trobriand child's "father figure").

How did Malinowski use his parallel descriptions of European and Trobriand practices in treating the topic of infantile sexuality? To begin with, he disputed Freud's emphasis on the virtually universal existence and importance of anal erotism. He maintained that a baby's intense interest in its excretory function was produced by the severe toilet-training procedures employed by upper-class European families, and by the accompanying attitude that excrement and activities related to it were "dirty" or "indecent." Because Freud's neurotic patients were primarily products of wealthy families, it was not surprising that their analyses often revealed a causal relation between their early experiences with excretory control and their adult behavior and attitudes. But neither the European peasantry nor the Trobrianders made such a large issue of early toilet-training, and for the latter the notions of decency or indecency simply did not apply to a child's bowel movements. The result of this, Malinowski maintained, was a total lack of any stage of development that could be called anal-erotic.

Malinowski aimed his second criticism at the concept of the latency period. His argument followed the same pattern: it was primarily the upper classes that suppressed genital activity and interest in four- and five-year-old children. These children normally had no opportunity to observe copulation, and they were likely to be reproved for inquiring about it and punished for experimenting with it. Prohibition on sexual discussion or expression remained until middle or late adolescence. Quite naturally, therefore, these children typically exhibited little in the way of genital sexuality from the time it became prohibited.

Again, what was true for the group from which Freud drew his patients was less so for other strata of European society, and was clearly false for the Trobrianders. Children of the European working class, living in crowded conditions with their parents and other adults, had the opportunity to hear about and observe the varieties of adult sexual behavior. Their vocabularies were early enriched with scatalogical and sexual slang. The children of farmers and peasants had an additional avenue of sexual enlightenment: they observed and discussed the matings and births of their domestic animals. For these European children, as well as for the Trobriand children there was no sexual latency period. Their knowledge and expression of genital sexuality started early and continued steadily throughout childhood and into adolescence.

Malinowski's third and major contention was that Freud had incorrectly emphasized the universality of the Oedipus complex. The student will probably be able to surmise the crux of the argument. The Oedipal conflict was produced in European children because of the early suppression of all sensual contact with the mother, and the always stern, sometimes brutal authority of the father. These were relatively invariant attributes of European families, to be found among the peasants, the working class, and the wealthy. However, as we have seen, neither of these attributes pertained to the relations between a Trobriand child and its parents. And Malinowski could find no evidence, whether through in-depth conversations with the natives, including discussions of the content of their dreams, or in an analysis of their myths and lore, to indicate that the Oedipal conflict existed for them.[12] Freud's contention that it was a universal human occurrence was incorrect.

It is important to notice that Malinowski did not attempt to dispose entirely of the Oedipus complex; he simply placed it in a cross-cultural perspective. Having done this, he went on to show that male Trobrianders possessed an analogous "system of sentiments," which he originally labeled the *family* or *nuclear complex*.[13] Unlike the Oedipus complex, the nuclear complex was not developed fully until puberty. The reason for this can be understood by reviewing the salient events of the childhood years.

Our earlier discussion of Trobriand sex life emphasized both the general freedom granted to erotic play and the strict prohibition on virtually any form of contact between a boy and his sister. We have also seen that the maternal uncle was directly responsible for training his nephew in the ways of the clan and tribe. Throughout childhood and into puberty the boy became progressively more involved with the *kada*, until he came to see himself as his uncle's successor. He became resentful of the uncle's demands and jealous of his possessions and authority. At the same time the break between himself and his sister was maintained strictly; even to be seen with her in public would have been a disgrace. Any feelings of affection or tenderness for her had always been immediately repressed. She was an "indecent" object, never to be thought about in sexual terms. And yet the boy could not forget her, because as her brother he would eventually become responsible for a large measure of her material welfare. He would also be the *kada* of her children. He would become for her what his *kada* was for his mother.

The "artificial and premature" repression of affection for the sister, coupled with the resentment of the *kada*, formed the basis of the nuclear complex. Malinowski summarized the matter this way: "Applying to each society a terse, though somewhat crude formula, we might say that in the

[12]For a reason that will soon become clear, Malinowski particularly emphasized the lack of Oedipal conflict in Trobriand boys. The girl's attitudes toward her parents was in all major respects not unlike that of a European. Recall that Freud believed the Oedipus complex to be a relatively minor feature of normal female sexual development.

[13]In a later paper (1927, p. 153) Malinowski described the phrase *nuclear family complex* as a generic term to cover the relevant circumstances in any culture. In the same paragraph, however, he argued that the term *complex* should be discarded altogether because of its misleading surplus meaning.

Oedipus complex there is the repressed desire to kill the father and marry the mother, while in the matrilineal society of the Trobriands the wish is to marry the sister and to kill the maternal uncle" (1927). Malinowski documented his assertion with references to the myths, lore, dreams, and day-to-day conduct of the islanders.

CONCLUSIONS

Psychodynamic theory, particularly as elaborated by Freud and his immediate followers, has been the major force in twentieth-century psychology. So much praise, paraphrase, condemnation, and plain nonsense have been written about psychoanalysis that the student is likely to get lost in the commentary without ever discovering its source. The reason for the enormous critical literature on Freud's life and thought is that the ideas he elaborated spoke and continue to speak to a number of fundamental human problems. As Ellenberger (1970) emphasized, many ideas currently labeled as Freudian have their origins within the development of dynamic psychiatry during the nineteenth century. Freud's insistence on the fundamental importance of sexuality in the determination of behavior and psychological illness has been questioned and denied many times. In fact it was partly a disagreement on this issue that led Jung and Adler to separate themselves from Freud during the early part of the second decade of this century. Moreover, as we have seen, the cross-cultural perspective forces us to limit the generality of at least some basic psychoanalytic ideas about human sexuality. Nevertheless, Freud remains without peer as a bold originator and systematizer of the foundations and vicissitudes of human sexual behavior. The currently entrenched position of psychoanalytic therapy as a medical subspecialty, which has created an aura of scientific exactitude around psychoanalytic theory, should not cause us to forget that one of Freud's major objectives was to bring the insights of literature and poetry into closer harmony with psychiatric practice. Our purpose here has been neither to attack nor defend the psychoanalytic theory of sexual behavior, but to present the basic material and to suggest the reader investigate further and come to his own conclusions.

References

Adams-Smith, W. N. (1967) The ovary and sexual maturation of the brain. *Journal of Embryology and Experimental Morphology* 17:1–10.

Adler, N. (1969) Effects of the male's copulatory behavior on successful pregnancy of the female rat. *Journal of Comparative Physiology and Psychology* 69:613–622.

Adler, N., and Bermant, G. (1966) Sexual behavior of male rats: Effects of reduced sensory feedback. *Journal of Comparative and Physiological Psychology* 61:240–243.

Adler, N., and Zoloth, S. (1970) Copulatory behavior can inhibit pregnancy in female rats. *Science* 168:1480–1482.

Alleva, J., Waleski, M., Alleva, R., and Umberger, E. (1968) Synchronizing effect of photoperiodicity on ovulation in hamsters. *Endocrinology* 82:1227–1235.

Amoroso, E., and Marshall, A. (1960) External factors in sexual periodicity. In A. Parkes (ed.), *Marshall's Physiology of Reproduction*, vol. I, part 2, pp. 707–831. London: Longmans.

Anand, B., Chhina, G., and Singh, B. (1962) Effect of glucose on the activity of hypothalamic "feeding centers." *Science* 138:597–598.

Anderson, C. O., Zarrow, M. X., and Denenberg, V. H. (1970) Maternal behavior in the rabbit: Effects of androgen treatment during gestation upon the nest building behavior of the mother and her offspring. *Hormones and Behavior* 1:337–346.

Ardrey, R. (1966) *The Territorial Imperative*. New York: Atheneum.

Armstrong, C., and Marshall, A. (eds.). (1964) *Intersexuality in Vertebrates Including Man*. New York: Academic.

Aron, C., Asch, G., and Roos, J. (1966) Triggering of ovulation by coitus in the rat. *International Review of Cytology* 201:139–172.

Aron, C., Roos, J., and Asch, G. (1968) New facts concerning the afferent stimuli that trigger ovulation by coitus in the rat. *Neuroendocrinology* 3:47–54.

Aronson, L. R. (1959) Hormones and reproductive behavior: Some phylogenetic considerations. In A. Gorbman (ed.), *Comparative Endocrinology*, pp. 98–120. New York: Wiley.

Aronson, L. R. (1965) Environmental stimuli altering the physiological condition of the individual among lower vertebrates. In F. Beach (ed.), *Sex and Behavior*, pp. 290–318. New York: Wiley.

Aronson, L. R. and Cooper, M. (1968) Desensitization of the glans penis and sexual behavior in cats. In M. Diamond (ed.), *Perspectives in Reproduction and Sexual Behavior*, pp. 51–82. Bloomington: University of Indiana Press.

Aronson, L. R. and Cooper, M. (1969) Mating behavior in sexually inexperienced cats after desensitization of the glans penis. *Animal Behavior* 17:208–212.

Arvidsson, T., and Larsson, K. (1967) Seminal discharge and mating behavior in the male rat. *Physiology and Behavior* 2:341–343.

Asdell, S. (1964) *Patterns of Mammalian Reproduction*, 2d ed. Ithaca: Comstock.

Atz, J. (1964) Intersexuality in fishes. In C. Armstrong and A. Marshall (eds.), *Intersexuality in Vertebrates Including Man*, pp. 145–232. New York: Academic.

Baldwin, J. (1960) *Another Country*. New York: Dial.

Ball, J. A. (1934) Demonstration of quantitative relations between stimulus and response in pseudopregnancy in the rat. *American Journal of Physiology* 107:698–703.

Ball, J. A. (1941) Effect of progesterone upon sexual excitability in the female monkey. *Psychological Bulletin* (abstract) 38:533.

Bard, P. (1940) The hypothalamus and sexual behavior. *Research Publication of the Association for Research in Nervous Mental Diseases* 20:551–579.

Barfield, R. (1969) Activation of copulatory behavior by androgen implanted into preoptic area of male fowl. *Hormones and Behavior* 1: 37–52.

Barfield, R., and Sachs, B. (1968) Sexual behavior: Stimulation by painful electrical shock to skin in male rats. *Science* 161:392–395.

Barr, M. (1959) Sex chromatin and phenotype in man. *Science* 130:679–685.

Barraclough, C. A. (1966) Hormones and development. Modifications in the CNS regulation of reproduction after exposure of prepuberal rats to steroid hormones. *Recent Progress in Hormone Research* 22:503–539.

Barraclough, C. A., and Gorski, R. A. (1962) Studies on mating behavior in the androgen-sterilized rat. *Journal Endocrinology* 25:175–182.

Barrington, E. (1971) Evolution of hormones. In E. Schoffeniels (ed.), *Biochemical Evolution and the Origin of Life*, pp. 174–190. Amsterdam: North-Holland.

Bartke, A., and Wolff, G. (1966) Influence of the lethal yellow (A) gene on estrous synchrony in mice. *Science* 153:79–80.

Bastock, M. (1956) A gene mutation which changes a behavior pattern. *Evolution* 10:421–439.

Bastock, M. (1967) *Courtship*. Chicago: Aldine.

Beach, F. (1938) Sex reversals in the mating pattern of the rat. *Journal of Genetic Psychology* 53:329.

Beach, F. (1940) Effects of cortical lesions upon the copulatory behavior of male rats. *Journal of Comparative Psychology* 29:193–244.

Beach, F. (1941) Female mating behavior shown by male rats after administration of testosterone propionate. *Endocrinology* 29:409–412.

Beach, F. (1942a) Execution of the complete masculine copulatory pattern by sexually receptive female rats. *Journal of Genetic Psychology* 60:137–142.

Beach, F. (1942b) Comparison of copulatory behavior of male rats raised in isolation, cohabitation, and segregation. *Journal of Genetic Psychology* 60:121–136.

Beach, F. (1944) Effects of injury to the cerebral cortex upon sexually receptive behavior in the female rat. *Psychosomatic Medicine* 6:40–55.

Beach, F. (1945a) Bisexual mating behavior in the male rat: Effects of castration and hormone administration. *Physiology* 18:390–402.

Beach, F. (1945b) Hormonal induction of mating responses in a rat with congenital absence of gonadal tissue. *Anatomical Record* 92:289–292.

Beach, F. (1947) A review of physiological and psychological studies of sexual behavior in mammals. *Physiological Review* 27:240–307.

Beach, F. (1956) Characteristics of masculine "sex drive." In M. Jones (ed.), *Nebraska Symposium on Motivation*: 1956, 1–31. Lincoln: University of Nebraska Press.

Beach F. (1958) Normal sexual behavior in male rats isolated at fourteen days of age. *Journal of Comparative and Physiological Psychology* 51:37–38.

Beach, F. (1967) Cerebral and hormonal control of reflexive mechanisms involved in copulatory behavior. *Physiological Review* 47:289–316.

Beach, F. (1968a) Coital behavior in dogs. III. Effects of early isolation on mating in males. *Behavior* 30:218–238.

Beach, F. (1968b) Factors involved in the control of mounting behavior by female mammals. In M. Diamond (ed.), *Reproduction and Sexual Behavior*, pp. 83–131. Bloomington: University of Indiana Press.

Beach, F. (1970) Coital behavior in dogs. VI. Long-term effects of castration upon mating in the male. *Journal of Comparative and Physiological Psychology* 70:1–32.

Beach, F. (1971) Hormonal factors affecting the differentiation, development and display of copulatory behavior in the ramstergig and related species. In E. Tobach (ed.), *Biopsychology of Development*, pp. 249–296. New York: Academic.

Beach, F., Conovitz, M., Steinberg, F., and Goldstein, A. (1956) Experimental inhibition and restoration of mating behavior in male rats. *Journal of Genetic Psychology*, 89:165–181.

Beach, F., and Inman, N. (1965) Effects of castration and androgen replacement on mating in male quail. *Proceeding of the National Academy of Science* 54:1426–1431.

Beach, F., and Jordan, T. (1956) Sexual exhaustion and recovery in the male rat. *Quarterly Journal of Experimental Psychology* 8:121–133.

Beach, F., and Kuehn, R. (1970) Coital behavior in dogs: X. Effects of androgenic stimulation during development on feminine mating responses in females and males. *Hormones and Behavior* 1:347–367.

Beach, F., and Levinson, G. (1950) Effects of androgen on the glans penis and mating behavior of castrated male rats. *Journal of Experimental Zoology* 114:159–170.

Beach, F., and Merari, A. (1968) Coital behavior in dogs: IV. Effects of progesterone in the bitch. *Proceedings of the National Academy of Science* 61:442–446.

Beach, F., and Merari, A. (1970) Coital behavior in dogs: V. Effects of estrogen and progesterone on mating and other forms of social behavior in the bitch. *Journal of Comparative and Physiological Psychology Monograph* 70:(1), Part 2, 1–22.

Beach, F., Noble, R. G., and Orndoff, R. K. (1969) Effects of perinatal androgen treatment on responses of male rats to gonadal hormones in adulthood. *Journal of Comparative and Physiological Psychology* 68: 490–497.

Beach, F., and Rasquin, P. L. (1942) Masculine copulatory behavior in intact and castrated female rats. *Endocrinology* 31:393–409.

Beach, F., Rogers, C. M., and LeBoeuf, B. J. (1968) Coital behavior in dogs: II. Effects of estrogen on mounting by females. *Journal of Comparative Physiological Psychology* 66:296–307.

Beach, F., Westbrook, W. H., and Clemens, L. G. (1966) Comparisons of the ejaculatory responses in men and animals. *Psychosomatic Medicine* 28:749–763.

Beach, F., and Whalen, R. (1959) Effects of intromission and ejaculation upon sexual behavior in male rats. *Journal of Comparative and Physiological Psychology* 52:476–481.

Beach, F., Zitrin, A., and Jaynes, J. (1955) Neural mediation of mating in male cats: II. Contribution of the frontal cortex. *Journal of Experimental Zoology* 130:381–401.

Beamer, W., Bermant, G., and Clegg, M. (1969) Copulatory behavior of the ram, *Ovis aries* II: Factors affecting copulatory satiation. *Animal Behavior* 17:706–711.

Beatty, R. (1967) Parthenogenesis in vertebrates. In C. Metz and A. Monroy (eds.), *Fertilization*, pp. 413–440. New York: Academic.

Benedek, T., and Rubinstein, B. B. (1942) The sexual cycle in women. *Psychosomatic Medicine Monograph. 3*, nos. 1 and 2.

Bermant, G. (1961a) Response latencies of female rats during sexual intercourse. *Science* 133:1771–1773.

Bermant, G. (1961b) Regulation of sexual contact by female rats. Unpublished Ph.D. dissertation, Harvard University.

Bermant, G. (1964) Effects of single and multiple enforced intercopulatory intervals on the sexual behavior of male rats. *Journal of Comparative and Physiological Psychology* 57:398–403.

Bermant, G. (1965) Rat sexual behavior: Photographic analysis of the intromission response. *Psychonomic Science* 2:65–66.

Bermant, G. (1967) Copulation in rats. *Psychology Today* (July) 1:52–60.

Bermant, G., Beamer, W., and Clegg, M. (1969) Copulatory behavior of the ram, *Ovis aries* I: A normative study. *Animal Behavior* 17:700–705.

Bermant, G., Glickman, S., and Davidson, J. (1968) Effects of limbic lesions on copulatory behavior of male rats. *Journal of Comparative and Physiological Psychology* 65:118–125.

Bermant, G., Lott, D., and Anderson, L. (1968) Temporal characteristics of the Coolidge Effect in male rat copulatory behavior. *Journal of Comparative Physiological Psychology* 65:447–452.

Bermant, G., Parkinson, S., and Anderson, T. (1969) Copulation in rats: Relations among intromission frequency, duration, and pacing. *Psychonomic Science* 17:293–294.

Bermant, G., and Taylor, L. (1969) Interactive effects of experience and olfactory bulb lesions in male rat copulation. *Physiology and Behavior* 4:13–18.

Bermant, G., and Westbrook, W. (1966) Peripheral factors in the regulation of sexual contact by female rats. *Journal of Comparative and Physiological Psychology* 61:244–250.

Bertolini, A., Vergoni, W., Gessa, G. L., and Ferrari, W. (1969) Induction of sexual excitement by the action of adrenocorticotrophic hormone in brain. *Nature* 221:667–669.

Beyer, C., McDonald, P., and Vidal, N. (1970) Failure of 5 α-dihydrotestosterone to elicit estrous behavior in the ovariectomized rabbit. *Endocrinology* 86:939–941.

Bieber, I., Dain, H., Dince, P., Drellick, M., Grand, H., Grundlack, R., Kremer, M., Rifkin, A., Wilbur, C., and Bieber, T. (1965) *Homosexuality—A Psychoanalytical Study*. New York: Basic Books.

Bingham, H. C. (1928) Sex development in apes. *Comparative Psychology Monograph*, 5:1.

Blair, W. F. (1955) Mating call and stage of speciation in the *Microhyla olivacea-M. carolinensis* complex. *Evolution* 9:469–480.

Blaudau, R. J. (1945) On the factors involved in sperm transport through the cervix uteri of the albino rat. *American Journal of Anatomy* 77: 253–272.

Bloch, G. J., and Davidson, J. M. (1968) Effects of adrenalectomy and prior experience on postcastrational sex behavior in the male rat. *Physiology and Behavior* 3:461.

Blumer, D. (1970) Hypersexual episodes in temporal lobe epilepsy. *American Journal of Psychiatry* 126:1099–1106.

Bolles, R. (1970) Species-specific defense reactions and avoidance learning. *Psychological Review* 77:32–48.

Borradaile, L., and Potts, F. (1935) *The Invertebrata*, 2d ed. New York: MacMillan.

Brazier, M. (1959) The historical development of neurophysiology. In J. Field (ed.), *Handbook of Physiology. Section I: Neurophysiology*, vol. I, pp. 1–58. Washington: The American Physiological Society.

Bremer, J. (1959) *Asexualization*. New York: MacMillan.

Brien, P. (1960) The fresh-water hydra. *American Scientist* 48:461–475.

Bronson, F. (1968) Pheromonal influences on mammalian reproduction. In M. Diamond (ed.), *Perspectives in Reproduction and Sexual Behavior*, pp. 341–362. Bloomington: Indiana University Press.

Bronson, F., and Whitten, W. (1968) Estrus accelerating pheromone of mice: Assay, androgen-dependency, and presence in bladder urine. *Journal of Reproduction and Fertility* 15:131–134.

Brooks, C. (1935) Studies on the neural basis of ovulation in the rabbit. *American Journal of Physiology* 113:18–19.

Brower, L. P., Brower, J., and Cranston, F. (1965) Courtship behavior of the queen butterfly, *Danaus gilippus berenice* (Cramer). *Zoologica* 50:1–54.

Browman, E. (1937) Light in its relation to activity and oestrus rhythms in the albino rat. *Journal of Experimental Zoology* 75:375–388.

Brown, W. L., Jr., and Wilson, E. O. (1956) Character displacement. *Systematic Zoology* 5:49–64.

Brown-Grant, K. (1969) The effects of progesterone and pentobarbitone administered at the dioestrous stage of the ovarian cycle of the rat. *Journal of Endocrinology* 43:539–552.

Brown-Grant, K., Davidson, J. M., and Greig, F. (1973) Induced ovulation in albino rats exposed to constant light. *Journal of Endocrinology* 57: 7–22.

Brown-Grant, K., and Sherwood, M. R. (1971) The early androgen syndrome in the guinea pig. *Journal of Endocrinology* 49:277–291.

Bruce, H. (1967) Effects of olfactory stimuli on reproduction in mammals. In G. Wolstenhome (ed.), *Effects of External Stimuli on Repro-*

duction, pp. 29–38. (Ciba Foundation Study Group No. 26), Boston: Little, Brown.

Bruce, H. M. (1969) Pheromones and behavior in mice. *Acta Neurologica Belgica* 69:529–538.

Buchsbaum, R. (1938) *Animals Without Backbones*. Chicago: University of Chicago Press.

Bullock, D. W. (1970) Induction of heat in ovariectomized guinea pigs by brief exposure to estrogen and progesterone. *Hormones and Behavior* 1:137–143.

Bullough, W. (1961) *Vertebrate Reproductive Cycles*, 2d ed. London: Methuen.

Burnett, A., and Diehl, N. (1964) The nervous system of hydra III. The initiation of sexuality with special reference to the nervous system. *Journal of Experimental Zoology* 157:237–250.

Burton, M. (1967) Mammals. In *The Larousse Encyclopedia of Animal Life*. New York: McGraw Hill, pp. 468–613.

Butler, C. (1967) Insect pheromones. *Biological Review* 42:42–87.

Caggiula, A. (1970) Analysis of the copulation-reward properties of posterior hypothalamic stimulation in male rats. *Journal of Comparative and Physiological Psychology* 70:399–412.

Caggiula, A., and Hoebel, B. (1966) Copulation-reward site in the posterior hypothalamus. *Science* 153:1284–1285.

Campbell, H. J. (1965) Effects of neonatal injection of hormones on sexual behavior and reproduction in the rabbit. *Journal of Physiology* 181:568–571.

Carlsson, S., and Larsson, K. (1964) Mating in male rats after local anesthetization of the glans penis. *Zeitschrift für Tierpsychologie* 21:854–856.

Carr, W. J., and Caul, W. F. (1962) The effect of castration in rats upon the dissemination of sex odors. *Animal Behavior* 10:20–27.

Caspari, E. (1965) The evolutionary importance of sexual processes and of sexual behavior. In F. A. Beach (ed.), *Sex and Behavior*, pp. 34–52. New York: Wiley.

Chace, L. (1952) Aerial mating of the great slug. *Discovery* 13:356–359.

Cheng, P., Ulberg, L. C., Christian, R. E., and Casida, L. E. (1950) Different intensities of sexual activity in relation to the effect of testosterone propionate in the male rabbit. *Endocrinology* 46:447–452.

Cherney, E., and Bermant, G. (1970) The role of stimulus female novelty in the rearousal of copulation in male laboratory rats. *Animal Behavior*, 18:567–574.

Chester, R. V., and Zucker, I. (1970) Influence of male copulatory behavior on sperm transport, pregnancy and pseudopregnancy in female rats. *Physiology and Behavior* 5:35–43.

Clark, E. (1959) Functional hermaphroditism and self-fertilization in a serranid fish. *Science* 129:215–216.

Clark, J. H., and Zarrow, M. W. (1971) Influence of copulation on time of ovulation in women. *American Journal of Obstetrics and Gynecology* 109:1083–1085.

Clayton, R. B., Kogura, J., and Kraemer, H. C. (1970) Sexual differentiation of the brain: Effects of testosterone on brain RNA metabolism in newborn female rats. *Nature* 226:810–812.

Clemens, L. G., Wallen, K., and Gorski, R. (1967) Mating behavior: Facilitation in the female rat after cortical application of potassium chloride. *Science* 157:1208–1209.

Clemens, L. G., Hiroi, M., and Gorski, R. A. (1969) Induction and facilitation of female mating behavior in rats treated neonatally with low doses of testosterone propionate. *Endocrinology* 84:1430–1438.

Cofer, C., and Apley, M. (1964) *Motivation: Theory and Research*. New York: Wiley.

Comfort, A. (1971) Likelihood of human pheromones. *Nature* 230:432–433.

Conner, R. L., and Levine, S. (1969) Hormonal influences on aggressive behavior. In S. Garattini and E. B. Siff (eds.), *Aggressive Behavior*, pp. 150–163. Amsterdam: Excerpta Medica.

Cowgill, U. (1966) The season of birth in man. *Man* 1:232–239.

Crane, J. (1941) Eastern pacific expeditions of the New York Zoological Society: Crabs of the genus *Uca* from the west coast of Central America. *Zoologica* 26:145–207.

Crane, J. (1957) Basic patterns of display in fiddler crabs *Ocypodidae* (genus *Uca*). *Zoologica* 42:69–82.

Crossley, D. A., and Swanson, H. H. (1968) Modification of sexual behavior of hamsters by neonatal administration of testosterone propionate. *Journal of Endocrinology* 41:xiii–xiv.

Dantchakoff, V. (1937) Sur l'obtention experimentale des freemartins chez le cobaye et sur la nature du facteur conditionnant leur histogenese sexuelle. *Computes Rendue Academie Scientifique* 204:195–197.

Darwin, C. (1859) *The Origin of Species*. New York: Modern Library (1936 ed.).

Davenport, W. (1965) Sexual patterns and their regulation in a society of the southwest Pacific. In F. Beach (ed.), *Sex and Behavior*, pp. 164–207. New York: Wiley.

Davidson, J. M. (1966a) Activation of male rats' sexual behavior by intracerebral implantation of androgen. *Endocrinology* 79:783–794.

Davidson, J. M. (1966b) Characteristics of sex behavior in male rats following castration. *Animal Behavior* 14:266–272.

Davidson, J. M. (1969a) Effects of estrogen on the sexual behavior of male rats. *Endocrinology* 84:1365.

Davidson, J. M. (1969b) Hormonal control of sexual behavior in adult rats. In G. Raspé (ed.), *Advances in Bioscience 1*, pp. 119–169. New York: Pergamon.

Davidson, J. M. (1972) Hormones and reproductive behavior. In H. Balin and S. Glasser (eds.), *Reproductive Biology*, pp. 877–918. Amsterdam: Excerpta Medica.

Davidson, J. M. and Bloch, G. (1969) Neuroendocrine aspects of male reproduction. *Biology of Reproduction Supplement* 1:67–92.

Davidson, J. M., and Levine, S. (1969) Progesterone and heterotypical sexual behavior in male rats. *Journal of Endocrinology* 44:129–130.

Davidson, J. M., Rodgers, C. H., Smith, E. R., and Bloch, G. J. (1968a) Relative thresholds of behavioral and somatic responses to estrogen. *Physiology and Behavior* 3:227–229.

Davidson, J. M., Rodgers, C. H., Smith, E. R., and Bloch, G. J. (1968b) Stimulation of female sex behavior in adrenalectomized rats with estrogen alone. *Endocrinology* 82:193–195.

Davidson, J. M., Weick, R. F., Smith, E. R., and Dominguez, R. D. (1970) Feedback mechanisms in relation to ovulation. (Symposium on control of ovulation). *Federation Proceedings*, 29:1900–1906.

Davis, D. E. (1964) The physiological analysis of aggressive behavior. In W. Etkin (ed.), *Social Behavior and Organization Among Vertebrates*, p. 307. Chicago: University of Chicago Press.

DeGroot, J. (1959) *The Rat Forebrain in Stereotoxic Coordinates*. Amsterdam: North Holland.

Dempsey, E. W., Herz, R., and Young, W. C. (1936) The experimental induction of estrus (sexual receptivity) in the normal and ovariectomized guinea pig. *American Journal of Physiology* 116:201–210.

Dempsey, E. W., and Searles, H. F. (1943) Environmental modification of certain endocrine phenomena. *Endocrinology* 32:119–128.

Dethier, V., and Stellar, E. (1964) *Animal Behavior*, 2d ed. Englewood Cliffs, N.J.: Prentice-Hall.

DeVore, I. (1963) Mother-infant relations in free-ranging baboons. In H. Rheingold (ed.), *Maternal Behavior in Mammals*, pp. 305–335. New York: Wiley.

DeVore, I., and Hall, K. (1965) Baboon ecology. In E. DeVore (ed.), *Primate Behavior—Field Studies of Monkeys and Apes*, pp. 20–52. New York: Holt, Rinehart & Winston.

Dewsbury, D. (1968) Copulatory behavior of rats—variations within the dark phase of the diurnal cycle. *Communications in Behavioral Biology* 1:373–377.

Dewsbury, D., Goodman, E., Salis, P., and Burnell, B. (1968) Effects of hippocampal lesions on the copulatory behavior of male rats. *Physiology and Behavior* 3:651–656.

Diamond, M. (1967) Androgen induced masculinization in the ovariectomized and hysterectomized guinea pig. *Anatomical Record* 157:47–52.

Diamond, M. (1970) Intromission pattern and species vaginal code in relation to induction of pseudopregnancy. *Science* 168:995–997.

Diamond, I. and Hall, W. (1969) Evolution of neocortex. *Science* 164: 251–262.

Diamond, M., and Young, W. C. (1963) Different responsiveness of pregnant and non-pregnant guinea pigs to the masculinizing action of testosterone propionate. *Endocrinology* 72:429–438.

Dobzhansky, T. (1955) *Evolution, Genetics, and Man.* New York: Wiley.

Dondey, M., Albe-Fessard, D., and Le Beau, J. (1962) Premières applications neurophysiologiques d'une methode permettant le blocage electif et reversible de structures centrales par refrigeration localisee. *Electroencephalography and Clinical Neurophysiology* 14:758–763.

Dougherty, E. C. (1955) The origin of sexuality. *Systematic Zoology* 4: 145–169.

Drori, D., and Folman, Y. (1964) Effects of cohabitation on the reproductive system, kidneys and body composition of male rats. *Journal of Reproduction and Fertility* 8:351–359.

Dumortier, B. (1963) The physical characteristics of sound emissions in Arthropoda. In R. G. Busnel (ed.), *Acoustic Behavior of Animals*, pp. 346–373. Amsterdam: Elsevier.

Eaton, G. (1970) Effect of a single prepubertal injection of testosterone propionate on adult bisexual behavior of male hamsters castrated at birth. *Endocrinology* 87:934–940.

Eayrs, J. T. (1968) Developmental relationships between brain and thyroid. In R. P. Michael (ed.), *Endocrinology and Human Behavior*. London: Oxford University Press, pp. 239–255.

Eccles, J. (1966) *Brain and Conscious Experience.* Berlin: Springer.

Eccles, J. (1970) *Facing Reality.* Berlin: Springer.

Eckert, J.E., and Shaw, F. (1960) *Bee Keeping.* New York: Macmillan.

Edwards, D. A. (1969) Early androgen stimulation and aggressive behavior in male and female mice. *Physiology and Behavior* 4:333.

Edwards, D. A. (1970) Induction of estrus in female mice: Estrogen-progesterone interactions. *Hormones and Behavior* (in press).

Edwards, D. A., and Burge, K. G. (1971) Early androgen treatment and male and female sex behavior in mice. *Hormones and Behavior* (in press).

Edwards, D. A., Whalen, R. E., and Nadler, R. D. (1968) Induction of estrus: Estrogen-progesterone interactions. *Physiology and Behavior* 3:29–33.

Eleftherion, B., Bronson, F., and Zarrow, M. (1962) Interaction of olfactory and other environmental stimuli on implantation in the deer mouse. *Science* 137:764.

Ellenberger, H. (1970) *The Discovery of the Unconscious.* New York: Basic Books.

Ellis, A. (1963) Constitutional factors in homosexuality: A re-examination of the evidence. In H. Beigel (ed.), *Advances in Sex Research*, pp. 161–186. New York: Harper & Row.

Ellis, H. (1905) *Sexual Selection in Man*. New York: Davis.

Endroczi, E., and Lissak, K. (1962) Role of reflexogenic factors in testicular hormone secretion. Effect of copulation on the testicular hormone production in the rabbit. *Acta Physiologica Acadamique* 21:203–206.

Engel, P. (1942) Female mating behavior shown by male mice after treatment with different substances. *Endocrinology* 30:623.

Ericsson, R. J., and Baker, V. F. (1966) Sexual receptivity and fertility of female rats that are in androgen induced persistent vaginal estrus. *Proceedings of the Society of Experimental Biology and Medicine* 122: 88.

Evans, Ray B. (1972) Physical and biochemical characteristics of homosexual men. *Journal of Consulting and Clinical Psychology* 39:140–147.

Everett, J. W. (1951) *Anatomical Record* 109:291–292.

Everett, J. W. (1961) The mammalian female reproductive cycle and its controlling mechanisms. In W. C. Young (ed.), *Sex and Internal Secretions*, vol. I, pp. 497–555. Baltimore: Williams & Wilkins.

Everitt, B. J., and Herbert, J. (1969) Adrenal glands and sexual receptivity in Rhesus female monkeys. *Nature* 222:1065–1066.

Falconer, D. (1960) *Introduction to Quantitative Genetics*. New York: Ronald.

Farber, L. (1964) I'm sorry dear. *Commentary* (November) 38:47–54.

Farberow, N.L. (ed.) (1966) *Taboo Topics*. New York: Atherton.

Farris, H. E. (1967) Classical conditioning of courting behavior in the Japanese quail, *Coturnix citurnix japonica*. *Journal of Experimental Analysis of Behavior* 10:213–217.

Feder, H. H., Goy, R. W., and Resko, J. A. (1967) Progesterone concentrations in the peripheral plasma of cyclic rats. *Journal of Physiology* 191:136–137.

Feder, H. H., Resko, J. A., and Goy, R. W. (1966) Progestin levels in peripheral plasma of guinea pigs during the estrous cycle. *American Zoologist* (abstract) 6:597.

Feder, H. H., and Ruf, K. B. (1969) Stimulation of progesterone release and estrous behavior by ACTH in ovariectomized rodents. *Endocrinology* 69:171–174.

Ferin, M., Tempone, A., Zimmering, P. E., and Vande Wiele, R. L. (1969) Effect of antibodies to 17β-estradiol and progesterone on the estrus cycle of the rat. *Endocrinology* 85:1070.

Finger, F. (1969) Estrus and general activity in the rat. *Journal of Comparative and Physiological Psychology* 68:461–466.

Firestone, S. (1970) *The Dialectic of Sex: The Case for Feminist Revolution*. New York: Morrow.

Fisher, A. E. (1956) Maternal and sexual behavior induced by intracerebral chemical stimulation. *Science* 124:228–229.

Fletcher, R. (1957) *Instinct in Man*. London: Allen and Unwin.

Folman, Y., and Drori, D. (1965) Normal and aberrant copulatory behavior in male rats (*R. Norvegicus*) raised in isolation. *Animal Behavior* 13:427–429.

Folman, Y., and Drori, D. (1966) Effects of social isolation and female odours on the reproductive system, kidneys and adrenals of unmated male rats. *Journal of Reproduction and Fertility* 11:43–50.

Ford, C., and Beach, F. (1951) *Patterns of Sexual Behavior.* New York: Harper & Row.

Fowler, H., and Whalen, R. (1961) Variation in incentive stimulus and sexual behavior in the male rat. *Journal of Comparative and Physiological Psychology* 54:68–71.

Freedman, L. Z., and Rosvold, H. E. (1962) Sexual, aggressive, and anxious behavior in the laboratory *Macaque. Journal of Nervous and Mental Diseases* 134:18–27.

Freud, S. (1900) The interpretation of dreams. (Trans. A. Brill) In A. Brill (ed.), *The Basic Writings of Sigmund Freud.* New York: Random House, 1938.

Freud, S. (1905) Three essays on the theory of sexuality. In T. Strachey (ed.), *The Standard Edition of the Complete Psychological Works of Sigmund Freud,* vol. 7, pp. 135–243. London: Hogarth.

Freud, S. (1908) Character and anal erotism. (Trans. R. McWalters.) In P. Rieff (ed.), *The Collected Papers of Sigmund Freud.* New York: Collier Books, 1963.

Freud, S. (1915) Instincts and their vicissitudes. (Trans. C. Baines.) In P. Rieff (ed.), *The Collected Papers of Sigmund Freud.* New York: Collier Books, 1963.

Freud, S. (1924a) The economic problem in masochism. (Trans. J. Riviere.) In P. Rieff (ed.), *The Collected Papers of Sigmund Freud,* New York: Collier Books, 1963.

Freud, S. (1924b) *A General Introduction to Psychoanalysis.* (Trans. J. Riviere.) New York: Washington Square, 1960.

Freud, S. (1924c) The passing of the Oedipus-complex. (Trans. J. Riviere.) In P. Rieff (ed.), *The Collected Papers of Sigmund Freud.* New York: Collier Books, 1963.

Freud, S. (1925) Some psychological consequences of the anatomical distinction between the sexes. (Trans. J. Strachey.) In P. Rieff (ed.), *The Collected Papers of Sigmund Freud.* New York: Collier Books, 1963.

Freud, S. (1931) Female sexuality. (Trans. J. Riviere.) In P. Rieff (ed.), *The Collected Papers of Sigmund Freud.* New York: Collier Books, 1963.

Freund, K. (1967) Diagnosing homo- and heterosexuality and erotic age preference by means of a psychophysiological test. *Behavior Research Therapy* 5:209–228.

Friedan, B. (1963) *The Feminine Mystique.* New York: Norton.

Frink, H. (1918) *Morbid Fears and Compulsions.* New York: Moffat, Yard.

Fuller, J., and Thompson, W. (1960) *Behavior Genetics.* New York: Wiley.

276   References

Ganong, W. F. (1967) *Review of Medical Physiology*, 3d ed. Los Altos, Calif.: Lange.

Gary, N. (1961) Chemical mating attractants in the queen honey bee. *Science* 136:773–774.

Gebhard, P. H. (1965) *The Sexual Offender.* New York: Harper & Row.

Gerall, A. A., and Kenney, A. M. (1970) Neonatally androgenized females' responsiveness to estrogen and progesterone. *Endocrinology* 87: 560–566.

Gerall, H., Ward, I., and Gerall, A. (1967) Disruption of the male rat's sexual behavior induced by social isolation. *Animal Behavior* 15:54–58.

Glasser, G. (1963) *EEG and Behavior.* New York: Basic Books.

Glick, I. D. (1966) Mood and behavioral changes associated with the use of the oral contraceptive agents. *Psychopharmacologia* 10:363–374.

Glickman, S., and Schiff, B. (1967) A biological theory of reinforcement. *Psychological Review* 74:81–109.

Goldfoot, D. A., Feder, H. H., and Goy, R. W. (1969) Development of biosexuality in the male rat treated neonatally with androstenedione. *Journal of Comparative and Physiological Psychology* 67:41–45.

Goodman, E., Jansen, P., and Dewsbury, D. (1971) Midbrain reticular formation lesions: Habituation to stimulation and copulatory behavior in male rats. *Physiology and Behavior* 6:151–156.

Gorski, R. A. (1966) Localization and sexual differentiation of the nervous structures which regulate ovulation. *Journal of Reproduction and Fertility*, suppl. 1:67–88.

Gorski, R. A., and Wagner, J. W. (1965) Gonadal activity and sexual differentiation of the hypothalamus. *Endocrinology* 76:226–239.

Gowen, J. (1961) Genetic and cytologic foundations for sex. In W. Young (ed.), *Sex and Internal Secretions*, vol. I, pp. 3–75. Baltimore: Williams & Wilkins.

Goy, R. W. (1968) Organizing effect of androgen on the behavior of rhesus monkeys. In R. P. Michael (ed.), *Endocrinology and Human Behavior.* London: Oxford University Press.

Goy, R. W., and Phoenix, C. (1963) Hypothalamic regulation of female sexual behavior: Establishment of behavioral oestrus in spayed guinea pigs following hypothalamic lesions. *Journal of Reproduction and Fertility* 5:23–40.

Goy, R. W., and Resko, J. A. (1969) Stress and dexamethasone on lordosis. Paper presented at meetings of the West Coast Sex Society.

Goy, R. W., and Young, W. C. (1957) Strain differences in the behavioral responses of female guinea pigs to α-estradiol benzoate and progesterone. *Behavior* 10:340–354.

Grant, E. C. G., and Mears, E. (1967) Mental effects of oral contraceptives. *Lancet* 2, 945–946.

Gray, D. A. (1971) Sex differences in emotional behavior in mammals including man. (in press).

Green, E. (1966) Breeding systems. In E. Green (ed.), *The Biology of the Laboratory Mouse*, 2d ed., pp. 11–22. New York: McGraw-Hill.

Green, J., Clemente, C., and DeGroot, J. (1957) Rhinencephalic lesions and behavior in cats. *Journal of Comparative Neurology* 108:505–545.

Greer, G. (1971) *The Female Eunuch.* New York: McGraw-Hill.

Greulich, W. W. (1934) Artificially induced ovulation in the cat (*Felis domestica*). *Anatomical Record* 58:217.

Gruendel, A., and Arnold, W. (1969) Effects of early social deprivation on reproductive behavior of male rats. *Journal of Comparative and Physiological Psychology* 67:123–128.

Grunt, J. A. and Young, W. C. (1953) Consistency of sexual behavior patterns in individual male guinea pigs following castration and androgen therapy. *Journal of Comparative and Physiological Psychology* 46:138–144.

Guttmacher, A. (1971) *Pregnancy and Birth.* New York: Signet.

Hafez, E. (1968) Environment and reproduction in domesticated species. In E. Hafez (ed.), *Adaptation of Domestic Animals*, pp. 113–164. Philadelphia: Lea & Febiger.

Hafez, E.S.E., and Evans, T.N. (Eds.) (1973) *Human Reproduction: Conception and Contraception.* New York: Harper & Row.

Hall, C. (1954) *A Primer of Freudian Psychology.* Cleveland: World Publishing.

Hall, K., and DeVore, I. (1965) Baboon social behavior. In I. DeVore (ed.), *Primate Behavior—Field Studies of Monkeys and Apes*, pp. 53–110. New York: Holt, Rinehart & Winston.

Hamilton, J. B. (1943) Demonstrable ability of penile erection in castrate men with markedly low titers of urinary androgen. *Proceedings of the Society Experimental Biology and Medicine* 54:309.

Hamilton, J., Walter, R., Daniel, R., and Mestler, G. (1969) Competition for mating between ordinary and supermale Japanese medaka fish. *Animal Behavior* 17:168–176.

Hampson, J. L. (1965) Determinants of psychosexual orientation. In F. Beach (ed.), *Sex and Behavior*, pp. 108–132. New York: Wiley.

Hampson, J. L., and Hampson, J. G. (1961) The ontogenesis of sexual behavior in man. In W. Young (ed.), *Sex and Internal Secretions*, vol. II, pp. 1401–1432. Baltimore: Williams & Wilkins.

Harlow, H. F. (1965) Sexual behavior in the Rhesus monkey. In F. Beach (ed.), *Sex and Behavior*, pp. 234–265. New York: Wiley.

Harrington, F. E., Eggert, R. G., Wilbur, R. D., and Linkenheimer, W. H. (1966) Effect of coitus on chlorpromzaine inhibition of ovulation in the rat. *Endocrinology* 79:1130–1134.

Harrington, R. W., Jr. (1961) Oviparous hermaphroditic fish with internal self-fertilization. *Science* 134:1749–1750.

Harris, G. W. (1964) Sex hormones, brain development and brain function. *Endocrinology* 74:627–648.

Harris, G. W., and Jacobsohn, D. (1953) Functional grafts of the anterior pituitary gland. *Proceedings Royal Society (London) Series B.*, 139: 263–276.

Harris, G. W., and Levine, S. (1965) Sexual differentiation of the brain and its experimental control. *Journal of Physiology* 181:379–400.

Harris, G. W., and Michael, R. P. (1964) The activation of sexual behavior by hypothalamic implants of oestrogen. *Journal of Physiology* 171: 275–301.

Harris, V. T. (1952) An experimental study of habitat selection by prairie and forest races of the deermouse, *Peromyscus maniculatus. Contributions from the laboratory of Vertebrate Biology,* No. 56. Ann Arbor: University of Michigan Press.

Hart, B. (1967a) Sexual reflexes and mating behavior in the male dog. *Journal of Comparative and Physiological Psychology* 64:388–399.

Hart, B. (1967b) Testosterone regulation of sexual reflexes in spinal rats. *Science* 155:1283–1284.

Hart, B. (1968a) Alteration of quantitative aspects of sexual reflexes in spinal male dogs by testosterone. *Journal of Comparative Physiological Psychology* 66:726–730.

Hart, B. (1968b) Neonatal castration: Influence on neural organization of sexual reflexes in male rats. *Science* 160:1135–1136.

Hart, B. (1968c) Role of prior experience in the effects of castration on sexual behavior of male dogs. *Journal of Comparative and Physiological Psychology* 66:719–725.

Hart, B. (1968d) Sexual reflexes and mating behavior in the male rat. *Journal of Comparative and Physiological Psychology* 65:453–460.

Hart, B. (1969) Gonadal hormones and sexual reflexes in the female rat. *Hormones and Behavior* 1:65–71.

Hart, B. (1970) Mating behavior in the female dog and the effects of estrogen on sexual reflexes. *Hormones and Behavior* 1:93–104.

Hart, B., and Haugen, C. M. (1968) Activation of sexual reflexes in male rats by spinal implantation of testosterone. *Physiology and Behavior* 3:735–738.

Hartman, C. G. (1962) *Science and the Safe Period.* Baltimore: Williams & Wilkins.

Hartmann, G., Endroczi, E., and Lissak, K. (1966) The effect of hypothalamic implantation of 17-β-oestardiol and systemic administration of prolactin (LTH) on sexual behavior in male rabbits. *Acta. Physiologica Academiae Scientiarum Hungaricae* 30:53–59.

Hartmann, H. (1959) Psychoanalysis as a scientific theory. In S. Hook (ed.), *Psychoanalysis, Scientific Method, and Philosophy,* pp. 3–35. New York: New York University Press.

Harvey, E., Yanagimachi, R., and Chang, M. (1961) Onset of estrus and ovulation in the golden hamster. *Journal of Experimental Zoology* 146: 231–235.

Hastings, D. W. (1963) *Impotence and Frigidity.* New York: Dell.

Hediger, H. (1965) Environmental factors influencing the reproduction of zoo animals. In F. Beach (ed.), *Sex and Behavior*, pp. 319–354. New York: Wiley.

Heimer, L., and Larsson, K. (1964a) Drastic changes in the mating behavior of male rats following lesions in the junction of diencephalon and mesencephalon. *Experientia* 20:460–461.

Heimer, L., and Larsson, K. (1964b) Mating behavior in male rats after destruction of the mammillary bodies. *Acta Neurologica Scandinavia* 40:353.

Heimer, L., and Larsson, K. (1966/1967) Impairment of mating behavior in male rats following lesions in the preoptic-anterior hypothalamic continuum. *Brain Research* 3:248–263.

Herberg, L. (1963) Seminal ejaculation following positively reinforcing electrical stimulation of the rat hypothalamus. *Journal of Comparative and Physiological Psychology* 56:679–685.

Herbert, J. (1966) The effect of estrogen applied directly to the genitalia upon the sexual attractiveness of the female rhesus monkey. *Excerpta. Medica International Congress Series* 3:212.

Herbert, J. (1970) Hormones and reproductive behavior in rhesus and tatapoin monkeys. *Journal of Reproduction and Fertility* (suppl.) 11:119–140.

Herschberger, R. (1948) *Adam's Rib: A Defense of Modern Women*. New York: Harper & Row.

Hertz, R., and Meyer, R. K. (1937) The effect of testosterone, testosterone propionate and dehydroandrosterone on the secretion of the gonadotropic complex as evidenced in parabiotic rats. *Endocrinology* 21:756–761.

Hess, E., (1962) Ethology: An approach toward the complete analysis of behavior. In R. Brown, E. Galanter, E. Hess, and G. Mandler (eds.), *New Directions in Psychology*. New York: Holt, Rinehart & Winston.

Hinde, R. (1960) Energy models of motivation. In *Models and Analogues in Biology. Symposia of the Society for Experimental Biology*, No. 14. New York: Academic.

Hinde, R. (1966) *Animal Behavior*. New York: McGraw-Hill.

Hirsch, J. (ed.) (1967) *Behavior—Genetic Analysis*. New York: McGraw-Hill.

Hitt, J., Hendericks, S., Ginsberg, S., and Lewis, J. (1970) Disruption of male but not female sexual behavior in rats by medial forebrain bundle lesions. *Journal of Comparative and Physiological Psychology* 73:377–384.

Hoebel, B., and Teitelbaum, P. (1962) Hypothalamic control of feeding and self-stimulation. *Science* 135:375–376.

Hooker, E. (1965a) Male homosexuals and their "worlds." In J. Marmor (ed.), *Sexual Inversion*, pp. 83–107. New York: Basic Books.

Hooker, E. (1965b) An empirical study of some relations between sexual patterns and gender identity in male homosexuals. In J. Money (ed.), *Sex Research—New Developments*, pp. 24–52. New York: Holt, Rinehart & Winston.

Hubbs, C. L., and Hubbs, L. C. (1932) Apparent parthenogenesis in nature in a form of fish of hybrid origin. Science 76:628–630.

Hulet, C. V. (1966) Behavioral, social and psychological factors affecting mating time and breeding efficiency in sheep. Journal of Animal Science (suppl.), 25:5–20.

Hull, C. D., Garcia, J., and Cracchiolo, E. (1965) Cerebral temperature changes accompanying sexual activity in the male rat. Science 149:89–90.

Hutchison, J. B. (1967) Initiation of courtship by hypothalamic implants of testosterone propionate in castrated doves (Streptopelia risoria). Nature 216:591–592.

Huxley, J. (1963) The evolutionary process. In J. Huxley, A. C. Hardy, and E. B. Ford (eds.), Evolution as a Process. New York: Collier Books.

Hyman, L. (1967) The Invertebrates, vol. VI. Mollusca I. New York: McGraw-Hill.

Jenkins, M. (1927) The effects of segregation on the sex behavior of the white rat as measured by the destruction method. Genetic Psychology Monographs 3:455–571.

Johnston, P. J., and Davidson, J. M. (1972) Intracerebral androgens and sexual behavior in the male rat. Hormones and Behavior, in press.

Jones, E. (1953–1957) The Life and Work of Sigmund Freud, 3 vols. New York: Basic Books.

Jost, A., Jones, H. W., and Scott, W. W. (eds.)(1969)Hermaphroditism, Genital Anomalies and Related Endocrine Disorders, 2d ed. Baltimore: Williams & Wilkins.

Kagan, J., and Beach, F. (1953) Effects of early experience on mating behavior in male rats. Journal of Comparative and Physiological Psychology 46:204–208.

Kallman, F. (1952) Comparative twin study on the genetic aspects of male homosexuality. Journal of Nervous Mental Disease 115:283–298.

Kallman, K. D. (1962) Gynogenesis in the Teleost, Mollienisia formosa (Girard) with discussion of detection of parthenogenesis in vertebrates by tissue transplantation. Journal of Genetics 58 (1):7–24.

Komisarauk, B.R., Adler, N.T., and Hutchison, J. (1972) Genital sensory fields: enlargement by estrogen treatment in female rats. Science 178:1295–1298.

Kane, F. J. (1968) Psychiatric reactions to oral contraceptives. American Journal of Obstetrics and Gynecology 102:1053–1063.

Karlson, P., and Butenandt, A. (1959) Pheromones (ectohormones) in insects. Annual Review of Entomology 4:39–58.

Katchadourian, H. A., and Lunde, D. T. (1972) Fundamentals of Human Sexuality. New York: Holt, Rinehart & Winston.

Kenniston, K. (1967) The physiological fallacy. Contemporary Psychology 12:113–115.

Kim, C. (1960) Sexual activity of male rats following ablation of hippocampus. Journal of Comparative Physiological Psychology 53:553–557.

Kincl, F. A., and Maqueo, M. (1965) Prevention by progesterone of steroid-induced sterility in neonatal male and female rats. *Endocrinology* 77:859–862.

Kinsey, A. Pomeroy, W., and Martin, C. (1948) *Sexual Behavior in the Human Male*. Philadelphia: Saunders.

Kinsey, A., Pomeroy, W., Martin, C., and Gebhard, P. (1953) *Sexual Behavior in the Human Female*. Philadelphia: Saunders.

Klein, M. (1952) Administration of sex hormones and sexual behavior. *Ciba Foundation Colloquium on Endocrinology* 3:323–337.

Kloek, J. (1961) The smell of some steroid sex hormone and their metabolites. Reflections and experiments concerning the significance of smell for the mutual relation of the sexes. *Psychiatria, Neurologia, Neurochirurgia* 64:301.

Klopfer, P. H. (1962) *Behavioral Aspects of Ecology*. Englewood Cliffs, N.J.: Prentice-Hall.

Kollberg, S., Petersen, I., and Stener, I. (1962) Preliminary results of an electromyographic study of ejaculation. *Acta Chirurgica Scandinavica* 123:478–483.

Kolodny, Robert C., Masters, William H., Hendryx, Julie, and Toro, Gelson (1971) Plasma testostcrone and semen analysis in male homosexuals. *New England Journal of Medicine* 285:1170–1174.

Kolodny, Robert C., Jacobs, Laurence S., Masters, William H., Toro, Gelson, and Daughaday, William II. (1972) Plasma gonadotrophins and prolactin in male homosexuals. *Lancet* 2:18–20.

Koranyi, L., Endroczi, E., and Tarnok, F. (1966) Sexual behavior in the course of avoidance conditioning in male rabbits. *Neuroendocrinology* 1:144–157.

Krafft-Ebing, R. (1892) *Psychopathia Sexualis*. Philadelphia: Davis.

Kuehn, R. E., and Beach, F. (1963) Quantitative measurement of sexual receptivity in female rats. *Behavior* 21:282–299.

Kuehn, R. E., and Zucker, I. (1968) Reproductive behavior of the Mongolian gerbil (*Meriones ungutculatus*). *Journal of Comparative Physiological Psychology* 66:747–752.

Kummer, H. (1968) *Social Organization of Hamadrayas Baboons*. Chicago: University of Chicago Press.

Lancaster, J., and Lee, R. (1965) The annual reproductive cycle in monkeys and apes. In I. DeVore (ed.), *Primate Behavior—Field Studies of Monkeys and Apes*, pp. 486–513. New York: Holt, Rinehart & Winston.

Larsson, K. (1956) *Conditioning and Sexual Behavior in the Male Albino Rat*. Stockholm: Almquist & Wiksell.

Larsson, K. (1959a) The effect of restraint upon copulatory behavior in the rat. *Animal Behavior* 7:23–25.

Larsson, K. (1959b) Experience and maturation in the development of sexual behavior in male rats. *Behaviour* 14:101–107.

Larsson, K. (1962a) Mating behavior in male rats after cerebral cortex ablation: I. Effects of lesions in the dorsolateral and the median cortex. *Journal of Experimental Zoology* 151:167–176.

Larsson, K. (1962b) Spreading cortical depression and the mating behavior in male and female rats. *Zeitschrift für Tierpsychologie* 19:321–331.

Larsson, K. (1963) Non-specific stimulation and sexual behavior in the male rat. *Behaviour* 20:110–114.

Larsson, K. (1964) Mating behavior in male rats after cerebral cortex ablation: II. Effects of lesions in the frontal lobes compared to lesions in the posterior half of the hemispheres. *Journal of Experimental Zoology* 155:203–214.

Larsson, K. (1966) Individual differences in reactivity to androgen in male rats. *Physiology and Behavior* 1:255–258.

Law, D. T., and Meagher, W. (1958) Hypothalamic lesions and sexual behavior in the female rat. *Science* 128:1626–1627.

Law, O. T., and Sackett, G. (1965/1966) Hypothalamic potentials in the female rat evoked by hormones and by vaginal stimulation. *Neuroendocrinology* 1:31–44.

Laws, D., and Rubin, H. (1969) Instructional control of an autonomic sexual response. *Journal of Applied Behavioral Analysis* 2:93–99.

LeBoeuf, B. J., and Peterson, R. S. (1969) Social status and mating activity in elephant seals. *Science* 163:91–93.

Lee, S. Van der, and Boot, L. (1955) Spontaneous pseudopregnancy in mice. *Acta Physiol. Pharmacol. Neerl.* 4.442–443.

Lee, S. Van der, and Boot, L. (1956) Spontaneous pseudopregnancy in mice. II. *Acta. Physiol. Pharmacol. Neerl.* 5:213–214.

Lehrman, D. S. (1961) Gonadal hormones and parental behavior in birds and infrahuman mammals. In W. C. Young (ed.) *Sex and Internal Secretions*, vol. II. Baltimore: Williams & Wilkins.

LeMagnen, J. (1952) Les phenomenes olfactosexuals chez le rat blanc. *Archives Scientifique Physiologique* 6:295–332.

Lentz, T. H. (1966) *The Cell Biology of Hydra.* Amsterdam: North Holland.

Levine, S. (1968) Hormones and conditioning. *Nebraska Symposium on Motivation: 1968*, pp. 85–201. Lincoln: University of Nebraska Press.

Levine, S., and Mullins, R. F., Jr. (1964) Estrogen administered neonatally affects adult sex behavior in male and female rats. *Science* 144:185–187.

Lincoln, D. W., and Cross, B. A. (1967) Effect of oestrogen on the responsiveness of neurones in the hypothalamus, septum and preoptic area of rats with light-induced persistent oestrus. *Journal of Endocrinology* 37:191–203.

Lisk, R. D. (1960) A comparison of the effectiveness of intravenous, as opposed to sub-cutaneous, injection of progesterone for the induction of estrous behavior in the rat. *Canadian Journal of Biochemistry and Physiology* 38:1381–1383.

Lisk, R. D. (1962) Diencephalic placement of estradiol and sexual receptivity in the female rat. *American Journal of Physiology* 203:493–496.

Lisk, R. D. (1966) Increased sexual behavior in the male rat following lesions in the mammillary region. *Journal of Experimental Zoology* 161:129–136.

Lisk, R. D. (1967) Neural localization for androgen activation of copulatory behavior in the rat. *Endocrinology* 80:754–761.

Lisk, R. D. (1969) Progesterone: Biphasic effects on the lordosis response in adult or neonatally gonadectomized rats. *Neuroendocrinology* 5: 149–160.

Lloyd, C. W. (1968) The influence of hormones on human sexual behavior. In *Clinical Endocrinology*, vol. 2, pp. 665–674. New York: Grune & Stratton.

Lloyd, J. E. (1966) Studies on the flash communication system in photinus fireflies. *Mis. Publ. Mus. Zool.*, No. 130. Ann Arbor: University of Michigan.

Loomis, W. F. (1959) The sex gas of hydra. *Scientific American* 200:145–156.

Loraine, J. A., Ismail, A. A. A., Adamopolous, D. A., and Dove, G. A. (1970) Endocrine function in male and female homosexuals. *British Medical Journal* 4:406–409.

Lorenz, K. (1957) Companionship in bird life—fellow members of the species as releasers of social behavior. In C. Schiller (ed.), *Instinctive Behavior*, pp. 83–128. New York: International Universities.

Lorenz, K. (1964) Ritualized fighting. In J. D. Carthy, and F. J. Ebling (eds.), *The Natural History of Aggression*, pp. 39–50. New York: Academic.

Lorenz, K. (1965) *Evolution and Modification of Behavior*. Chicago: University of Chicago Press.

Malinowski, B. (1927) *Sex and Repression in Savage Society*. Cleveland: World Publishing.

Malinowski, B. (1929) *The Sexual Life of Savages in North-Western Melanesia*. New York: Harcourt Brace Jovanovich.

Manning, A. (1961) The effects of artificial selection for mating speed in *Drosophila melanogaster*. *Animal Behavior* 9:82–92.

Manning, A. (1963) Selection for mating speed in *Drosophila melanogaster* based on the behavior of one sex. *Animal Behavior* 11:116–120.

Manning, A. (1965) Drosophila and the evolution of behavior. *Viewpoints in Biology* 4, pp. 125–169. London: Butterworth.

Margoles, S. (1970) Homosexuality: A new endocrine correlate. *Hormones and Behavior* 1:151–156.

Marler, P. (1968) Aggregation and dispersal: Two functions in primate communication. In P. Jay (ed.), *Primates: Studies in Adaptation and Variability*, pp. 420–438. New York: Holt, Rinehart & Winston.

Marmor, J. (1965) Introduction. In J. Marmor (ed.), *Sexual Inversion*, pp. 1–24. New York: Basic Books.

Marsden, H., and Bronson, F. (1964) Estrous synchrony in mice: Alteration by exposure to male urine. *Science* 144:1469.

Masters, W. H. (1960) The sexual response cycle of the human female. *West. J. Surg., Obst. & Gynec.* 68:57–72.

Masters, W. H. and Johnson, V. (1966) *Human Sexual Response.* Boston: Little, Brown.

Matthews, L. H. (1939) Visual stimulation and ovulation in pigeons. *Proceedings of the Royal Society (London),* 126B:557–580.

Mayr, E. (1950) The role of antennae in the mating behavior of female Drosophila. *Evolution* 4:149–154.

Mayr, E. (1957) The species problem. *American Association for the Advancement of Science,* Pub. 50:1–22.

Mayr, E. (1963) *Animal species and evolution.* Cambridge, Mass.: Harvard University Press.

Mead, M. (1961) Cultural determinants of sexual behavior. In W. Young (ed.), *Sex and Internal Secretions,* vol. II, pp. 1433–1479. Baltimore: Williams & Wilkins.

Meinwald, J., Meinwald, Y., and Mazzocchi, P. (1969) Sex pheromone of the queen butterfly: Chemistry. *Science* 164:1174–1175.

Merari, A. (1970) Personal communication.

Meyer, H. (1938) Investigation concerning the reproductive behavior of *Molienisia formosa. Journal of Genetics* 36 (3):329–366.

Michael, R. P. (1969) Behavioral effects of gonadal hormones and contraceptive steroids in primates. In Salhanick, H. A., Kipnis, D. M., and Vandewiele, R. (eds.), *Metabolic Effects of Gonadal Hormones and Contraceptive Steroids,* pp. 706–721. New York: Plenum.

Michael, R. P., Herbert, J., and Welegalla, J. (1967) Ovarian hormones and the sexual behavior of the male rhesus monkey (*Macaca mulatta*). *Journal of Endocrinology* 39:81–93.

Michael, R. P., and Keverne, E. B. (1968) Pheromones in the communication of sexual status in primates. *Nature* 218:746–749.

Michael, R. P., Keverne, E. B., and Bonsall, R. W. (1971) Pheromones: Isolation of male sex attractants from a female primate. *Science* 172:964–966.

Michael, R. P., and Zumpe, D. (1970) Rhythmic changes in the copulatory frequency of Rhesus monkeys (*Macaca mulatta*) in relation to the menstrual cycle and a comparison with the human cycle. *Journal of Reproduction and Fertility* 21:119–201.

Miller, N., Hubert, G., and Hamilton, J. B. (1938) Mental and behavioral changes following male hormone treatment of adult castration hypogonadism and psychic impotence. *Proceedings of the Society for Experimental Biology and Medicine* 38:538–540.

Missakian, E. (1969) Reproductive behavior of socially deprived male rhesus monkeys (*Macaca mulatta*). *Journal of Comparative Physiological Psychology* 69:403–407.

Mitchell, G. (1973) Comparative development of social and emotional behavior. In G. Bermant (ed.), *Perspectives in Animal Behavior,* pp. 102–128. Chicago: Scott, Foresman.

Moment, G. (1958) *General Zoology.* Cambridge: Houghton Mifflin.

Money, J. (1961a) Sex hormones and other variables in human eroticism. In W. Young (ed.), *Sex and Internal Secretions*, vol. II, pp. 1383–1400. Baltimore: Williams & Wilkins.

Money, J. (1961b) Hermaphroditism. In A. Ellis and A. Abarbanel (eds.), *The Encyclopedia of Sexual Behavior*, vol. I., pp. 472–484. New York: Hawthorn.

Money, J. (1965a) Influence of hormones on sexual behavior. *Annual Review of Medicine* 16:67–82.

Money, J. (1965b) *Sex Research: New Developments*. New York: Holt, Rinehart & Winston.

Money, J., and Ehrhardt, A. A. (1968) Prenatal hormonal exposure: Possible effects on behavior in man. In R. P. Michael (ed.), *Endocrinology and Human Behavior*. London: Oxford University Press.

Money, J., and Ehrhardt, A. A. (1972) *Man and Woman, Boy and Girl*. Baltimore: Johns Hopkins Press.

Morgan, R. (ed.) (1970) *Sisterhood Is Powerful*. New York: Random House.

Muller-Schwarze, D. (1969) Complexity and relative specificity in a mammalian pheromone. *Nature* 223:525–526.

Mullins, R. F., Jr., and Levine, S. (1968a) Hormonal determinants during infancy of adult sexual behavior in the female rat. *Physiology and Behavior* 3:333–338.

Mullins, R. F., Jr., and Levine, S. (1968b) Hormonal determinants during infancy of adult sexual behavior in the male rat. *Physiology and Behavior* 3:339–343.

Mullins, R. F., Jr., and Levine, S. (1969) Differential sensitivity of penile tissues by sexual hormones in infancy. *Communications in Behavioral Biology* 3:1–4.

MacKinnon, P. C. B. (1970) A comparison of protein synthesis in the brains of mice before and after puberty. *Journal of Physiology* 210: 10–11.

McCary, J. (1967) *Human Sexuality*. New York: Van Nostrand.

McClearn, G. (1967) Genes, generality, and behavioral research. In J. Hirsch (ed.), *Behavior–Genetic Analysis*, pp. 307–321. New York: McGraw-Hill.

McClintock, M. K. (1971) Menstrual synchrony and suppression. *Nature, London* 229:244–245.

McDonald, P. G., Vidal, N., and Beyer, C. (1970) Sexual behavior in the ovariectomized rabbit after treatment with different amounts of gonadal hormones. *Hormones and Behavior* 1:161–172.

McGill, T. E. (1962) Sexual behavior in three inbred strains of mice. *Behaviour* 19:341–350.

McGill, T. E. (1970) Induction of luteal activity in female house mice. *Hormones and Behavior* 1:211–222.

McGill, T. E., and Blight, W. (1963) The sexual behavior of hybrid male mice compared with the sexual behavior of males of the inbred parent strains. *Animal Behavior* 11:480–483.

McGill, T. E., and Ransom, T. (1968) Genotypic change affecting conclusions regarding the mode of inheritance of elements of behavior. *Animal Behavior* 16:88–91.

Nadler, R. D. (1968) Masculinization of female rats by intracranial implantation of androgen in infancy. *Journal of Comparative Physiological Psychology* 66:157–167.

Nalbandov, A. (1964) *Reproductive Physiology*, 2d ed. San Francisco: Freeman.

Napier, J., and Napier, P. (1967) *A Handbook of Living Primates*. New York: Academic.

Netter, F. H. (1953) *The Ciba Collection of Medical Illustrations, vol. I, Nervous System*. Summit, N.J.: Ciba Pharmaceutical Products Co.

Niemi, M., and Ikonen, M. (1963) Histochemistry of the Leydig cells in the postnatal prepubertal testis of the rat. *Endocrinology* 72:443–448.

Noble, G. K., and Greenberg, B. (1941) Induction of female behavior in male *Anolis carolinensis* with testosterone propionate. *Proceedings of the Society for Experimental Biology and Medicine* 47:32–37.

Odell, W. D., and Moyer, D. L. (1971) *Physiology of Reproduction*. St. Louis: Mosby.

Olsen, M. W. (1962) The occurrence and possible significance of parthenogenesis in eggs of mated turkeys. *Journal of Genetics* 58:1–6.

Olsen, M. W., and Marsden, S. J. (1954) Development of unfertilized turkey eggs. *Journal of Experimental Zoology* 126:337–347.

Orbach, J. (1961) Spontaneous ejaculation in the rat. *Science* 134:1072–1073.

Orchard, W. H. (1969) Psychiatric aspects of oral contraceptives. *Medical Journal of Australia* 1:872–876.

Palka, Y. S., and Sawyer, C. H. (1966) Induction of estrous behavior in rabbits by hypothalamic implants of testosterone. *American Journal of Physiology* 211:225–228.

Parkes, A., and Bruce, H. (1961) Olfactory stimuli in mammalian reproduction. *Science* 134:1049–1054.

Parsons, P. (1967) *The Genetic Analysis of Behavior*. London: Methuen.

Patterson, R. L. S. (1968) Identification of 3-hydroxy-5-androst-16-ene as the mask odour component of boar submaxillary salivary gland and its relationship to the sex odour taint in pork meat. *Journal of the Science of Food and Agriculture* 19:434.

Pelligrino, L., and Cushman, A. (1967) *A Stereotoxic Atlas of the Rat Brain*. New York: Appleton.

Pepelko, W., and Clegg, M. (1965) Studies of mating behavior and some factors influencing the sexual response in the male sheep, *Ovis aries*. *Animal Behavior* 13:249–259.

Perdeck, A. C. (1958) The isolating value of specific song patterns in two sibling species of grasshoppers (*Chorthippus brunneus* Thunb. and *C. bigguttulus* L.). *Behaviour* 12:1–75.

Pfeiffer, C. A. (1936) Sexual differences of hypophyses and their determination by the gonads. *American Journal of Anatomy* 58:195–225.

Phoenix, C. H., Goy, R. W., Gerall, A. A., and Young, W. C. (1959) Organizing action of prenatally administered testosterone propionate on the tissues mediating mating behavior in the female guinea pig. *Endocrinology* 65:369–382.

Pierce, J., and Nuttall, R. (1961) Self-paced sexual behavior in the female rat. *Journal of Comparative Physiological Psychology* 54:310–313.

Pincus, G. (1965) *The Control of Fertility.* New York: Academic.

Pliske, T., and Eisner, T. (1969) Sex pheromone of the queen butterfly: Biology. *Science* 164:1170–1172.

Powers, J. B. (1970) Hormonal control of sexual receptivity during the estrus cycle of the rat. *Physiology and Behavior* 5:831.

Powers, J. B., and Zucker, I. (1969) Sexual receptivity in pregnant and pseudopregnant rats. *Endocrinology* 84:820–827.

Price, D., Ortiz, E., and Zaaijer, J. J. P. (1963) Secretion of androgenic hormone by testes and adrenal glands of fetal guinea pigs. *American Zoology* 3:553–554.

Pumpian-Mindlin, E. (ed.) (1952) *Psychoanalysis as Science.* Stanford, Calif.: Stanford University Press.

Rado, S. (1940) A critical examination of the theory of bisexuality. *Psychosomatic Medicine* 2:459–467.

Rado, S. (1949) An adaptational view of sexual behavior. In P. Hoch and J. Zubin (eds.), *Psychosexual Development in Health and Disease*, pp. 159–189. New York: Grune & Stratton.

Ramirez, V., Komisaruk, B., Whitmoyer, D., and Sawyer, C. (1967) Effects of hormones and vaginal stimulation on the EEG and hypothalamic units in rats. *American Journal of Physiology* 212:1376–1384.

Rapaport, D. (1959) The structure of psychoanalytic theory: A systematizing attempt. In S. Koch (ed.), *Psychology: A Study of a Science*, vol. 3, pp. 55–183. New York: McGraw-Hill.

Rapaport, D. (1960) On the psychoanalytic theory of motivation. In M. Jones (ed.), *The Nebraska Symposium on Motivation: 1960*, pp. 173–247. Lincoln: Nebraska University Press.

Rechy, J. (1963) *City of Night.* New York: Grove.

Reichlin, S. (1968) Neuroendocrinology. In R. H. Williams (ed.), *Textbook of Endocrinology*, pp. 967–1016. Philadelphia: Saunders.

Report of the Commission on Obscenity and Pornography. (1970) Washington, D.C.: GPO.

Resko, J. A., Feder, H. H., and Goy, R. W. (1968) Androgen concentrations in plasma and testes of developing rats. *Journal of Endocrinology* 40:485–491.

Reuben, D. (1970) *Everything You Always Wanted to Know About Sex but Were Afraid to Ask.* New York: McKay.

Ribbands, C. (1964) *The Behaviour and Social Life of Honeybees.* New York: Dover.

Riddiford, L. M. (1967) Trans-2-Hexenal: Mating stimulant for Polyphemus moths. *Science* 158:139–140.

Riddiford, L. M., and Williams, C. M. (1967) Volatile principle from oak leaves: Role in sex life of the Polyphemus moth. *Science* 155:589–590.

Roberts, R. (1967) Some concepts and methods in quantitative genetics. In J. Hirsch (ed.), *Behavior–Genetic Analysis*, pp. 214–286. New York: McGraw-Hill.

Roeder, K. (1963) *Nerve Cells and Insect Behavior*. Cambridge, Mass.: Harvard University Press.

Roelofs, W., and Comeau, A. (1969) Sexpheromone specificity: Taxonomic and evolutionary aspects in *Lepidoptera. Science* 165:398–399.

Romm, M. (1965) Sexuality and homosexuality in women. In J. Marmor (ed.), *Sexual Inversion*, pp. 282–301. New York: Basic Books.

Rose, Robert M., Bourne, Peter G., Poe, Richard O., Mougey, Edward H., Collins, David R., and Mason, John W. (1969) Androgen responses to stress. II. Excretion of testosterone, epitestosterone, androsterone, and etiocholanolone during basic combat training and under threat of attack. *Psychosomatic Medicine* 31:418–436.

Rosenblatt, J. (1965) Effects of experience on sexual behavior in cats. In F. Beach (ed.), pp. 416–439. *Sex and Behavior*. New York: Wiley.

Rosenblatt, J., and Aronson, L. R. (1958) The decline of sexual behavior in male cats after castration with special reference to the role of prior sexual experience. *Behaviour* 12:285–338.

Ross, J., Claybough, C., Clemens, L. G., and Gorski, R. A. (1971) Short latency induction of estrous behavior with intra-cerebral gonadal hormones in ovariectomized rats. *Endocrinology* 81:32–38.

Rowan, W. (1938) Light and seasonal reproduction in animals. *Biological Review* 13:374–402.

Rowell, T. (1963) Behavior and female reproductive cycles of rhesus macaques. *Journal of Reproduction and Fertility* 6:193–203.

Rowell, T. (1967) Female reproductive cycles and behavior of baboons and rhesus macaques. In S. Altmann (ed.), *Social Communication Among Primates*, pp. 13–52. Chicago: University of Chicago Press.

Rozin, P., and Kalat, J. (1971) Specific hungers and poison avoidance as adaptive specializations of learning. *Psychological Review* 78:459–486.

Saayman, G. (1970) The menstrual cycle and sexual behavior in a troop of free ranging chacma baboons (*Papio ursinus*). *Folia Primatologica* 12:81–110.

Saginor, M., and Horton, R. (1968) Reflex release of gonadotropin and increased plasma testosterone concentration in male rabbits during copulation. *Endocrinology* 82:627–628.

Salhanick, H. A., Kipnis, D. M., and Vande Wiele, R. L. (eds.). (1969) *Metabolic Effects of Gonadal Hormones and Contraceptive Steroids.* New York: Plenum.

Salmon, U. J., and Geist, S. H. (1943) Effect of androgens upon libido in women. *Journal of Clinical Endocrinology and Metabolism* 3:235–238.

Sawyer, C. H. (1957) Endocrine malfunctioning and homosexuality. *Practitioner* 172:374–377.

Sawyer, C. H. (1959) Nervous control of ovulation. In C. W. Lloyd (ed.), *Recent Progress in the Endocrinology of Reproduction.* New York: Academic.

Sawyer, C. H. (1960) Reproductive behavior. In J. Field (ed.), *Handbook of Physiology. Section 1: Neurophysiology,* vol. II, pp. 1225–1240. Washington, D.C.: The American Physiological Society.

Sawyer, C. H., and Everett, J. W. (1959) Stimulatory and inhibitory effects of progesterone on the release of pituitary ovulating hormone in the rabbit. *Endocrinology* 65:644–651.

Sawyer, C. H., and Markee, J. E. (1959) Estrogen facilitation of release of pituitary ovulating hormone in the rabbit in response to vaginal stimulation. *Endocrinology* 65:614–621.

Schaller, G. (1965) The behavior of the mountain gorilla. In I. DeVore (ed.), *Primate Behavior—Field Studies of Monkeys and Apes,* pp. 324–367. New York: Holt, Rinehart & Winston.

Schein, M. W. (1965) The physical environment and behavior. In E. Hafez (ed.), *The Behavior of Domestic Animals,* pp. 82–95. Baltimore: Williams & Wilkins.

Schinckel, P. G. (1954) The effect of the ram on the incidence and occurrence of oestrus in ewes. *Australian Veterinary Journal* 30:189.

Schneider, D., and Seiht, U. (1969) Sex pheromone of the Queen butterfly: Electroantennogram responses. *Science* 164:1172–1173.

Schon, M., and Sutherland, A. M. (1960) The role of hormones in human behavior. III. Changes in female sexuality after hypophysectomy. *Journal of Clinical Endocrinology and Metabolism* 20:833–841.

Schreiner, L., and Kling, A. (1956) Rhinencephalon and behavior. *American Journal of Physiology* 184:486–490.

Schwartz, N. B. (1969) A model for control of the rat estrous cycle. *Journal of Basic Engineering* 91:381.

Schwartz, N. B., and Talley, W. L. (1965) Effects of acute ovariectomy on mating in the cyclic rat. *Journal of Reproduction and Fertility* 10: 463–466.

Scott, J. P. (1958) *Animal Behavior.* Chicago: University of Chicago Press.

Selby, H. (1957) *Last Exit to Brooklyn.* New York: Grove.

Seligman, M. (1970) On generality of the laws of learning. *Psychological Review* 77:406–418.

Selinger, H., and Bermant, G. (1967) Hormonal control of aggressive behavior in Japanese quail (*Coturnix coturnix japonica*). *Behaviour* 28: 255–268.

Sherrington, C. (1947) *The Integrative Action of the Nervous System,* 2d ed. New Haven: Yale University Press.

Signoret, J. P. (1970) Reproductive behavior in pigs. *Journal of Reproduction and Fertility Suppl.,* 11:105–117.

Simpson, G. G. (1949) *The Meaning of Evolution.* New Haven: Yale University Press.

Singer, J. (1968) Hypothalamic control of male and female sexual behavior in female rats. *Journal of Comparative Physiological Psychology* 66:738–742.

Sluckin, W. (1965) *Imprinting and Early Learning*. Chicago: Aldine.

Smith, E. R., and Davidson, J. M. (1968) Role of estrogen in the cerebral control of puberty in female rats. *Endocrinology* 82:100–108.

Sokal, R. R., and Sneath, P. H. A. (1963) *Principles of Numerical Taxonomy*. San Francisco: Freeman.

Sperry, R. (1969) A modified concept of consciousness. *Psychological Review* 76:532–536.

Spiess, E., and Langer, B. (1961) Chromosomal adaptive polymorphism in *Drosophila persimilis*. III. Mating propensity of homokanyotypes. *Evolution* 15:535–544.

Spiess, E., and Langer, B. (1964) Mating speed control by gene arrangement carriers in *Drosophila persimilis*. *Evolution* 18:430–444.

Spiess, E., and Spiess, L. (1967) Mating propensity, chromosomal polymorphism, and dependent conditions in *Drosophila persimilis*. *Evolution* 21:672–678.

Spieth, H. (1952) Mating behavior within the genus *Drosophila* (Dystern). *Bulletin of the American Museum of Natural History* 99:395–474.

Spieth, H. T. (1958) Behavior and isolating mechanisms. In A. Roe and G. G. Simpson (eds.), *Behavior and Evolution*. New Haven: Yale University Press, pp. 363–389.

Srb, A. M., Owen, R. D., and Edgar, R. S. (1965) *General Genetics*, 2d ed. San Francisco: Freeman.

Staats, J. (1966a) The laboratory mouse. In E. Green (ed.), *The Biology of the Laboratory Mouse*, 2d ed., pp. 1–10. New York: McGraw-Hill.

Staats, J. (1966b) Nomenclature. In E. Green (ed.), *The Biology of the Laboratory Mouse*, 2d ed., pp. 45–50. New York: McGraw-Hill.

Staples, R. E. (1967) Behavioral induction of ovulation in the oestrous rabbit. *Journal of Reproduction and Fertility* 13:429–435.

Stebbins, G. (1966) *Processes of Organic Evolution*. Englewood Cliffs, N.J.: Prentice-Hall.

Stern, J. J. (1969) Neonatal castration, androstenedione, and the mating behavior of the male rat. *Journal of Comparative Physiological Psychology* 69:608–612.

Stone, C. P. (1924) A note on "female" behavior in adult male rats. *American Journal of Physiology* 68:39.

Strashimirov, D., Bohus, B., and Kovacs, S. (1969) Pituitary-thyroid function of the rat after suppression of ACTH release by dexamethosone implants in the median eminence. *Acta Physilogica Acad. Scientiarum Hungarieve* 35:335–344.

Svomalainen, E. (1950) Parthenogenesis in vertebrates. *Advances in Genetics* 3:193–253.

Taleisnik, S., Caligaris, L., and Astrada, J. J. (1966) Effect of copulation on the release of pituitary gonadotropins in male and female rats. *Endocrinology* 79:49–54.

Tauber, E. S. (1940) Effects of castration upon the sexuality of the adult male. *Psychosomatic Medicine* 2:74.

Teitelbaum, P. (1967) The biology of drive. In G. Quarton, T. Melnechulz, and F. Schimitt (eds.), *The Neurosciences*, pp. 557–567. New York: Rockefeller University Press.

Terzian, H., and Dall Ore, G. (1955) Syndrome of Kluver and Bucy reproduced in man by bilateral removal of the temporal lobes. *Neurology* 5:373–380.

Thomas, T. R., and Neiman, C. N. (1968) Aspects of copulatory behavior preventing atrophy in male rats' reproductive system. *Endocrinology* 83:633–635.

Thompson, R. (1967) *Foundations of Physiological Psychology*. New York: Harper & Row.

Thorpe, D. (1967) Basic parameters in the reaction of ferrets to light. In G. Wolstenholme, and M. O'Connor (eds.), *Effects of External Stimuli on Reproduction*, pp. 53–66. Boston: Little, Brown.

Tiefer, L. (1970) Gonadal hormones and mating behavior in the adult golden hamster. *Hormones and Behavior* 1:189–202.

Tinbergen, N. (1951) *The Study of Instinct*. London: Oxford University Press.

Tinbergen, N. (1960) *The Herring Gull's World*. New York: Basic Books.

Tinbergen, N. (1965) Some recent studies of the evolution of sexual behavior. In F. Beach (ed.), *Sex and Behavior*, pp. 1–33. New York: Wiley.

Tourney, Garfield, and Hatfield, Lon M. (1973) Androgen metabolism in schizophrenics, homosexuals, and normal controls. *Biological Psychiatry* 6:23–36.

Tyler, A. (1967) Introduction: Problems and procedures of comparative gametology and syngamy. In Metz, C., and Monroy, A. (eds.), *Fertilization*, pp. 1–26. New York: Academic.

Udry, J. R., and Morris, N. M. (1968) Distribution of coitus in the menstrual cycle. *Nature* 220:593–596.

Uexkull, J. von (1921) *Umwelt und Innenwelt der Tiere*. Berlin: Springer.

Ulrich, R., and Azrin, N. (1962) Reflexive fighting in response to aversive stimulation. *Journal of Experimental Analysis of Behavior* 5:511–520.

Valenstein, E., Riss, W., and Young, W. (1955) Experiential and genetic factors in the organization of sexual behavior in male guinea pigs. *Journal of Comparative Physiological Psychology* 48:397–403.

Vandenbergh, J., and Vessey, S. (1968) Seasonal breeding of free-ranging rhesus monkeys and related ecological factors. *Journal of Reproduction and Fertility* 15:71–79.

Van Dis, H., and Larsson, K. (1971) Induction of sexual arousal in the castrated male rat by intracranial stimulation. *Physiological Behavior* 6:85–86.

Van Lawick-Goodall, J. (1968) The behavior of free-living chimpanzees in the Gombe Stream reserve. *Animal Behavior Monograph* 1 (3) 161–311.

Vaughn, E., and Fisher, A. (1962) Male sexual behavior induced by intracranial electrical stimulation. *Science* 137:758–760.

Walker, P. (1967) Protozoans. In *The Larousse Encyclopedia of Animal Life*, pp. 11–31. New York: McGraw-Hill.

Wallace, B., and Srb, A. (1961) *Adaptation*. Englewood Cliffs, N.J.: Prentice-Hall.

Wang, G. (1923) The relation between "spontaneous" activity and oestrus cycle in the white rat. *Comparative Psychology Monographs* 2 (whole No. 6).

Ward, I. L. (1969) Differential effects of pre- and postnatal androgen on the sexual behavior of intact and spayed female rats. *Hormones and Behavior* 1:25–36.

Watson, J. D. (1968) *The Double Helix*. New York: Atheneum.

Watson, J. D., and Crick, F. H. C. (1953) A structure for deoxyribose nucleic acid. *Nature* 171:737–738.

Waxenberg, S. E., Drellich, M. G., and Sutherland, A. M. (1959) The role of hormones in human behavior: I. Changes in female sexuality after adrenalectomy. *Journal of Clinical Endocrinology and Metabolism* 19:193–202.

Weick, R. F., Smith, E. R., Dominguez, D., Dhariwal, A. P. S., and Davidson, J. M. (1971) Mechanism of stimulatory feedback effect of estradiol benzoate on the pituitary. *Endocrinology* 88:293–301.

West, C. D., Bamast, B. L., Sarro, S. D., and Pearson, O. H. (1956) Conversion of testosterone to estrogens in castrated, adrenalectomized human females. *Journal of Biological Chemistry* 218:409–418.

Wetherbee, D. (1961) Investigations in the life history of the Common Coturnix. *American Midland Naturalist* 65:168–186.

Whalen, R. E. (1966) Sexual motivation. *Psycholo. Rev.* 73:151–163.

Whalen, R.E. (1967) *Hormones and Behavior*. New York: Van Nostrand.

Whalen, R. E., and Edwards, D. A. (1967) Hormonal determinants of the development of masculine and feminine behavior in male and female rats. *Anatomical Record* 157:173–180.

Whalen, R. E., and Robertson, R. T. (1968) Sexual exhaustion and recovery of masculine copulatory behavior in virilized female rats. *Psychonomic Science* 11:319–320.

White, M. (1954) *Animal Cytology and Evolution*. New York: Cambridge University Press.

Whitten, W. K. (1958) Modification of the oestrous cycle of the mouse by external stimuli associated with the male. *Journal of Endocrinology* 17:307–313.

Whitten, W. K. (1966) Pheromones and mammalian reproduction. In A. McLaren (ed.), *Advances in Reproductive Physiology*. London: Logos (Academic) Press, 155–178.

Whitten, W. K., Bronson, F., and Greenstein, J. (1968) Estrus-inducing pheromone of male mice: Transport by movement of air. *Science* 161: 584–585.

Wickler, W. (1967) Socio-sexual signals and their intra-specific imitation among primates. In D. Morris (ed.), *Primate Ethology*, pp. 69–147. Chicago: Aldine.

Wiener, H. (1966) External chemical messengers 1. Emission and reception in man. *N.Y. State Journal of Medicine* 66:3153–3170.

Wiener, H. (1967) *N.Y. State Journal of Medicine* 67:1144.

Wilbur, C. (1965) Clinical aspects of female homosexuality. In J. Marmor (ed.), *Sexual Inversion*, pp. 268–281. New York: Basic Books.

Wilcock, J. (1969) Gene action and behavior: An evaluation of major gene plerotropism. *Psychological Bulletin* 72:1–29.

Wilson, E. O. (1965) Chemical communication in the social insects. *Science* 149:1064–1071.

Wilson, J. Adler, N., and LeBoeuf, B. (1965) The effects of intromission frequency on successful pregnancy in the female rat. *Proceedings of the National Academy of Science* 53:1392–1395.

Wilson, J., Kuehn, R., and Beach, F. (1963) Modifications in the sexual behavior of male rats produced by changing the stimulus female. *Journal of Comparative Physiological Psychology* 56:636–644.

Wilson, M., and Bermant, G. (1972) An analysis of social interactions in the Japanese quail (*Coturnix coturnix japonica*). *Animal Behavior* 20: 252–258.

Winick, C. (1968) *The New People-Desexualization in American Life*. New York: Pegasus Press.

Winton, F. R., and Bayliss, L. E. (1968) *Human Physiology*, 6th ed. London: Churchill Livingstone.

Woodard, A., Abplanalp, H., and Wilson, W. (1965) *Japanese Quail Husbandry in the Laboratory. (Coturnix coturnix japonica)*. Davis, Calif.: Department of Poultry Husbandry, University of California.

Yamamoto, J., and Seeman, W. (1960) The psychological study of castrated male rats. *Psychiatric Research Reports* 12:97–103.

Ying, S., and Meyer, R. K. (1969) Effect of coitus on barbiturate-blocked ovulation in immature rats. *Fertility and Sterility* 20:772–778.

Young, W. C. (1961) The hormones and mating behavior. In W. C. Young (ed.), *Sex and Internal Secretions*, vol. II, pp. 1173–1239. Baltimore: Williams & Wilkins.

Young, W. C., and Orbison, W. D. (1944) Changes in selected features of behavior in pairs of oppositely sexed chimpanzees during the sexual cycle after ovariectomy. *Journal of Comparative Physiological Psychology* 37:107–143.

Zarrow, M. X., Brody, P. N., and Denenberg, V. H. (1968) The role of progesterone in behavior. In M. Diamond (ed.), *Reproduction and Sexual Behavior*, pp. 341–361. Bloomington: Indiana University Press.

Zarrow, M. X., Campbell, P. S., and Clark, J. H. (1968) Pregnancy following coital-induced ovulation in a spontaneous ovulator. *Science* 159:329–330.

Zarrow, M. X., and Clark, J. H. (1968) Ovulation following vaginal stimulation in a spontaneous ovulator and its implications. *Journal of Endocrinology* 40:343–352.

Zarrow, M. X., Naqvi, R. H., and Denenberg, V. H. (1969) Androgen-induced precocious puberty in the female rat and its inhibition by hippocampal lesions. *Endocrinology* 84:14–19.

Zitrin, A., Jaynes, J., and Beach, F. (1956) Neural mediation of mating in male cats: III. Contribution of the occipital, parietal, and temporal cortex. *Journal of Comparative Neurology* 105:111–125.

Zucker, I. (1966) Facilitatory and inhibitory effects of progesterone on sexual responses of spayed guinea pigs. *Journal of Comparative Physiological Psychology* 62:376–381.

Zucker, I. (1967) Actions of progesterone in the control of sexual receptivity of the spayed female rat. *Journal of Comparative Physiological Psychology* 63:313–316.

Zuckerman, M. (1971) Physiological measures of sexual arousal in the human. *Psychological Bulletin* 75:297–329.

Index of Names

Abplanalp, H., 77
Adams-Smith, W. N., 195
Adler, A., 263
Adler, N., 62, 80, 107, 171
Albe-Fessard, D., 100
Alleva, J., 64
Amoroso, E., 56, 58, 59
Anand, B., 102
Anderson, C. O., 201
Anderson, L., 91
Anderson, T., 112, 122
Aphrodite, 48
Apley, M., 254
Ardrey, R., 7, 94
Aristotle, 134
Armstrong, C., 48
Arnold, W., 96
Aron, C., 139, 167, 169
Aronson, L. R., 58, 59, 80, 150, 236
Arvidsson, T., 116
Asdell, S., 56, 57, 61, 166
Atz, J., 52
Azrin, N., 118

Baker, V. F., 167
Ball, J. A., 171, 240
Bard, P., 106, 152
Barfield, R., 117, 160
Barr, M., 246
Barraclough, C. A., 193, 200
Bartke, A., 65
Bastock, M., 28, 29, 30, 31
Bayliss, L. E., 249
Beach, F., 63, 78, 92, 95, 96, 106, 107, 112, 115, 119, 135, 143, 148, 150, 152,

161, 164, 175, 180, 181, 182, 195, 197, 201, 202, 220, 227, 228
Beamer, W., 89, 90, 116
Beatty, R., 44, 48
Benedek, T., 242
Berger, H., 101
Bermant, G., 78, 79, 80, 81, 82, 89, 91, 92, 112, 113, 115, 116, 117, 118, 120, 121, 122
Bertolini, A., 124
Beyer, C., 183
Bieber, I., 224, 225, 228, 230, 231
Bingham, H. C., 180
Blair, W. F., 21, 22, 23
Blandau, R. J., 172
Blight, W., 34, 35, 36
Bloch, G. J., 145, 236
Blumer, D., 121
Bolles, R., 4
Boot, L., 64
Borradaile, L., 49
Brazier, M., 99, 100, 101
Bremer, J., 233, 234, 239
Brien, P., 42
Bronson, F., 65, 66, 174, 175, 176
Brooks, C., 166
Brower, J., 69
Brower, L. P., 69, 70
Browman, E., 64
Brown, W. L., Jr., 22
Brown-Grant, K., 132, 137, 168, 169, 200
Bruce, H., 65, 174
Buchsbaum, R., 49, 50
Bullock, D. W., 136
Bullough, W., 56, 58

Bunnell, B., 117
Burge, K. G., 200
Burnett, A., 42, 43
Burton, M., 58
Butenandt, A., 66
Butler, C., 68, 69

Caggiula, A., 115, 117, 118, 119
Campbell, H. J., 201
Carr, W. J., 175
Caspari, E., 41
Caul, W. F., 175
Chace, L., 51
Chang, M., 62
Cheng, P., 148
Cherney, E., 91
Chester, R. V., 171, 172
Chhina, G., 102
Clark, E., 49
Clark, J. H., 168, 169
Clark, R. H., 99
Clayton, R. B., 203
Clegg, M., 89, 116
Clemens, L. G., 107, 115, 120, 159, 203, 204
Clemente, C., 121
Cofer, C., 254
Comeau, A., 68
Comfort, A., 177
Conner, R. L., 201
Conovitz, M., 117
Cooper, M., 80
Cowgill, U., 60, 61
Cracchiolo, E., 102
Crane, J., 17–18
Cranston, F., 69
Crick, F. H. C., 13
Cross, B. A., 102
Crossley, D. A., 184, 200
Cushman, A., 100

Dalle Ore, G., 121
Dantchakoff, V., 199
Darwin, C., 12–13, 17
Davenport, W., 217, 218, 220
Davidson, J. M., 117, 121, 135, 144, 145, 146, 150, 160, 161, 163, 168, 169, 184, 185, 196, 204, 236
Davis, D. E., 179
DeGroot, J., 100, 115, 121, 161
Dempsey, E. W., 136
Dethier, V., 98
DeVore, I., 85, 86
Dewsbury, D., 64, 117
Diamond, I., 98
Diamond, M., 172, 173, 199
Diehl, N., 42, 43
Dobzhansky, T., 15, 33
Dondey, M., 100
Dougherty, E. C., 41

Drori, D., 96, 169, 170, 174
DuMortier, B., 24

Eaton, G., 200
Eayrs, J. T., 124
Eccles, J., 98
Eckert, J. E., 44
Edgar, R. S., 26, 39, 41
Edwards, D. A., 137, 196, 200, 201
Ehrhardt, A. A., 229, 245, 250
Eisner, T., 70
Eleftheriou, B., 66
Ellenberger, H., 263
Ellis, A., 228
Ellis, H., 177
Endroczi, E., 169
Engel, P., 182
Ericsson, R. J., 167
Evans, R., 229
Evans, T. N., 128
Everett, J. W., 135, 137, 170
Everitt, B. J., 241, 242

Falconer, D., 8
Farber, L., 4
Farris, H. E., 78
Fechner, G., 252
Feder, H. H., 138, 141
Ferin, M., 139
Finger, F., 63
Firestone, S., 250
Fisher, A. E., 115, 116; 117, 182
Fletcher, R., 251, 252, 253˝
Folman, Y., 96, 169, 170, 174
Ford, C., 220, 227, 228
Fowler, H., 92
Freedman, L. Z., 240
Freud, S., 207, 251, 252, 253, 254, 255, 256, 257, 258, 259, 261, 263
Freund, K., 214
Friedan, B., 250
Frink, H., 254
Fuller, J., 27

Garcia, J., 102
Gary, N., 68
Gebhard, P. H., 224
Geist, S. H., 240
Gerall, A. A., 96, 204
Gerall, H., 96
Glasser, G., 102
Glick, I. D., 244
Glickman, S., 117, 121
Goldfoot, D. A., 197
Goldstein, A., 117
Goodman, E., 115, 117
Gorski, R. A., 107, 138, 193, 194, 200
Gowen, J., 246
Goy, R. W., 107, 115, 119, 138, 141, 147, 202

Grant, E. C. G., 244
Gray, D. A., 201, 202
Green, E., 34
Green, J., 121, 185
Greenberg, B., 182
Greenstein, J., 65
Greer, G., 250
Greig, F., 168, 169
Greulich, W. W., 167
Gruendel, A., 96
Grumbach, M. M., 188, 189, 190
Grunt, J. A., 147
Guttmacher, A., 60

Hafez, E., 58, 59, 61, 128
Hale, E., 70, 71, 72, 89, 91, 92
Hall, C., 251
Hall, K., 85, 86
Hall, W., 98
Hamilton, J. B., 33, 233, 234
Hampson, J. G., 232, 242, 245, 246, 247, 248
Hampson, J. L., 232, 242, 245, 246, 247, 248
Harlow, H. F., 96
Harrington, F. E., 167
Harrington, R. W., Jr., 49
Harris, G. W., 157, 158, 192, 195, 196
Harris, V. T., 15
Hart, B., 106, 107, 108, 109, 110, 111, 114, 150, 152, 162, 163, 236
Hartman, C. G., 168
Hartmann, G., 124
Hartmann, H., 251
Harvey, E., 62
Hastings, D. W., 235
Hatfield, L., 229
Haugen, C. M., 162, 163
Hebb, D. O., 8
Hediger, H., 94
Heimer, L., 115, 116, 117, 118, 119, 154
Herberg, L., 115, 117, 119
Herbert, J., 176, 240, 241, 242
Hercules, 41, 125
Hermes, 48
Herschberger, R., 250
Hertz, R., 136, 186
Hess, E., 251
Hinde, R., 4, 74, 76, 252
Hirsch, J., 27
Hitt, J., 115, 119
Hoebel, B., 100, 115, 117
Hooker, E., 224, 225, 226, 230, 231
Horsley, V., 99
Horton, R., 169
Hubbs, C. L., 47
Hubbs, L. C., 47
Hubert, G., 234
Hulet, C. V., 174
Hull, C. D., 102

Hutchison, J. B., 160, 167
Huxley, J., 13
Hyman, L., 48, 49

Ikonen, M., 195
Ingle, D., 149
Inman, N., 78

Jacobsohn, D., 192
Jansen, P., 117
Jaynes, J., 119
Jocasta, 258
Johnson, V., 81, 173, 206, 208, 210, 236, 239, 240, 257
Johnston, P. J., 160
Jones, E., 251
Jordan, T., 112
Jost, A., 187
Jung, C., 263

Kagan, J., 95, 96, 148
Kalat, J., 4
Kallman, F., 228
Kallman, K. D., 47
Kane, F. J., 244
Karlson, P., 66
Keller, H., 176
Kenney, A. M., 204
Kenniston, K., 4
Keverne, E. B., 175
Kim, C., 117
Kincl, F. A., 199
Kinsey, A., 207, 222, 223, 227, 229, 237, 238, 239, 242
Klein, M., 183
Kling, A., 121
Klock, J., 175
Klopfer, P. H., 14
Kollberg, S., 207, 209
Kolodny, R., 229
Komisaruk, B., 102, 167
Koranyi, L., 124
Krafft-Ebing, R., 253
Kuehn, R. E., 63, 92, 141, 201
Kummer, H., 67, 85, 86, 87, 88

Lancaster, J., 60
Langer, B., 31–32
Larsson, K., 64, 84, 112, 113, 115, 116, 117, 118, 119, 120, 150, 154, 236
Law, D. T., 107, 115
Laws, D., 214
Le Beau, J., 100
Le Boeuf, B. J., 62, 92
Lee, R., 60
Lee, S. Van der, 64
Lehrman, D. S., 124
LeMagnen, J., 175, 177
Lentz, T. H., 42, 43
Levine, S., 124, 195, 196, 197, 201, 204

Levinson, G., 161
Lincoln, D. W., 102
Lisk, R. D., 116, 137, 139, 158, 160
Lissak, K., 169
Lloyd, C. W., 236
Lloyd, J. E., 18, 19, 20, 21
Loomis, W. F., 43
Loraine, J. A., 229
Lorenz, K., 73, 94, 95, 251
Lott, D., 91
Louis XVI, 235

McClearn, G., 28
McClintock, M. K., 177
McDonald, P. G., 135, 183
McGill, T. E., 34, 35, 36, 172
Malinowski, B., 207, 214, 215, 217, 219, 259, 260, 261, 262, 263
Manning, A., 29, 36, 37, 38
Maqueo, M., 199
Margolese, S., 229
Marie Antoinette, 235
Markee, J. E., 167
Marler, P., 67
Marmor, J., 224
Marsden, H., 48, 65
Marshall, A., 48, 56, 58, 59
Masters, W. H., 81, 173, 206, 208, 210, 212, 236, 239, 240, 257
Matthews, L. H., 167
Mayr, E., 13, 14, 15, 16, 23, 26, 30, 41, 48
Mazzocchi, P., 70
Meagher, W., 107, 115
Mears, E., 244
Meinwald, J., 70
Meinwald, Y., 70
Mendel, G., 12, 13
Merari, A., 135, 175
Meyer, H., 47
Meyer, R. K., 167, 186
Michael, R. P., 158, 175, 176, 177, 240, 243
Miller, N., 234
Missakian, E., 96
Mitchell, G., 95
Money, J., 124, 229, 236, 241, 245, 246, 247, 250
Moment, G., 39, 41, 48, 49, 50
Morgan, R., 250
Morris, N. M., 242, 243
Moyer, D. L., 140
Muller-Schwarze, D., 175
Mullins, R. F., 195, 196, 197

Nadler, R. D., 203
Nalbandov, A., 80
Napier, J., 59
Napier, P., 59
Neiman, C. N., 170, 174

Netter, F. H., 129
Niemi, M., 195
Noble, G. K., 182
Nuttall, R., 81

Odell, W. D., 140
Oedipus, 257
Olsen, M. W., 48
Orbach, J., 114
Orbison, W. D., 240
Orchard, W. H., 244
Owen, R. D., 26, 39, 41

Palka, Y. S., 159, 183
Parkes, A., 65
Parkinson, S., 112, 122
Parsons, P., 27
Patterson, R. L. S., 175
Pellegrino, L., 100
Pepelko, W., 116
Perdeck, A. C., 23, 24, 25, 26
Petersen, I., 209
Peterson, R. S., 92
Pfeiffer, C. A., 192, 193
Phoenix, C., 107, 115, 119, 196, 199
Pierce, J., 81
Pincus, G., 244
Pliske, T., 70
Potts, F., 49
Powers, J. B., 137, 141, 142
Price, D., 199
Pumpian-Mindlin, E., 251

Rado, S., 230
Ramirez, V., 102
Ransom, T., 34, 35, 36
Rapaport, D., 251, 253
Rasquin, P. L., 183
Reichlin, S., 238
Resko, J. A., 138, 141, 195
Riddiford, L. M., 68
Riss, W., 95
Roberts, R., 28
Robertson, R. T., 197
Rodgers, C. H., 135, 146
Roeder, K., 103, 106
Roelofs, W., 68
Romm, M., 227
Rose, R., 229
Rosenblatt. J., 96, 150, 236
Ross, J.. 159, 164
Rosvold, E., 240
Rowan, W., 58
Rowell, T., 67
Rozin, P., 4
Rubenstein, B. B., 242
Rubin, H., 214
Ruf, K. B., 141

Saayman, G., 86

Sachs, B., 117
Saginor, M., 169
Salhanick, H. A., 244
Salis, P., 117
Salmon, U. J., 240
Sawyer, C. H., 98, 102, 118, 119, 135,
  137, 153, 154, 156, 157, 159, 167, 183
Schaller, G., 61
Schein, M. W., 59, 70, 71, 72, 89, 91, 92
Schiff, B., 117
Schinckel, P. G., 174
Schneider, D., 70
Schon, M., 240
Schreiner, L., 121
Schwartz, N. B., 139, 141
Scott, J. P., 15
Seeman, W., 234
Seibt, U., 70
Seligman, M., 4
Selinger, H., 78
Shaw, F., 44
Sherrington, C., 103
Sherwood, M. R., 200
Signoret, J. P., 175
Simpson, G. G., 13
Singer, J., 119, 154
Singh, B., 102
Sluckin, W., 94, 95
Smith, E. R., 135, 146, 163
Sneath, P. H. A., 13
Sokal, R. R., 13
Sonneborn, T. M., 40
Sophocles, 257, 258
Spalding, D. A., 94
Sperry, R., 98
Spiess, E., 31, 32
Spiess, L., 31
Spieth, H., 14, 26
Srb, A. M., 26, 39, 41
Staats, J., 34, 35
Staples, R. E., 166, 167
Stebbins, G., 13
Steinberg, F., 117
Stellar, E., 98
Stener, I., 209
Stern, J. J., 197
Stone, C. P., 180, 182
Suomalainen, E., 44
Sutherland, A. M., 240
Swanson, H. H., 184, 200

Taleisnik, S., 169
Talley, W. L., 139, 141
Tauber, E. S., 233
Taylor, L., 92, 120
Teitelbaum, P., 100
Terzian, H., 121
Thomas, T. R., 170, 174
Thompson, R., 102
Thompson, W., 27

Thorpe, D., 58
Tiefer, L., 183, 184
Tinbergen, N., 4, 26, 73, 74, 75, 76, 77,
  251
Tourney, G., 229

Udry, J. R., 242, 243
Uexkull, Jacob von, 3, 4
Ulrich, R., 118

Valenstein, E., 95
Vandenbergh, J., 59, 60
Van Dis, H., 116
Van Lawick-Goodall, J., 61
Van Wyk, J. J., 189, 190
Vaughn, E., 115, 116, 117
Vessey, S., 59, 60

Wagner, J. W., 194
Walker, P., 40
Wallen, K., 107
Wang, G., 63
Ward, I. L., 96, 197
Watson, J. D., 13
Waxenberg, S. E., 241, 242
Weick, R. F., 135, 163
Welegalla, J., 176, 240
West, C. D., 183
Westbrook, W. H., 82
Wetherbee, D., 77
Whalen, R., 92, 166, 196, 197
White, M., 44, 47, 48
Whitmoyer, D., 102
Whitten, W. K., 62, 64, 65
Wickler, W. K., 67
Wiener, H., 176
Wilbur, C., 227
Wilcock, J., 27, 30
Williams, C. M., 68
Wilson, E. O., 22
Wilson, J., 62, 92, 171
Wilson, M., 78, 79
Wilson, W., 77
Winick, C., 250
Winton, F. R., 249
Wolff, G., 65
Woodard, A., 77

Yamamoto, J., 234
Yanigimachi, R., 62
Ying, S., 167
Young, W. G., 95, 135, 136, 137, 147,
  148, 149, 150, 181, 183, 195, 199,
  240, 248

Zarrow, M. X., 66, 124, 167, 168, 169
Zitrin, A., 119
Zucker, I., 137, 141, 142, 171, 172
Zuckerman, M., 213
Zumpe, D., 243

Index of Subjects

Ablation, 132–133
Abnormal sexual behavior
  environmental determinants, 91–96
Accessory sex glands, 148–149, 160–161,
  169–170, 186
Adrenal cortex, 124, 126, 130, 143, 149,
  240–242
Adrenocortical androgen, 240–242
Adrenocorticotropin (ACTH), 124, 130,
  141, 241
Adrenogenital syndrome, 127
Adultery, 220
African lions, 180
Aging, 235, 236, 238
Aggression, 73–79, 87–88, 118, 124, 179,
  201, 202, 203
Amazon Molly fish (*Mollienisia formosa*),
  46, 47
Amphibians, 150
Anal eroticism, 255–256, 261
Androgen, 65, 71, 73, 78, 88, 130–131,
  142, 167, 174, 188, 197–198,
  229–230, 240–242. *See also*
  Hormones; Testosterone
Androstenedione, 142, 197, 198. *See
  also* Androgen
Anlage, 187, 198
Ano-genital intercourse, 216, 218, 224,
  227, 231
Apes, 66–67
Aphids (*Tetraneur ulmi*), 46
Aranda, 227
Arrhenotoky, 44–45
Artificial insemination, 89

Artificial selection, 28, 36–38
Asexual reproduction, 39. *See also*
  Binary fission
Auditory stimuli, 21–25
Autoerotic stages, 256
Autogamy, 41
Autonomic nervous system (ANS), 152,
  97–122, 208, 213

Baboons (genus *Papio*), 55, 67, 73, 85–88
Barbary dove, 160
Bats, 4
Beagles, 108–109. *See also* Dogs
Bears, 57–58, 61
Behavioral differentation, 202–205
Behavior-genetic analysis, 27. *See also*
  Genetic correlates of sexual
  behavior
Behavioral isolation
  chemical signals, 26–27
  mating calls, 21–25
  visual displays, 17–21
Berdache, 227
Bighorn sheep (*Ovis canadensis*), 57
Binary fission, 39, 41
Biological definitions of sex, 245–247
Biological psychology, 3, 7
Biological species concept, 9, 13, 14
Birds, 56, 73, 95, 124, 150, 165
Bisexuality, 52, 198. *See also*
  Hermaphroditism
Blacktailed deer, 175
Boyawa. *See* Trobriand Islanders
Breams (*Sparidae*), 52

Breasts, 210–211
Bruce effect, 65–66, 165, 173–174
Budding, 42–43
Buffalo, 73
Bukumatula, 217
Bulbocavernosus muscle, 211
Bulls, 131, 175. See also Cattle
Butterflies, 68, 69, 70

Canadian junco (Junco hyemalis), 58
Cannibalism, 104–106
Capacitation, 134
Capon, 160
Carnivores, 80, 236
Castration, 143, 145–146, 147, 149,
    161–162, 175, 182, 184, 187, 192,
    195, 196, 197, 201, 232–234, 238,
    239, 258. See also Ovariectomy
Castration complex, 257
Cats (Felidae), 57, 62, 79, 80, 93, 96, 102,
    106, 119–120, 121, 135, 142, 150,
    153, 154, 158, 161, 165, 166, 169,
    174, 175, 180, 185, 236
Cattle, 89, 90, 135, 180, 181
    wild, 62
Central hierarchy, 86. See also Baboons
Cervix, 169, 171, 173
Character displacement, 22
Chemical stimulation, 100–101
Chicks, 94
"Children's republic," 215
Chimpanzees, 60
Chipmunks, 15
Cholesterol, 160
Chromosomal sex, 245–246
Chuckchee, 227
Climacteric, 235–236, 238
Clitellum, 50
Clitoris, 188, 197, 198, 199, 210, 221,
    241, 247
Clone, 39
Cockroaches, 106
Comb-jellies (Ctenophora), 48, 49
Concubines, 220
Conjugation, 40–41
"Coolidge effect," 89, 110
Copulation, 14, 17, 19, 21, 25, 49–50,
    50–52, 64–66, 106, 108, 208, 213,
    216, 217, 254, 261
    female control over, 81–84
    and lesions, 118–119
    and pace, 117–118
    and pheromones, 68
Copulatory lock, 108, 110, 150
    simulated, 153
Copulatory sequence
    quantitative description, 34–35
Corona glandis, 208
Corpus luteum, 127, 128, 132, 170, 172,
    192–193

Cortex, 98, 100, 102, 107, 119–122, 150,
    153–155, 157–159, 160, 164. See
    also Tissue destruction
"Cortexin," 187
Cortisone, 126
Courtship-mating sequence, 71–72,
    104–106
    abnormal, 95–96
    definition, 14
    and fight/flee responses, 73–79
    genetic correlates, 27–38, 59
    as RIM, 17–27
    short-term regulatory stimuli, 79–82
    stimuli eliciting, 67–84
Courting posture, 76
Cross-cultural approach, 214–232,
    259–263
Cunnilingus, 219
Cypronodont fish (Rivulus marmoratus),
    49
Deer, 57, 61
Deer mouse (Peromyscus maniculatus),
    15, 66
Deuterotoky, 45–46
Diethyl stilbestrol, 158. See also
    Estrogen
Disinhibition, 103–122, 164
DNA, 13
Dogs (Canis familiaris), 4, 15, 62, 93, 96,
    107–111, 135, 150, 153, 175, 180,
    181, 182, 183, 201, 236
Dolphins, 4
Dominance, 36
Dominance hierarchy, 85–86
Donkey, 15
Dragonflies, 15
Ducklings, 94

Earthworms (Lumbricus terrestris), 49–50
East Bay Society, 217–221, 223, 227, 231
Ectohormone, 176. See also Pheromones
Ejaculation, 34, 35, 71, 82, 86, 89–91, 95,
    111, 114, 143, 147, 160, 169, 170,
    172, 180, 184, 196, 197, 200, 207,
    208, 216, 236
    short term regulatory stimuli, 79–83
Ejaculation latency (EL), 35, 112
Ejaculatory frequencies, 116, 122
Ejaculatory threshold, 113, 114, 116, 122
Electrical stimulation, 100–101, 114–119
Electrodermal response (EDR), 213
Electroencephalogram (EEG), 101
Elephant seals, 92
Elk (Cervus candensis), 57
Environmental determinants of sexual
    behavior
    abnormal sexual behavior, 91–96
    definition, 3
    in hydra, 43
    and light, 130–131

Environmental determinants (cont.)
    reproductive cycles, 56–67
    sign stimuli, 4
    social control of mating, 84–88
    and temperature, 131
    varieties, 54–55
Erection, 79–80, 89, 108, 110, 143, 162,
    208, 213, 214
    in human males, 233–236
Erotogenic zones, 255
Estradiol, 146, 159, 184. See also
    Estrogen
Estrogen, 33, 86, 107, 146, 147, 149, 153,
    154, 163, 167, 175, 191, 240, 241,
    242, 244. See also Hormones
    and behavioral differentiation, 193–197
    circulating levels, 124–128
    implantation, 157–159
    female mounting, 181–183
    female sexual behavior, 134–142
    in males, 184–186
    neuroendocrine mechanisms, 130–132
    and pheromones, 176
Estrus cycles, 61–66, 80, 88, 107, 134,
    138, 173, 180, 181, 192, 194, 200,
    243
Evolution, 12, 13
"Exaltolide phenomenon," 177
Excitation, 112, 113, 122
Excitement phase, 210
External genital appearance
    and sex determination, 247–248

Fear-related behavior, 202–203
"Feedback," 163, 166. See also "Negative
    feedback" mechanisms
Fellatio, 213, 218, 219, 224, 225
Ferrets, 135, 165
Fertility, 173
Fetishes, 253
Fiddler crabs (genus Uca), 17–18
Fireflies (Photinus), 18–21
Fish, 15, 73, 150
Flukes, 48
Follicle stimulating hormone (FSH),
    127–128, 130, 132, 140. See also
    Gonadotropins
Fowl, 238
Fragmentation, 52
Freudian theory, 94, 250–263
    and orgasm, 211–213
Frigidity, 241
Frogs, 16, 48
    genus Microhyla, 21
Fruit flies (Drosophila), 15
    D. melanogaster, 28–32, 36–38
    D. persimilis, 26, 28–33
    D. pseudoobscura, 26

Gamete transport, 138

Gametic mortality, 15, 24
Gametogenesis, 132, 191
Gay Liberation Movement, 227
Gender identity, 225, 230
Gender roles, 103–104, 248–249
Genetic correlates of sexual behavior,
    27–38
    artificial selection, 35–38
    differences in sex chromosome
        complement, 33–34
    in homosexuality, 228
    and hybridization, 34–36
    single gene effects, 28–31
    third chromosome pair arrangements,
        31–33
Genital sexuality, 261
Glucocorticoids, 126
Goats, 15, 61, 62, 135
Gonadal sex, 246
Gonadotropins, 119, 121, 127, 130–132,
    139, 142, 153, 159, 163–164,
    168–169, 170–171, 186, 191–192,
    195, 235, 238. See also names of
    individual hormones
Gonochorism, 48
Gorillas, 60
Goslings, 94
Grasshoppers (genus Chorthippus), 23
Great slug (Limax maximus), 49, 50–51
Ground squirrels, 59, 165
Guinea pigs, 62, 89, 95, 96, 107, 128, 135,
    136, 141, 147, 148, 153, 164, 174,
    180, 181, 182, 183, 195, 196, 199,
    200, 202
Gulls (Larus), 16, 74, 75–77
Gynecomastia, 127
Gynogenesis, 47

Habitat isolating mechanisms, 15, 24. See
    also Reproductive isolating
    mechanisms (RIM)
Hamsters, 62, 128, 135, 173, 178, 180,
    183, 200
Hares, 15
Heredity vs. environment controversy, 8, 9
Hermaphroditism, 48–52, 247–249
    and gender roles, 248–249
    in humans, 245
    successive, 51–52
Herring gull (Largus argentata), 74–77
Heterokaryotype, 31
Heterosis, 36
Heterozygosity, 34
Hippocampus, 117, 164
Homokaryotype, 31
Homosexuality, 237, 253
    causes, 228–231
    cross-cultural data, 225
    definition, 223–231
    in East Bay Society, 218

Homosexuality (cont.)
  endocrine correlates, 229–230
  female, 227
  and psychoanalytic theory, 230–231
  stereotypes, 223–224
  in Trobriand Islanders, 216
Homosexual mounting, 78–79
Homozygosity, 28, 34, 41
Honey bee (Hymenoptera), 4, 44–45, 68
Hormonal sex, 246–247
Hormones, 118, 123–150
  actions on nervous system, 151–164
  activational, 239–245
  and brain lesions, 153–155
  and evolution, 150
  gonadal, 78, 123–124, 125, 186
  in heterotypical behavior, 186, 203
  in homotypical and heterotypical
    behavior, 180–186
  human menstrual cycle, 242–245
  human sex behavior, 232–250
  implantation effects, 157–164
  and IQ, 124
  Kinsey study, 237–239
  male, 127, 142–150, 175
  neuroendocrine mechanisms, 130–132
  nonreproductive behavior, 201
  ovarian, 133–142, 175, 181 182, 201,
    203
  and pregnancy, 170–173
  regulatory feedback mechanisms, 131
  sites of action, 102, 151–164
  social stimuli, 165–176
  thyroid, 238
Hormone specificity, 184
Horses, 15, 61, 62, 175
Host specificity, 185
Human sexual behavior
  anatomy and physiology, 207–214
  anthropological and sociological
    perspectives, 214–231
  clinical-medical approach, 232–250
  and hormones, 232–245
  psychoanalytic theory, 250–263
  social control, 84–85
Hybrid inferiority, 15, 24
Hybridization, 34–36
Hybrid sterility, 15, 24
Hydra, 41–43, 55
Hyperadrenocorticism, 246
Hypogonadism, 235–236
Hypogonadotropic eunuchoidism,
  235–236
Hypophysectomy, 241. See also Pituitary
  gland
Hypothalamus, 100, 102, 107, 115–119,
  121, 125, 130, 154, 157–158, 160,
  162, 164, 183, 192–193, 200, 203

Impotence, 234–235

Imprinting, 94–95
Indirect erotic stimulation, 213–214
"Infant hercules" condition, 127
Infantile sexuality, 254, 258–259
Infant play, 203
Inhibition, 103–122, 152, 159, 164,
  173–174
Instincts, 251–263
Intense Ejaculatory Reaction (IER),108,
  110, 111, 153
Intercopulatory interval, 112–113
Intermediacy, 36
Internal accessory organs, 247
The Interpretation of Dreams, 257
Intromission, 34–35, 80–81, 82, 83, 92,
  95, 111, 113, 114, 116, 117, 143,
  146–147, 150, 167, 169, 171, 172,
  180, 184, 196, 197, 200
Intromission duration, 112
Intromission frequency, 111, 113, 122,
  146
Intromission latency (IL), 35, 111–112,
  143, 149–150, 160
Ischiocavernosus muscle, 211

Japanese macaque monkeys, 6
Japanese medaka fish (Oryzias latipes),
  33–34
Japanese quail (Coturnix coturnix
  japonica), 74, 77–79

Kada, 260–262. See also Trobriand
  Islanders
Katuyasi, 216–217
Keraki, 227
Kiwai, 227
Klinefelter's syndrome, 235, 246
Koniag, 227

Labia, 210, 247
Lactation, 128
Lango, 227
Latency period, 256, 258, 259, 261
Lee-Boot effect, 65, 165, 173–174, 177
Leeches, 48–49
Lesions, 116, 118–122. See also Tissue
  destruction; Cortex
Lesu, 221
Leydig cells, 127
Libido, 236
Light, 167–169
Lizards, 182
  Lacerta saxicola armeniaca, 46–47
Lordosis, 34, 63, 67, 107, 119–120, 135,
  139, 141, 146–147, 152, 159, 167,
  180–185, 194, 197, 198, 200, 203, 205
  in males, 180, 182–183, 184
  and social isolation, 96
  and testosterone, 183

Lordosis quotient, 63, 135, 184
Luteinizing hormone (LH), 127, 128,
    130, 131, 132, 138, 139, 142, 166,
    167, 168, 169, 170, 171

Marsupials, 61
Marten, 180
Masculine-feminine dichotomy, 225–226
Masturbation, 121, 216, 218, 221–223,
    231, 256
Maternal behavior, 201
Mechanical (anatomical) isolation, 15, 24.
    See also Reproductive isolating
    mechanisms
"Medullarin," 187
Menstrual cycle, 80, 86, 168, 177, 192
    and hormones, 243–245
    in primates, 66–67
    in women, 242–243
Menopause, 239
Mice (Mus musculus), 34–36, 62, 64–66,
    128, 135, 137, 141, 165, 172, 173,
    174, 177, 200, 201
Milk-ejection reflex, 131
Minks, 165
Mongolian gerbil, 141
Monkeys, 56, 66, 67, 141, 177. See also
    Rhesus monkey
Mount frequency, 111
Mounting, 79–83, 89, 96, 104, 111, 119,
    121, 143, 166–167, 169, 171–172,
    176, 180–185, 196, 201, 205, 239
    in females, 180–183
Mount latency (ML), 35
Mule, 15
Müllerian duct, 187
Müllerian system, 247
Musk, 177
Mutual masturbation, 221, 223, 224. See
    also Homosexuality

"Natural" behavior
    definition, 93–94
Natural selection, 12
Nature-nuture controversy. See
    Heredity vs. environment
    controversy
"Negative feedback" mechanisms, 131,
    142, 186
Neonatal period, 193–205
Neural determinants of sexual
    behavior, 97–122
    and sexual reflexes, 102–122
    social stimuli, 165–176
Neural psychology, 99–102
Neuroendocrine integration, 98, 103,
    150–177
Nipple erection, 210
Nonsexual and sexual reproduction,
    38–52

Nonsexual behavior differentiation, 205
Nuclear complex, 262

Oedipus complex, 216, 257–263
Olfactory stimuli, 25, 90, 111, 120,
    172–177
    in female mice, 64–66
    in humans, 176–177
    See also Bruce effect; Lee-Boot effect;
    Whitten effect
Oral contraceptives, 168–169, 177,
    244–245
Oral-genital intercourse, 216, 223. See
    also Fellatio; Cunnilingus
Oral sexuality, 255–256
Orangutans, 93
Orgasm, 79, 143, 207–208, 211–213, 216,
    221, 223, 233, 234, 238, 241, 243,
    253
Orgasmic phase, 211
"Orgasmic platform," 211
Ova, 39, 43, 44, 45, 49–51, 58–59, 74,
    128, 134, 138
Ovariectomy, 134, 159, 239
Ovaries, 49, 107, 125–128, 131, 134, 154,
    171, 183, 187, 191–195, 239–240
Oviduct, 71, 134
Ovulation, 62, 66, 67, 80, 86, 128, 130,
    132, 134, 139, 140, 156, 163, 164,
    167, 177, 192, 200, 242–243
Oxytocin, 172
Oyster, 13

Paramecium, 39–41
Parthenogenesis, 39, 44–48
Pathological behavior, 230
Pelvic thrusting, 34–35, 79, 80, 83, 86,
    108–111, 121, 172, 183
    in females, 180
Penis, 34, 80, 82, 108, 109, 111, 114, 122,
    161, 172, 186, 196, 197, 198, 207,
    213, 221, 254, 257, 258–259
    artificial, 172, 221
Phallic phase, 256–258
Phallus. See Penis
Pheromones, 66–69, 165, 173–177
Physiological isolation, 15. See also
    Reproductive isolating mechanism
    (RIM)
Physiology
    determinants of sexual behavior, 4–5
Pigeons, 167
Pigs, 125, 134, 175
Pituitary gland, 58, 99, 118, 121, 127,
    130–131, 153–154, 163–164, 166,
    171, 174, 191–193, 235, 238
Planarian flatworms, 48, 49, 52
Plateau phase, 210–211
Pleasure principle, 253
Pleiotropism, 30–31

Polyphemus moth (*Antheraea polyphemus*), 68
Porcupines, 180
Porgies (*Sparidae*), 52
Pornography, 213–214
Postejaculatory intervals, 84, 89, 91, 112, 113, 116, 143, 144
Posture, 74–79, 88, 160, 175. *See also* Lordosis
Prägung. See Imprinting
Praying mantis (*Heirodula tenuidentata*), 103–104
Preejaculatory mounting, 72
Pregenital stages, 256
Pregnancy, 61, 62, 64–66, 86, 88, 128, 132, 142, 167, 170–173, 199, 254
Presenting response, 67. *See also* Posture
Primates, 59–67, 80, 128, 180, 192, 239. *See also names of individual species*
Progesterone, 107, 125–128, 130–132, 138, 140, 142, 147, 153, 170–171, 174, 176–177, 181, 184, 192, 194–195, 199, 203, 241–244
and behavior, 123–124
facilitatory effects, 135, 143
inhibitory effects, 136–137, 141
and pheromones, 175
receptors, 159
Proglottids, 49
Prolactin, 124, 127, 128, 130, 170–173
Protandry, 52
Protogyny, 52
Pseudogamy, 47
Pseudohermaphroditism, 246
Pseudopregnancy, 61, 62, 64–66, 128, 142, 170–172
Psychoanalytic theory, 207, 230–231, 242, 250–263

Queen butterfly (*Danaus glippus berenice*), 69–70

Rabbits, 15, 62, 102, 119, 124, 134, 135, 137, 142, 148, 154, 159, 165, 166, 169, 174, 180, 183, 201
Rape, 134, 168
Rassenkreise, 16
Rats, 7, 62, 63, 79–80, 89, 91, 95–96, 100, 107, 111, 114, 116, 118–120, 122, 128, 130–132, 134, 135, 137, 141, 143, 147–148, 150, 152–153, 158–161, 163, 167, 169, 171, 173, 174, 178, 180–184, 186, 192–194, 196, 199–201, 203, 236, 241
Receptivity, 61, 62–63, 67, 70–71, 78, 81, 102, 119, 131, 134, 138, 140, 158, 167, 183, 192, 201, 239. *See also* Lordosis
Red deer (*Cervus elaphus*), 57
Reflex, definition of, 102–103. *See also* Sexual reflexes

Reflex ovulation, 62, 134, 142, 165–170
Relaxin, 172
Replacement therapy, 132–133, 145–147, 153–155. *See also* Hormones; Castration
Report of the President's Commission on Obscenity and Pornography, 213
Reproductive Biology Research Foundation, 208
Reproductive cycles, 128, 134, 163, 191
annual, 56–61
cultural factors in, 61
estrus cycles, 61–66
in humans, 60–61
and light, 58–59
in primates, 59–67
and rainfall, 59–60
and temperature, 59
testicular growth, 59
Reproductive isolating mechanisms (RIM)
behavioral, 16–27
classification, 14
in gynogenesis, 47
habitat, 15
physiological, 15
seasonal, 15
Reproductive isolation, advantages of, 13–14
Reproductive system
differentiation of, 187–190, 191–192
male, atrophy, 170
See also Müllerian system
Reserpine, 174
Resolution phase, 211
Rhesus monkeys (*Macaca mulatta*), 59, 67, 96, 121, 135, 176, 180, 202, 203, 239, 241, 243, 244
Rhinoceroses, 94
Roaches, 68
Rodents, 56, 61–62, 64–66, 80, 96, 124, 127, 128, 146, 170, 180, 192, 236, 239. *See also names of individual species*
Rosy flush, 208, 210
"Rough and tumble play," 202
Rwala bedouins, 227

Scrotum, 188, 202, 208
Sea-bass (*Serranellus subligarius*), 49
Sea lions, 57
Seals, 57
Seasonal isolation, 24. *See also* Reproductive isolating mechanism (RIM)
Sea urchins, 15
Semen, 89, 108, 114, 143, 207
Sex
biological characteristics, 245–247
definition, 9

Sexual behavior, definition of, 9
Sexual and nonsexual reproduction, 38–52
Sexual differences
  anomalies, 245
  behaviorial determinants, 193–205
  causes, 186–192
  heterotypical and homotypical
    responses, 178–180
Sexual differentiation, 178
*The Sexual Life of Savages in
  Northwestern Melanesia*, 215
Sexual reflexes, 107–122, 164
Sexual response cycle, 210–213
Sharp-tailed shrews, 180
Sheep, 15, 61–62, 89–91, 116, 141,
  174–175
Siwans, 227
Skoptsi sect, 232–233
Smegma, 177
Snails, 48
Social control
  of extended copulatory performance,
    88–92
  and human sexual behavior, 84–85
  of mating opportunities, 84–88
Social isolation, 95–96
Social stimuli
  and gender role, 248–250
  and homosexuality, 229–231
  and hormones, 165–176
  and pheromones, 173–174
  and pregnancy, 170–173
  and pseudopregnancy, 170–172
  and reflex ovulation, 165–169
Spawning, 14, 33, 58, 73
Sperm, 39, 43, 49–51, 58–59, 80, 104,
  106, 128, 134, 172–173, 192
Spinal animals, 106–111, 152–157,
  162–164
Sporulation, 52
Stereotaxic surgery, 99–100. *See also*
  Tissue destruction
Steroidogenesis, 132, 191
Steroids, 125, 175. *See also* Hormones
Stickleback (*Gasterosteus aculeatus*),
  72–74
Sucking response, 255
Swine, 180
Sympatric population, definition of, 14
Synchronous population, definition of, 14

Tanala, 227
Tapeworms, 49
Temporal lobe, 121. *See also* Cortex
Testes, 43, 49, 58–59, 126–128, 131–132,
  169, 187, 192, 195, 197, 208, 229,
  234, 235, 247
Testosterone, 119, 125–127, 169, 175,
  181, 187, 193, 196, 199–201, 203,
  205, 240–241
  in clinical disorders, 234–237
  in females, 183–186, 197
  in males, 142–150, 159–164, 182-183
  in neonatal period, 194–195
  in spinal animals, 153
  and uterine growth, 186
Thelytoky, 45–46
Threat posture, 74, 76. *See also* Posture
Three Essays on the Theory of Sexuality,
  254, 258
Tissue destruction, 99–100, 116, 119. *See
  also* Cortex
Toads, 14–16
Transverse perineal muscle, 207–209, 211
Trobriand Islanders, 207, 214–216, 219,
  221, 223, 259–263
Tse-tse flies, 15
Turkeys, 48, 71–72, 78, 89
Turner's syndrome, 246

Ulatile, 216–217
Umwelt, 4
Uterus, 66, 170–172, 187

Vagina, 34, 49, 80–81, 82, 108, 130, 167,
  176, 202, 210, 211
  artificial, 89
  stimulation, 169–173
Vaginal plug, 82–83, 171, 172
Visual stimuli, 111, 120, 167, 213–214
Vulva, 188, 210

"Walking stick" insect (*Saga pedo*), 46
Walruses, 57
Whitten effect, 65–66, 165, 173–174, 177
Wild canines, 57
Wish fulfillment, 253
Wolffian duct, 187, 247
Women's Liberation Movement, 250
Worms, 106

Zygotic mortality, 15, 24